Titles in This Series

Volume

1. **Markov random fields and their applications,** Ross Kindermann and J. Laurie Snell
2. **Proceedings of the conference on integration, topology, and geometry in linear spaces,** William H. Graves, Editor
3. **The closed graph and P-closed graph properties in general topology,** T. R. Hamlett and L. L. Herrington
4. **Problems of elastic stability and vibrations,** Vadim Komkov, Editor
5. **Rational constructions of modules for simple Lie algebras,** George B. Seligman
6. **Umbral calculus and Hopf algebras,** Robert Morris, Editor
7. **Complex contour integral representation of cardinal spline functions,** Walter Schempp
8. **Ordered fields and real algebraic geometry,** D. W. Dubois and T. Recio, Editors
9. **Papers in algebra, analysis and statistics,** R. Lidl, Editor
10. **Operator algebras and K-theory,** Ronald G. Douglas and Claude Schochet, Editors
11. **Plane ellipticity and related problems,** Robert P. Gilbert, Editor
12. **Symposium on algebraic topology in honor of José Adem,** Samuel Gitler, Editor
13. **Algebraists' homage: Papers in ring theory and related topics,** S. A. Amitsur, D. J. Saltman, and G. B. Seligman, Editors
14. **Lectures on Nielsen fixed point theory,** Boju Jiang
15. **Advanced analytic number theory. Part I: Ramification theoretic methods,** Carlos J. Moreno
16. **Complex representations of GL(2, K) for finite fields K,** Ilya Piatetski-Shapiro
17. **Nonlinear partial differential equations,** Joel A. Smoller, Editor
18. **Fixed points and nonexpansive mappings,** Robert C. Sine, Editor
19. **Proceedings of the Northwestern homotopy theory conference,** Haynes R. Miller and Stewart B. Priddy, Editors
20. **Low dimensional topology,** Samuel J. Lomonaco, Jr., Editor
21. **Topological methods in nonlinear functional analysis,** S. P. Singh, S. Thomeier, and B. Watson, Editors
22. **Factorizations of $b^n \pm 1$, $b = 2, 3, 5, 6, 7, 10, 11, 12$ up to high powers,** John Brillhart, D. H. Lehmer, J. L. Selfridge, Bryant Tuckerman, and S. S. Wagstaff, Jr.
23. **Chapter 9 of Ramanujan's second notebook—Infinite series identities, transformations, and evaluations,** Bruce C. Berndt and Padmini T. Joshi
24. **Central extensions, Galois groups, and ideal class groups of number fields,** A. Fröhlich
25. **Value distribution theory and its applications,** Chung-Chun Yang, Editor
26. **Conference in modern analysis and probability,** Richard Beals, Anatole Beck, Alexandra Bellow, and Arshag Hajian, Editors
27. **Microlocal analysis,** M. Salah Baouendi, Richard Beals, and Linda Preiss Rothschild, Editors
28. **Fluids and plasmas: geometry and dynamics,** Jerrold E. Marsden, Editor
29. **Automated theorem proving,** W. W. Bledsoe and Donald Loveland, Editors
30. **Mathematical applications of category theory,** J. W. Gray, Editor
31. **Axiomatic set theory,** James E. Baumgartner, Donald A. Martin, and Saharon Shelah, Editors
32. **Proceedings of the conference on Banach algebras and several complex variables,** F. Greenleaf and D. Gulick, Editors
33. **Contributions to group theory,** Kenneth I. Appel, John G. Ratcliffe, and Paul E. Schupp, Editors
34. **Combinatorics and algebra,** Curtis Greene, Editor

Titles in This Series

Volume

Titles in This Series

Volume

68 **Differential geometry: The interface between pure and applied mathematics,** Mladen Luksic, Clyde Martin, and William Shadwick, Editors

69 **Methods and applications of mathematical logic,** Walter A. Carnielli and Luiz Paulo de Alcantara, Editors

70 **Index theory of elliptic operators, foliations, and operator algebras,** Jerome Kaminker, Kenneth C. Millett, and Claude Schochet, Editors

71 **Mathematics and general relativity,** James A. Isenberg, Editor

72 **Fixed point theory and its applications,** R. F. Brown, Editor

73 **Geometry of random motion,** Rick Durrett and Mark A. Pinsky, Editors

74 **Geometry of group representations,** William M. Goldman and Andy R. Magid, Editors

75 **The finite calculus associated with Bessel functions,** Frank M. Cholewinski

76 **The structure of finite algebras,** David C. Hobby and Ralph Mckenzie

77 **Number theory and its applications in China,** Wang Yuan, Yang Chung-chun, and Pan Chengbiao, Editors

78 **Braids,** Joan S. Birman and Anatoly Libgober, Editors

79 **Regular differential forms,** Ernst Kunz and Rolf Waldi

80 **Statistical inference from stochastic processes,** N. U. Prabhu, Editor

81 **Hamiltonian dynamical systems,** Kenneth R. Meyer and Donald G. Saari, Editors

82 **Classical groups and related topics,** Alexander J. Hahn, Donald G. James, and Zhe-xian Wan, Editors

83 **Algebraic K-theory and algebraic number theory,** Michael R. Stein and R. Keith Dennis, Editors

84 **Partition problems in topology,** Stevo Todorcevic

85 **Banach space theory,** Bor-Luh Lin, Editor

86 **Representation theory and number theory in connection with the local Langlands conjecture,** J. Ritter, Editor

87 **Abelian group theory,** Laszlo Fuchs, Rüdiger Göbel, and Phillip Schultz, Editors

88 **Invariant theory,** R. Fossum, W. Haboush, M. Hochster, and V. Lakshmibai, Editors

89 **Graphs and algorithms,** R. Bruce Richter, Editor

90 **Singularities,** Richard Randell, Editor

91 **Commutative harmonic analysis,** David Colella, Editor

92 **Categories in computer science and logic,** John W. Gray and Andre Scedrov, Editors

93 **Representation theory, group rings, and coding theory,** M. Isaacs, A. Lichtman, D. Passman, S. Sehgal, N. J. A. Sloane, and H. Zassenhaus, Editors

94 **Measure and measurable dynamics,** R. Daniel Mauldin, R. M. Shortt, and Cesar E. Silva, Editors

95 **Infinite algebraic extensions of finite fields,** Joel V. Brawley and George E. Schnibben

96 **Algebraic topology,** Mark Mahowald and Stewart Priddy, Editors

97 **Dynamics and control of multibody systems,** J. E. Marsden, P. S. Krishnaprasad, and J. C. Simo, Editors

98 **Every planar map is four colorable,** Kenneth Appel and Wolfgang Haken

99 **The connection between infinite dimensional and finite dimensional dynamical systems,** Basil Nicolaenko, Ciprian Foias, and Roger Temam, Editors

100 **Current progress in hyperbolic systems: Riemann problems and computations,** W. Brent Lindquist, Editor

101 **Recent developments in geometry,** S.-Y. Cheng, H. Choi, and Robert E. Greene, Editors

102 **Primes associated to an ideal,** Stephen McAdam

103 **Coloring theories,** Steve Fisk

Titles in This Series

Volume

104 **Accessible categories: The foundations of categorical model theory,** Michael Makkai and Robert Paré

105 **Geometric and topological invariants of elliptic operators,** Jerome Kaminker, Editor

Geometric and Topological Invariants of Elliptic Operators

CONTEMPORARY MATHEMATICS

105

Geometric and Topological Invariants of Elliptic Operators

Proceedings of the AMS-IMS-SIAM
Joint Summer Research Conference
held July 23–29, 1988 with support
from the National Science Foundation
and the U. S. Army Research Office

Jerome Kaminker, Editor

AMERICAN MATHEMATICAL SOCIETY • PROVIDENCE, RHODE ISLAND

The AMS-IMS-SIAM Joint Summer Research Conference in the Mathematical Sciences on Geometric and Topological Invariants of Elliptic Operators was held at Bowdoin College, Brunswick, Maine on July 23–29, 1988 with support from the National Science Foundation, Grant DMS-8846813 and the U.S. Army Research Office, Grant 26725-MA-CF.

1980 *Mathematics Subject Classification* (1985 *Revision*). Primary 58G12; Secondary 46L80.

Library of Congress Cataloging-in-Publication Data

AMS-IMS-SIAM Joint Summer Research Conference in the Mathematical Sciences on Geometric and Topological Invariants of Elliptic Operators (1988: Bowdoin College)

Geometric and topological invariants of elliptic operators: proceedings of the AMS-IMS-SIAM joint summer research conference held July 23–29, 1988 with support from the National Science Foundation and the U.S. Army Research Office/Jerome Kaminker, editor.

p. cm.—(Contemporary mathematics, ISSN 0271-4132; v. 105)

Includes bibliographical references.

ISBN 0-8218-5112-8 (alk. paper)

1. Elliptic operators—Congresses. 2. Invariants (Mathematics)—Congresses. I. Kaminker, Jerome. II. American Mathematical Society. III. Institute of Mathematical Statistics. IV. Society for Industrial and Applied Mathematics. V. Title. VI. Series: Contemporary mathematics (American Mathematical Society); v. 105.

QA329.42.A47 1988 89-18660
515′.7242—dc20 CIP

Printed in the United States of America.
This volume was printed directly from author-prepared copy.
The paper used in this book is acid-free and falls within the guidelines established to ensure permanence and durability. ♾
Portions of this publication were typeset using $\mathcal{AMS}$-TEX, the American Mathematical Society's TEX macro system.

10 9 8 7 6 5 4 3 2 1 95 94 93 92 91 90

Contents

Preface

During the week of July 23–29, 1988, there was a Joint Summer Research Conference on "Geometric and topological invariants of elliptic operators" held at Bowdoin College. It was organized by Jeff Cheeger, Alain Connes, and Jerry Kaminker, along with additional members of the organizing committee, Ron Douglas and Ed Effros.

A broad range of topics was covered, and a selection of those is included in this volume. Some of the themes that were covered at the conference and, which appear here, are the use of more sophisticated asymptotic methods to obtain index theorems, the study of the eta invariant and analytic torsion, and index theory on open manifolds and foliated manifolds. Moreover, the current state of non-commutative differential geometry, as well as operator algebraic and K-theoretic methods are presented in several of the articles. There is also a discussion of the relations between the analytic and topological approaches to the Novikov conjecture.

While many striking results were presented at the conference, there was a feeling that the subject was still building momentum, particularly through the infusion of ideas from physics and the possibility of obtaining more refined invariants for elliptic operators. The present collection provides several excellent examples of this.

The papers in this volume are in final form, and no version of them will appear for publication elsewhere.

Jerome Kaminker

Contemporary Mathematics
Volume **105**, 1990

Asymptotic Pseudodifferential Operators And Index Theory

JONATHAN BLOCK
JEFFREY FOX

M.I.T.
University of Colorado-Boulder

Abstract. We introduce the notion of an asymptotic pseudodifferntial operator on spin bundles over a compact manifold and develop a calculus for these operators. We then use the the formula of Jaffe, Lesniewski, and Osterwalder for theta-summable Fredholm modules, along with the asymptotic calculus, to compute the cyclic cocycle that Connes attaches to the Dirac operator on a compact spin manifold.

INTRODUCTION

This paper had its origins in a seminar held at MSRI in the Fall of 1987. Our goal was to understand Connes's computation of the cyclic cocycle corresponding to the Dirac operator on a compact spin manifold ([**C-1**], [**C-2**]). In particular, we needed to understand the ideas in the paper by Getzler [**G**]. It soon became clear that a change of view would simplify the situtation. Of course, this change is implicit in the letter from Connes to Quillen.

Getzler introduced an algebra of pseudodifferential operators ($\psi D0$'s) on a compact spin manifold that incorporated the quantization map of Bokobza-Haggiag [**B**] with some supersymmetric ideas from quantum field theory. These ideas suggest that Clifford multiplication contributes to the degree of a differential operator. If p denotes a symbol, $\theta(p)$ will be the associated pseudodifferential operator. Given a ψDO Q, there is a symbol map, $\sigma(Q)$, which was first introduced by Widom ([**W1**], [**W2**]). Next, a scaling was introduced, which is essentially the traditional scaling of asymptotic analysis, but extended to take into account the Clifford variables [**MF**]. A family of compositions was defined by:

$$p \circ_t q = \sigma(\theta(p_t) \circ \theta(q_t))_{t^{-1}}.$$

This scaling had the important property that $\lim_{t \to 0} p \circ_t q$ was computable. While the $\theta(p)$'s generate an algebra of ψDO's, it is clear that $\theta(p_t) \circ \theta(q_t)$ is not

1980 *Mathematics Subject Classification* (1985 *Revision*). Primary 58G12, 46L80.
The second author was partially supported by NSF Grant DMS 8843757. Both authors were supported by the Mathematical Sciences Research Institute.

the scaled version of some other operator. It was this observation that led us to enlarge the class of symbols, moving closer to Widom's work and the classical asymptotic methods.

Let $p(x,\xi,t)$ be a symbol of degree n such that $p(x,\xi,t)$ has an asymptotic expansion in t with respect to the natural topology on the symbol space:

$$p(x,\xi,t) \sim \sum_{l=o}^{\infty} t^l p_l(x,\xi) \quad \text{with } p_l \text{ a symbol of degree } n-l.$$

We define an asymptotic operator to be (essentially) $\theta(p(x,\xi,t)_t) = P_t$. In this setting Getzler's rescaled operator is just an asymptotic operator whose symbol has the trivial asymptotic expansion $p(x,\xi,t) \sim p(x,\xi)$. Once the notion of an asymptotic pseudodifferential operator ($A\psi DO$) has been defined, the development of a calculus of $A\psi DO$'s paralleling the calculus of ψDO's is straightforward.

As we were working to understand Connes's work, we realized that a formula for the character of a Fredholm module in which all of the terms were asymptotic operators would be extremely useful. One of the main difficulties Connes needed to address was that $D_t = tD$ is not an $A\psi DO$ so D_t^{-1} is not an $A\psi DO$ either. We then received a preprint of Getzler and Szenes [**GS**], in which they presented such a formula, which is due to Jaffe, Lesniewski, and Osterwalder [**JLO**]. This formula has three beautiful attributes:

(1) one does not have to cup with the small complex to make D invertible;
(2) all terms are manifestly asymptotic operators; and
(3) it applies to the larger context of θ-summable Fredholm modules.

Using this formula for the Chern character, we write out a computation of the cyclic cocycle attached to the Dirac operator on a spin manifold.

We are grateful to MSRI for the financial support that enabled us to work together and to the organizers of MSRI's special year in representation theory–Wilfred Schmid, David Vogan, and Joseph Wolf–for organizing a program that was conducive to progress on this and other problems. Finally, we would like to thank Henri Moscovici for his constant encouragement and advice.

1. Review of Clifford Algebras and Spinors.

In this section a brief review of the construction of the Clifford algebra and spinors is presented. (For more details see [**Gi**], [**ABP**].) Let V be an even-dimensional vector space with an inner product, $\langle\cdot,\cdot\rangle$. Denote by V^n the n-fold tensor product of V with itself, so

(1.1.) $$V^n = V \otimes V \otimes \cdots \otimes V \text{ n-factors.}$$

(We set V^0 to be $\mathbf{R}$.) The tensor algebra of V is the graded algebra

(1.2) $$T(V) = \oplus_{n=0}^{\infty} V^n.$$

Let I be the ideal generated by the relations $v \otimes v - \langle v,v\rangle$. It should be noted that the ideal is not homogeneous under the Z grading but is homogeneous under the Z_2 grading. We have the following.

DEFINITION 1.1. *The Clifford algebra of the pair* $(V, \langle\cdot,\cdot\rangle)$ *is* $Cliff(V) = T(V)/I$.

Since the ideal I is homogeneous under the Z_2 grading, we have that $Cliff(V)$ is a Z_2 graded algebra. Let $e_1, e_2, \cdots, e_n$ be an orthonormal basis for V; then $Cliff(V)$ is the algebra generated by the e_i's subject to the relations:

$$e_ie_j + e_je_i = 0 \quad \text{if } i \neq j \qquad e_i^2 = 1. \tag{1.3}$$

Using the above relations we can write down a basis for $Cliff(V)$. Let $I = (i_1 < i_2 < \cdots < i_p)$ be a multi-index and set $e_I = e_{i_1} \cdot e_{i_2} \cdots e_{i_p}$; then $Cliff(V)$ has the 2^n monomials e_I as a basis. Denote by $\hat{\otimes}$ the graded tensor product. By considering the basis of $Cliff(V)$ as above we see that:

$$Cliff(V \oplus W) = Cliff(V)\hat{\otimes}Cliff(W). \tag{1.4}$$

Define the transpose map $a \rightarrow a^t$ for $a \in Cliff(V)$ by defining it on a basis and then extending it by linearity, for e_I we have:

$$(e_{i_1} \cdots e_{i_p})^t = e_{i_p} \cdots e_{i_1}. \tag{1.5}$$

A simple computation shows that $(e_{i_1} \cdots e_{i_p})^t = (-1)^{p(p-1)/2}(e_{i_1} \cdots e_{i_p})$. Next define $\tau = (i)^{n/2}e_1 \cdots e_n$ where $e_1, \ldots, e_n$ is an oriented orthonormal basis of V. Then we see that $\tau^2 = 1$.

Let $Cliff(V)_C$ denote the complexified Clifford algebra, $Cliff(V)_C = Cliff(V) \otimes_R C$. If V is a two-dimensional vector space, we can define the well-known isomorphism of $Cliff(V)_C$ with the two-by-two complex matrix algebra. This isomorphism is implemented by the Pauli matrices:

$$\begin{aligned} e_1 &\rightarrow \begin{pmatrix} 0 & 1 \\ 1 & 0 \end{pmatrix} \\ e_2 &\rightarrow \begin{pmatrix} 0 & i \\ -i & 0 \end{pmatrix} \\ \tau = ie_1e_2 &\rightarrow \begin{pmatrix} 1 & 0 \\ 0 & -1 \end{pmatrix}. \end{aligned} \tag{1.6}$$

If the dimension of the vector space V is $n = 2m$, then it follows from (1.4) and (1.6) that $Cliff(V)_C$ is isomorphic to the matrix algebra $M_{2^m}(C)$ of $2^m \times 2^m$ matrices. In particular, it is a simple algebra. Left multiplication of $Cliff(V)_C$ on itself decomposes into a direct sum of irreducible submodules. After choosing an isomorphism of $Cliff(V)_C$ with $M_{2^m}(C)$, each column of $M_{2^m}(C)$ yields an irreducible submodule. All of these are isomorphic, and in fact there is only one irreducible $Cliff(V)_C$ module up to isomorphism.

DEFINITION 1.2. *Let* Δ *be an irreducible submodule for* $Cliff(V)_C$ *acting on itself by left multiplication; then* Δ *is called the space of spinors.*

DEFINITION 1.3. *Define* $Spin(V)$ *by:*

$$Spin(V) = \{w \in Cliff(V) | w = v_1 \cdots v_{2j} \quad and |v_k| = 1\}. \tag{1.7}$$

Note that $Spin(V)$ is a group with $w^{-1} = w^t$.

Define a map $\rho : Cliff(V) \to End(Cliff(V))$ as follows. If $w \in Cliff(V)$, then:

$$\rho(w)(v) = wvw^t. \tag{1.8}$$

Suppose that $e_1, \ldots, e_n$ is an orthonormal basis for V. Then:

$$\begin{cases} \rho(e_1)(e_1) = e_1 \\ \rho(e_1)(e_i) = -e_i \end{cases} \Rightarrow -\rho(e_1) = \begin{cases} \text{reflection through the} \\ \text{hyperplane orthogonal to } e_1 \end{cases}. \tag{1.9}$$

We see that if $v \in V \subset Cliff(V)$, then $\rho(v)$ takes V to V. Furthermore, if $|v| = 1$, then $\rho(v) \in O(V)$, the orthogonal group of V. Consequently, ρ maps $Spin(V)$ into $SO(V)$, and since every element in $SO(V)$ can be written as a product of an even number of reflections, we see that ρ maps onto $SO(V)$. If $\rho(w) = 1$, then $wvw^t = v$ for all $v \in V$. It is easy to check that this implies that w is in the center of $Cliff(V)$. Since the center of $Cliff(V)$ consists of the scalars and $|w| = 1$, we have that $w = \pm 1$. Therefore, we have the short exact sequence of groups:

$$1 \to Z_2 \to Spin(V) \to SO(V) \to 1. \tag{1.10}$$

To show that $Spin(V)$ is connected, we only have to produce a path from -1 to 1 in $Spin(V)$. We simply write one down:

$$\gamma(\theta) = (cos(\theta)e_1 - sin(\theta)e_2)e_1. \tag{1.11}$$

Thus $\gamma(0) = 1$, $\gamma(\pi) = -1$, and we conclude that $Spin(V)$ is a two fold-cover of $SO(V)$.

Once we fix an orthonormal basis of V, $e_1, \cdots, e_n$, we can identify $SO(V)$ with $SO(n, R)$. The Lie algebra of $SO(n, R)$ and of $Spin(V)$ is then identified with the skew symmetric matrices of $M_n(R)$, which we denote by $so(n)$. If $A = (a_{ij}) \in so(n)$, then the action of A on Δ is given by Clifford multiplication by:

$$\frac{1}{4}\sum_{ij} a_{ij}e_ie_j. \tag{1.12}$$

For the details of this computation, see [**G**, p.173].

DEFINITION 1.4. *Let V be an oriented vector space and $e_1, \cdots, e_n$ be an oriented basis for V. If $\tau = (i)^{n/2}e_1 \cdots e_n$, then we define the half-spin representations of $Spin(V)$ by $\Delta^\pm = \{x \in \Delta | \tau \cdot x = \pm x\}$.*

Note that τ commutes with every element of $Spin(V)$, so $\Delta^\pm$ are invariant under the action of $Spin(V)$. If $v \in V$, let $c(v)$ be the operator on Δ which is Clifford multiplication by v. Then it is clear that:

$$c(v) : \Delta^\pm \to \Delta^\mp. \tag{1.13}$$

Let V be an oriented inner-product space with oriented orthonormal basis $e_1, \cdots, e_n$. Let $\Lambda(V)$ be the associated exterior algebra of V.

DEFINITION 1.5. *Define a linear isomorphism σ from $Cliff(V)$ to $\Lambda(V)$ as follows : for each multi-index $I = (i_1 < \cdots < i_p)$ we have*

$$\sigma(e_{i_1} \cdots e_{i_p}) = e_{i_1} \wedge \cdots \wedge e_{i_p}. \tag{1.14}$$

It is easy to check that this map is independent of the basis and that

$$\sigma(a \cdot b) = \sigma(a) \wedge \sigma(b) \quad \mod \Lambda^{deg(a)+deg(b)-1}. \tag{1.15}$$

Recall that a Z_2 graded vector space, $V = V^+ \oplus V^-$, is called a supervector space. Let ϵ be the operator that is $+1$ on V^+ and -1 on V^-; then ϵ is called the grading operator for V. If V is Z_2 graded, then $End(V)$ is also Z_2 graded with the grading operator $Ad(\epsilon)$. If V is finite dimensional, we define the supertrace of $a \in End(V)$ by;

$$tr_s(a) = tr(\epsilon a). \tag{1.16}$$

The fundamental property of a trace is that $tr([a, b]) = 0$, where $[a, b] = ab - ba$. There is an analogous property for the supertrace, but one has to replace the commutator $[a, b]$ by the graded commutator. If $a \in End(V)$, we say that $deg(a) = 0$ or 1 when $Ad(\epsilon)(a) = \pm a$, and we write $(-1)^a$ for $(-1)^{deg(a)}$. For $a, b \in End(V)$, we define the graded commutator as:

$$[a, b]_s = ab - (-1)^{deg(a)deg(b)} ba. \tag{1.17}$$

For the supertrace we have:

$$tr_s([a, b]_s) = 0. \tag{1.18}$$

Let V and W be two Z_2 graded vector spaces, and recall that $\hat{\otimes}$ represents the graded tensor product. We have:

$$End(V \hat{\otimes} W) = End(V) \hat{\otimes} End(W). \tag{1.19}$$

Under this isomorphism we have:

$$a \hat{\otimes} b(v \hat{\otimes} w) = (-1)^{deg(a)deg(v)} a(v) \hat{\otimes} b(w). \tag{1.20}$$

It follows from (1.20) that:

$$tr_s(a \hat{\otimes} b) = tr_s(a) tr_s(b). \tag{1.21}$$

The space of spinors Δ is naturally a super vector space with the grading operator τ, and $Cliff(V)_C = End(\Delta)$ is a super algebra. Equation (1.21) makes it easy to compute the value of the supertrace on elements of the Clifford algebra acting on the spinors. We start off with a two-dimensional vector space spanned by the orthonormal basis e_1 and e_2. Recall the isomorphism of equation (1.6) implemented by the Pauli matrices; then we see that $tr_s(e_1) = tr_s(e_2) = 0$ and $tr_s(e_1 e_2) = \frac{2}{i}$. More generally, for a real vector space of dimension $n =$

$2m$, $Cliff(V)_C$ is identified with $M_{2^m}(C) = \otimes^m(M_2(C))$ and the spinors are identified with 2^m-dimension column vectors. By equation (1.21) we have:

$$tr_s(e_I) = \begin{cases} 0 & I \neq \{1,\ldots,n\} \\ (\frac{2}{i})^m & I = \{1,\ldots,n\} \end{cases}. \tag{1.22}$$

2. Pseudodifferential Operators.

This section introduces the main ingredients for the asymptotic pseudodifferential operators. We first discuss the spin bundle and then talk about the standard symbol spaces and the ordinary (that is, non-asymptotic) pseudodifferential operators.

The Spin Bundle.

Let M be a compact, oriented, Riemannian manifold with tangent bundle $TM \xrightarrow{\pi} M$. Since M is oriented, we can find an open cover, $\{U_\alpha\}$, of M that trivializes the tangent bundle such that the transition functions $g_{\alpha\beta}$ are in $C^\infty(U_\alpha \cap U_\beta; SO(n))$. Assume that M is equipped with a Riemannian connection ∇. If $e_1, \ldots, e_n$ is a local orthonormal frame, then

$$\nabla_{e_i}(e_j) = \sum_k \Gamma_{ij}^k e_k. \tag{2.1}$$

Set $f^i =$ to the one-form dual to e_i and define $\omega_i^k = \sum_i \Gamma_{ji}^k f^j$, then

$$\nabla(e_i) = \sum_k \omega_i^k e_k. \tag{2.2}$$

The ω_i^k are called *the local connection one forms* defining the connection. From

$$0 = d\langle e_i, e_j\rangle = \langle \nabla(e_i), e_j\rangle + \langle e_i, \nabla(e_j)\rangle \tag{2.3}$$

we see that $\omega = (\omega_j^k)$ is an $so(n)$-valued one-form. The curvature of ∇ is defined by

$$\Omega = d\omega - \omega \wedge \omega. \tag{2.4}$$

If we evaluate $\Omega(e_i, e_j) \in Hom(TM)$, then

$$\Omega(e_i, e_j) = \sum R_{lij}^k e_k \otimes f^l. \tag{2.5}$$

The tensor R_{lij}^k is skew in the ij indices (that is, Ω is a two form) and also skew in the kl indices (that is, $\Omega(e_i, e_j) \in so(n)$) .

Recall that $\rho : Spin(n) \to SO(n)$ is a two-fold covering map. If the U_α are suitably chosen, then we can lift the transition functions $g_{\alpha\beta}$ to functions $\bar{g}_{\alpha\beta} : U_\alpha \cap U_\beta \to Spin(n)$ such that $\rho(\bar{g}_{\alpha\beta}) = g_{\alpha\beta}$. However, we may not be able to lift them so that the cocycle condition $\bar{g}_{\alpha\beta}\bar{g}_{\beta\gamma} = \bar{g}_{\alpha\gamma}$ holds. If we can, then we say that the manifold is spinable, and a choice of liftings is called a spin structure (see [**Gi**]). We will assume that a spin structure has been chosen for M; then we can form the bundle of spinors over M, which we denote by $\mathbb{S}$. Let us recall how $\mathbb{S}$ is constructed. First, form the disjoint union $S = \cup_\alpha U_\alpha \times \Delta$. Then define an equivalence relation on S as follows: if $(m,x) \in U_\alpha \times \Delta$ and $(n,y) \in U_\beta \times \Delta$, then $(m,x) \sim (n,y)$ iff $m = n$ and $\bar{g}_{\alpha\beta}(x) = y$. The bundle of spinors is then equal to $\mathbb{S} = S/\sim$. It is clear that $\mathbb{S}$ trivializes over U_α and that $\bar{g}_{\alpha\beta}$ are the transition functions.

Let $e_1, \ldots, e_n$ be an orthonormal frame for TM over U_α, and let ω be the associated connection one-form. We can define a local connection one-form on $\mathbb{S}$ restricted to U_α by using the homomorphism from $so(n)$ into $Cliff(R^n)$:

$$\bar{\omega} = \frac{1}{4}\sum_{ij} \omega^i_j e_i e_j \in Cliff(TM|_{U_\alpha}). \tag{2.6}$$

Denote by $\bar{\nabla}$ the corresponding connection on $\mathbb{S}$. The curvature of this connection is given by:

$$\bar{\Omega}(e_i, e_j) = \frac{1}{4}\sum_{kl} R^k_{lij} e_k e_l \in Cliff(TM|_{U_\alpha}). \tag{2.7}$$

Since M is assumed to be oriented, we have a decomposition of $\mathbb{S} = \mathbb{S}^+ \oplus \mathbb{S}^-$ corresponding to the decomposition of Δ into $\Delta^+ \oplus \Delta^-$. In fact, $\tau = (i)^{n/2} e_1 \cdots e_n$ defines a global section of $Cliff(TM)$ and $\Gamma(\mathbb{S}^\pm) = \{s \in \Gamma(\mathbb{S}) | \tau s = \pm s\}$. Since $\tau\omega = \omega\tau$ the connection $\bar{\nabla}$ restricts to a connection on $\mathbb{S}^\pm$, which we still denote by $\bar{\nabla}$.

If $\mathbb{E}$ is any vector bundle over M with connection $\nabla_{\mathbb{E}}$, then we will always use the tensor product connection on $\mathbb{E} \otimes \mathbb{S}$:

$$\nabla_{\mathbb{E}\otimes\mathbb{S}} = \nabla_{\mathbb{E}} \otimes Id + Id \otimes \nabla_{\mathbb{S}}. \tag{2.8}$$

Symbols and Their Calculus.

Recall the standard notion of a symbol for a vector bundle $\mathbb{E}$ over a compact manifold M. Let $\pi : T^*M \to M$ be the natural map and let $S = \pi^*(Hom(\mathbb{E}, \mathbb{E}))$ be the pull-back of the bundle $Hom(\mathbb{E}, \mathbb{E})$ to a bundle over T^*M.

DEFINITION 2.1. *A section $p \in S$ is called a symbol of order m if for every multi-index α and β we have the estimates:*

$$\| \partial^\alpha_x \partial^\beta_\xi p(x,\xi) \| \le C_{\alpha\beta}(1 + |\xi|)^{m-|\beta|}. \tag{2.9}$$

We denote by $\Sigma^m(\mathbb{E})$ the symbols of order m.

Note that $\Sigma^m(\mathbb{E})$ is a Fréchet space with respect to the semi-norms:

$$\rho_{\alpha\beta}(p) = inf\{\ C_{\alpha\beta} \mid \ \| \partial^\alpha_x \partial^\beta_\xi p(x,\xi) \| \le C_{\alpha\beta}(1 + |\xi|)^{m-\beta}\ \}. \tag{2.10}$$

Following Getzler [**G**], we will introduce a new filtration on the space of symbols of $\mathbb{E} \otimes \mathbb{S}$. Recall that we have a linear isomorphism of vector bundles $\sigma : Cliff(TM) \to \Lambda(TM) \cong \Lambda(T^*M)$, where we use the inner product to implement the isomorphism between TM and T^*M. Consequently, we have:

$$\begin{aligned} Hom(\mathbb{E} \otimes \mathbb{S}) &\cong Hom(\mathbb{E}) \otimes Hom(\mathbb{S}) \\ &\cong Hom(\mathbb{E}) \otimes Cliff(TM) \cong Hom(\mathbb{E}) \otimes \Lambda(T^*M). \end{aligned} \tag{2.11}$$

We will denote by $\bar{\sigma}$ the linear isomorphism from $Hom(\mathbb{E}) \otimes Cliff(TM)$ to $Hom(\mathbb{E}) \otimes \Lambda(T^*M)$ and by $\bar{\theta}$ its inverse. We denote by $\Omega^j(M)$ the smooth forms of degree j on M. Let $\mathbb{L}$ be the pull-back to T^*M of the bundle $Hom(\mathbb{E}) \otimes \Lambda(T^*M)$ via the map $\pi : T^*M \to M$.

DEFINITION 2.2. *A section $p \in \Gamma(\mathbb{L})$ will be called an s-symbol of order l if*

$$p = \sum_{j=0}^{m} p_j(x,\xi) \otimes \omega_j \quad \textit{with } p_j \in \Sigma^{l-j} \textit{ and } \omega_j \in \Omega^j(M). \tag{2.12}$$

We will denote the collection of s-symbols of order l by $S\Sigma^l(\mathbb{E})$.

It should be noted that $S\Sigma^l$ is also a Fréchet space when we give it the projective tensor product topology, since $\Omega(M)$ is a Fréchet space.

Remark. This definition of an s-symbol seem a bit strange and its ultimate justification will be the resulting asymptotic calculus that we develop. However, if one looks at the paper by the physicist Alvarez-Gaumé [**AG**] one sees that one is led to this definition in a relatively straightforward manner from the study of supersymmetric quantum field theory.

Given a s-symbol $p \in S\Sigma^l$, we want to define an operator $P = \theta(p)$. To do this we follow Bokobza-Haggiag's definition. First, we need to recall the definition of normal coordinates.

Let m be a point in the manifold M and T_mM be the tangent space to M at m. Since we have a connection on TM, there is an exponential map $Exp_m : T_mM \to M$. If $X \in T_mM$, let $\gamma_X(t)$ be the unique geodesic through m such that $\gamma'_X(0) = X$; then $Exp_m(X) = \gamma_X(1)$. Choose an orthonormal basis, $X_1, \ldots, X_n$, for T_mM, then $Exp(x_1X_1 + \cdots + x_nX_n) = (x_1, \cdots, x_n)$ defines a system of normal coordinates in a neighborhood of m_o. Exp can also be viewed as a map $Exp : TM \to M \times M$ which is a diffeomorphism of a neighborhood of the zero section onto a neighborhood of the diagonal in $M \times M$:

$$Exp(m, X_m) = (m, Exp_m(X_m)). \tag{2.13}$$

Let α be a function that is identically one in a neighborhood of the diagonal of $M \times M$ such that the exponential map is a diffeomorphism on the support of α.

Let $(m,x) \in Supp(\alpha)$. There is a unique geodesic from m to x; if $x = Exp_m(X)$ then that geodesic is $Exp_m(tX)$. Let $\tau(m,x) : (\mathbb{E} \otimes \mathbb{S})_m \to (\mathbb{E} \otimes \mathbb{S})_x$ be parallel translation along the unique geodesic from m to x. If $s \in \Gamma(\mathbb{E} \otimes \mathbb{S})$, then we define:

$$\hat{s}_m(x) = \alpha(m,x)\tau(x,m)s(x). \tag{2.14}$$

Notice that $\hat{s}_m(x)$ is really a function on $T_m(M)$, since if x is not in the image of the exponential map, $\alpha(x,m) = 0$ and $\hat{s}_m(x) = 0$. We will write $\hat{s}_m(X)$ for $\hat{s}_m(Exp_m(X))$ when there is no chance of confusion.

DEFINITION 2.3. *Let $p \in S\Sigma(\mathbb{E})$ and $s \in \Gamma(\mathbb{E} \otimes \mathbb{S})$; then we define:*

$$\theta(p)(s)(m) = \int_{T_m(M)\times T_m^*(M)} e^{-i\langle X,\xi\rangle}\bar{\theta}(p)(m,\xi)\hat{s}_m(X)\,dX\,d\xi. \tag{2.15}$$

Remark. The existence of a Riemannian structure on M yields a smooth choice of Lebesque measure on $T_m(M)$. We choose the measure $d\xi$ to be dual to dX in the sense that Fourier inversion holds:

$$\phi(o) = \int_{T_m^*(M)}\int_{T_m(M)} e^{-i\langle X,\xi\rangle}\phi(X)\,dX\,d\xi. \tag{2.16}$$

Remark. The operator $\theta(p)$ depends on the choice of the cut-off function α; however, when we introduce the notion of an asymptotic operator, we will see that the asymptotic operator is independent of the choice of α.

The set of all such operators, along with all infinitely smoothing operators, will be denoted by $Op^a(\mathbb{E})$, and we set $Op(\mathbb{E}) = \cup_a Op^a(\mathbb{E})$.

DEFINITION 2.4. *Given $s \in \Gamma(\mathbb{E}\otimes\mathbb{S})$, define $\bar{s}_m(x) = \alpha(m,x)\tau(m,x)s(m)$. Let $P \in Op(\mathbb{E})$ and $s \in \Gamma(\mathbb{E}\otimes S)$; define $\mu(P) \in End(\mathbb{E})_m \otimes End(\mathbb{S})_m$ by:*

$$\mu(P)(m,\xi)(s(m)) = P_y(e^{i\langle Exp_m^{-1}(y),\xi\rangle}\bar{s}_m(y))|_{y=m}. \tag{2.17}$$

Recall the map $\bar{\sigma} : Cliff(TM) \to \Omega(M)$ from equation 2.10. We define the s-symbol of P by:

$$\sigma(P)(m,\xi) = \bar{\sigma}(\mu(P)(m,\xi)). \tag{2.18}$$

Thus, $\sigma(P)(m,\xi)$ is a form-valued endomorphism of $\mathbb{E}_m$.

Remark. The two maps θ and σ are not inverses of each other; however, at the level of asymptotic operators, they will be inverses.

The following two examples will be key steps in obtaining the formula for the s-symbol of the composition of two operators.

Fix a point, $m \in M$, and choose an orthonormal basis, $X_1,\dots,X_n$, of $T_m(M)$; then we have normal coordinates, $(x_1,\dots,x_n) \to Exp_m(x_1X_1+\cdots+x_nX_n)$, in a neighboorhood of m. We also have that $(x_1,\dots,x_n)$ provide coordinates on $T_m(M)$. Let $(\xi_1,\dots,\xi_n)$ be dual coordinates to $(x_1,\dots,x_n)$ on $T_m^*(M)$, so $\xi = \sum \xi_i X_i^*$. Let $\mathbf{s} = (\mathbf{s}_1,\dots,\mathbf{s}_p)$ be a basis for $(\mathbb{E}\otimes\mathbb{S})_m$ and extend this to a sychronous frame around m. Thus, $\mathbf{s}(x)$ is parallel translation of $\mathbf{s}$ along the unique geodesic from m to x. If $s(m) = \sum s_i(m)\cdot\mathbf{s}_i(m)$, then $s_m(x) = \sum s_i(m)\cdot\mathbf{s}_i(x)$; that is, the coeffients of $\bar{s}_m(x)$ with respect to the frame $\mathbf{s}$ are constant.

Example 1.

Let $X = \sum c_i \frac{\partial}{\partial x_i}$ with $c_i \in \mathbb{R}$. (This is only valid in a neighborhood of m.) We can then compute:

$$\begin{aligned}\sigma(\nabla_X)(s)(m) &= \nabla_X(e^{i\langle Exp_m^{-1}(y),\xi\rangle}\bar{s}_m(y))|_{y=m} \\ &= i\langle X,\xi\rangle_m s(m) + \sum_{ij} s_i(m)\omega_i^j(X)_m(\mathbf{s}_j)_m\, .\end{aligned} \tag{2.19}$$

In a synchronous frame around m, $\omega_i^j(X)_m = 0$; so we have:

$$\sigma(\nabla_X)(m,\xi) = i\langle X,\xi\rangle_m. \tag{2.20}$$

Example 2.

Next, we let $X = \sum c_i \frac{\partial}{\partial x_i}$ and $Y = \sum d_i \frac{\partial}{\partial x_i}$ with $d_i, c_i \in \mathbb{R}$. We have:

$$\begin{aligned}&\sigma(\nabla_X\nabla_Y)(m,\xi)s(m) = \nabla_X\nabla_Y(e^{i\langle Exp_m^{-1}(y),\xi\rangle}\bar{s}_m(y))|_{y=m} \;= \\ &\nabla_X(i\langle Y.\xi\rangle e^{i\langle Exp_m^{-1}(y),\xi\rangle}\bar{s}_m(y) + e^{i\langle Exp_m^{-1}(y),\xi\rangle}(\nabla_Y\bar{s}_m)(y))|_{y=m} = \\ &(i\langle Y,\xi\rangle i\langle X,\xi\rangle e^{i\langle Exp_m^{-1}(y),\xi\rangle}\bar{s}_m(y) + i\langle Y,\xi\rangle e^{i\langle Exp_m^{-1}(y),\xi\rangle}\nabla_X\bar{s}_m(y) \\ &\quad + i\langle X,\xi\rangle e^{i\langle Exp_m^{-1}(y),\xi\rangle}(\nabla_Y\bar{s}_m(y) + e^{i\langle Exp_m^{-1}(y),\xi\rangle}\nabla_X\nabla_Y\bar{s}_m(y))|_{y=m}.\end{aligned} \tag{2.21}$$

As we saw above, $(\nabla_X\bar{s}_m(y))|_{y=m} = 0$, and it follows from appendix II of [**ABP**] that in this coordinate system, $\nabla_X\nabla_Y\bar{s}_m(y)|_{y=m} = \Omega(X,Y)s(m)$, where Ω is the curvature of the bundle $\mathbb{E}\otimes\mathbb{S}$. We can break Ω into two pieces, $\Omega_{\mathbb{E}} + \Omega_{\mathbb{S}}$. As we saw before, $\Omega_{\mathbb{S}}(X,Y) = \frac{1}{4}\sum R_{lij}^k c_i d_j e_k e_l$; thus we have:

$$\sigma(\nabla_X\nabla_Y)(m,\xi) = i\langle X,\xi\rangle i\langle Y,\xi\rangle + \frac{1}{4}\sum R_{lij}^k c_i d_j\, f_k\wedge f_l + \Omega_{\mathbb{E}}(X,Y). \tag{2.22}$$

Note that the first two terms are of order two while the last term, $\Omega_{\mathbb{E}}$, is of order zero.

Given two polynomial s-symbols, p, q, with associated differential operators, $\theta(p)$ and $\theta(q)$, we want to compute $\sigma(\theta(p)\circ\theta(q))$. The idea behind the computation is very simple. Recall that on the real line $\frac{d}{dx}$ generates a one-parameter group, $e^{t\frac{d}{dx}}$ and $(e^{t\frac{d}{dx}}\cdot f)(x) = f(t+x)$. Clearly, we have:

$$e^{t\frac{d}{dx}}(f\cdot g) = (e^{t\frac{d}{dx}}f)(e^{t\frac{d}{dx}}g). \tag{2.23}$$

There is an analogous construction when we have a vector bundle with a connection over a compact manifold. Suppose that $\mathbb{F}\to M$ is a vector bundle over M with connection ∇. If $X\in\Gamma(TM)$, denote by ψ_t^X the one-parameter group of diffeomorphisms generated by X. Given a point, $m\in M$, the map $t\to\psi_t^X(m)$ is a curve in M. Let $s\in\Gamma(\mathbb{F})$: then we have $e^{t\nabla_X}s$ is parallel translation of

s along the integral curve $t \to \psi_t^X(m)$. If we have $\phi \in C^\infty(M)$ and $s \in \Gamma(\mathbb{F})$, then:

$$e^{t\nabla_X}(\phi \cdot s)(m) = \phi(\psi_t^X(m))(e^{t\nabla_X}s)(m). \tag{2.24}$$

LEMMA 2.1. *In a normal coordinate system centered at m, let $X = \sum c_i \frac{\partial}{\partial x_i}$; then $\sigma(e^{t\nabla_X})(x,\xi) = e^{t\sigma(\nabla_X)(m,\xi)}(1 + r(t))$, where $r(t) \in \Omega(M)$. If we write $r(t) = \sum t^i r_i, r_i \in \Omega(M)$, then $deg(r_i) < i$.*

PROOF: This is just an application of Taylor's theorem with remainder as well as the Campbell-Baker-Hausdorff (CBH) formula. For $s \in \Gamma(\mathbb{E} \otimes \mathbb{S})$, we have:

$$\begin{aligned}\sigma(e^{t\nabla_X})(m,\xi)s(m) &= e^{t\nabla_X}(e^{i\langle Exp_m^{-1}(y),\xi\rangle} \cdot \bar{s}_m(y))|_{y=m} = \\ & e^{i\langle Exp_m^{-1}(\psi_t^X(y)),\xi\rangle} \cdot e^{t\nabla_X}(\bar{s}_m(y))|_{y=m} = \\ & e^{it\langle X,\xi\rangle_m}(\bar{s}_m + t\nabla_X\bar{s}_m + \frac{t^2}{2!}\nabla_X^2\bar{s}_m + r(t))|_{y=m}.\end{aligned} \tag{2.25}$$

Now we saw before that $\nabla_X\bar{s}_m|_{y=m} = 0$ and $(\nabla_X^2\bar{s}_m)|_{y=m} = \Omega(X,X)s(m)$; but Ω is skew so we have $\Omega(X,X) = 0$. One should note that $r(t)$ has no ξ dependence, $r(t) \in \Gamma(\bigwedge^{even}(M) \otimes End(\mathbb{E}))$.

Next, we need to recall the CBH formula. This formula says that if $c_i(A,B)$ are defined by:

$$exp(A)exp(B) = exp\left(\sum_{j=1}^{\infty} c_j(A,B)\right), \tag{2.26}$$

then if $c_1(A,B) = A + B$, the c_j's are uniquely determined by the following recursion formula:

$$\begin{aligned}(n+1)c_{n+1}(A:B) &= \frac{1}{2}[A - B, c_n(A,B)] + \\ \sum_{p\geq 1, 2p\leq n} K_{2p} & \sum_{\substack{k_1,\ldots,k_{2p}>0 \\ k_1+\cdots+k_{2p}=n}} [c_{k_1}(A,B),[\cdots[c_{k_{2p}}(A,B), A+B]\cdots]\cdot\end{aligned} \tag{2.27}$$

Let $\mathbf{s}$ be the synchronous frame for $\mathbb{E}\otimes\mathbb{S}$ associated with our coordinate system. Thus we have $\bar{s}_m(y) = \sum s_i(m)\mathbf{s}_i(y)$, the point being that the coefficients in this frame for $\bar{s}_m$ are constants.

In this frame we have $\nabla_X = \partial_X + \omega(X)$, where $\omega(X)$ is the connection one-form. Thus we have:

$$e^{t\nabla_X} = e^{t\partial_X + t\omega(X)} = e^{t(\partial_X+\omega(X))}e^{-t\partial_X}e^{-t\omega(X)}e^{t\omega(X)}e^{t\partial_X}. \tag{2.28}$$

(1) Set $A = \partial_X$;
(2) Set $B = \omega(X)$.

If we apply the CBH formula to equation 2.28, we get:

$$e^{t(\partial_X+\omega(X))} = e^{t\omega(X)+\{tA,tB\}}e^{-t\omega(X)}e^{t\omega(X)}e^{t\partial_X}. \tag{2.29}$$

The term in the brackets, $\{tA,\, tB\}$, is of the form $[E_1[E_2,\ldots,[E_l,F],\ldots,]]$, where the E_i's are either tA or tB. Now tB is in $so(n)$ and $[tA,\, tB] = t^2\partial_X(B) \in so(n)$. Since $so(n)$ is closed under brackets, we see that $\{tA, tB\}$ is an $so(n)$-valued section of $Cliff$; hence, as a differential operator, it is of degree two. Now assign to t the weight -1 and assign to a Clifford operator the weight that is its degree. We see then that the weight of the term $\{tA,\, tB\}$ is less than or equal to zero. If we apply CBH once more, we have:

$$e^{t(\partial_X+\omega(X))} = e^{G(tA,tB)}e^{t\omega(X)}e^{t\partial_X}. \tag{2.30}$$

Again, the explicit form of the CBH formula tells us that now the weight of $G(tA, tB)$ is strictly less than zero. Finally, if we evaluate this at the point m, the term $e^{t\omega(X)}|_m = 1$, since $\omega(X)|m = 0$. Thus we see that:

$$\left(e^{t\nabla_X}\bar{s}_m\right)|_{y=m} = e^{G(tA,tB)}\bar{s}(m). \tag{2.31}$$

Since the weight of $G(tA,\, tB)$ is strictly less than zero, we have that the weight of $e^{G(tA,tB)}$ is also strictly less than zero. If we write $e^{G(tA,tB)} = \sum t^i r_i$, then this says that $deg(r_i) < i$.

One should also note that $\psi_t^X(m) = exp_m(tX)$ for small t, since solving the flow equation is purely a local matter. ∎

COROLLARY 2.2. *Let $m \in M$ and X be the vector field defined in a neighborhood of m as in Lemma 2.1. Then in that neighborhood, we have:*

$$\sigma(\nabla_X^n) = (i\langle X,\xi\rangle)^n + p(m,\xi) \tag{2.32}$$

where $p(m,\xi) = \sum_{i\geq 1}\omega^i \otimes p_i(m,\xi)$ and $p_i \in \Sigma^{n-i}(\mathbb{E})$.

PROOF: One has that $\sigma(\nabla_X^n)$ is equal to $n!$ times the coefficient of t^n in the expansion of $\sigma(e^{t\nabla_X})$. This now follows from lemma 2.1. ∎

Recall that an s-symbol is a section of the vector bundle over $T^*(M)$ defined by
$\pi^*(\Omega(M) \otimes Hom(\mathbb{E}))$ with certain growth properties. In [**G**] a proof of the following theorem was outlined. The following proof is similar to the one in [**G**] but relies on Taylor's theorem instead of the algebra of formal symbols.

Let $\mathbb{L}$ be the bundle over $T^*(M)$ given by:

$$\mathbb{L} = \pi^*(End(\mathbb{E}) \otimes Cliff(T^*(M))). \tag{2.33}$$

The natural map $c : Cliff(T^*(M)) \otimes Cliff(T^*(M)) \to Cliff(T^*(M))$ given by the Clifford product induces a map from $\mathbb{L} \otimes \mathbb{L} \to \mathbb{L}$, which we will also call c.

THEOREM 2.1. *Let $p \in S\Sigma^l(\mathbb{E})$ and $q \in S\Sigma^k(\mathbb{E})$ be two symbols whose dependence on the ξ variable is polynomial. There exist differential operators $\bar{a}_n : \Gamma(\mathbb{L} \otimes \mathbb{L}) \to \Gamma(\mathbb{L})$ such that if $a_n(p,q) = \sigma(\bar{a}_n(p,q))$, then:*

(1) $a_n(p,q) \in S\Sigma^{l+k-n}(\mathbb{E})$;
(2) $p \circ q = \sum_{n=0}^{\infty} a_n(p,q)$ *(finite sum); and*
(3) $a_0(p,q) = e^{-\frac{1}{4}R(\frac{\partial}{\partial \xi},\frac{\partial}{\partial \eta})} p(x,\xi) \wedge q(x,\eta)|_{\xi=\eta}$.

We have used the notation:

$$-\frac{1}{4}R(\frac{\partial}{\partial \xi},\frac{\partial}{\partial \eta})(p,q) = -\frac{1}{4}\sum_{ijkl} R^k_{lij}\, f_k \wedge f_l \wedge \frac{\partial p}{\partial \xi_i}(x,\xi) \wedge \frac{\partial q}{\partial \eta_j}(x,\eta). \tag{2.34}$$

Remark. There are two important parts to this theorem. First, the order of the symbol of the composition of two operators is less than or equal to the sum of the orders of the operators. While this is true for ordinary differential operators, the presence of the Clifford variables makes this far from trivial for supersymmetric differential operators. It will be this fact that is crucial in developing the calculus of asymptotic operators. The second part will be the formula for $a_0(p,q)$.

PROOF: We will work locally; thus let $m \in M$ and let X and Y be the vector fields defined in a neighborhood of m as in example 2. Let $\wp^n_X = \sigma(\nabla^n_X)$; then by corollary 2.2 we have $\wp^n_X = (i\langle X,\xi\rangle)^n$ plus lower order terms in ξ. An easy induction shows that if $p \in S\Sigma(\mathbb{E})$ is an s-symbol with polynomial dependence in ξ, then we can write p as :

$$p(x,\xi) = \sum_i h_i(x)\wp^{n_i}_{X_i}, \quad h_i \in End(\mathbb{E}) \otimes \Omega(M). \tag{2.35}$$

Since the map $(p,q) \to \sigma(\theta(p)\circ\theta(q))$ is bilinear, it suffices to prove the theorem for symbols of the form $h(x)\wp^n_X$.

Let $h, g \in End(\mathbb{E}) \otimes \Omega(M)$; then we have:

$$\begin{aligned} &\sigma(\theta(h(\wp_X)^n\theta(g(\wp_Y)^m)) = \sigma(\bar\theta(h)(\nabla_X)^n\bar\theta(g)(\nabla_Y)^m) = \\ &\sigma(\bar\theta(h)\left(\sum_k \bar\theta(\nabla^{n-k}_X)(g)(\nabla^k_X\nabla^m_Y)\right)) = \\ &\sum_k \sigma(\bar\theta(h)\bar\theta(\nabla^{n-k}_X g))\sigma\left((\nabla_X)^k(\nabla_Y)^m\right) \quad \text{mod lower order terms.} \end{aligned} \tag{2.36}$$

In the last line we have used equation 1.15. Thus, we are reduced to proving the theorem for the operators $(\nabla_X)^n$ and $(\nabla_Y)^m$.

Next, we need to recall the Campbell-Baker-Hausdorff formula again.

Note that it follows from 2.27 that $c_n(tA,tB) = t^n c_n(A,B)$, so $c_n(A,B)$ is a homogeneous polynomial in A and B of degree n. If we apply 2.26 with $A = t\nabla_X$ and $B = s\nabla_Y$, then we have:

(1) $c_1(t\nabla_X, s\nabla_Y) = t\nabla_X + s\nabla_Y$; and
(2) $c_2(t\nabla_X, s\nabla_Y) = \frac{1}{2}ts[\nabla_X,\nabla_Y] = \frac{1}{2}ts\Omega(X,Y)$.

Recall that $\Omega = \Omega_{\mathbb{S}} + \Omega_{\mathbb{E}}$, where $\Omega_{\mathbb{S}}$ is the curvature of the spin bundle and $\Omega_{\mathbb{E}}$ is the curvature of $\nabla_{\mathbb{E}}$. As a ψDO, $\Omega_{\mathbb{S}}(X,Y)$ is of order two, while $\Omega_{\mathbb{E}}(X,Y)$ is of order zero.

Suppose that ϕ is a section of $Cliff(T^*(M))$ such that $\phi(x) \in so(n) \subset Cliff(T^*(M)_x)$ for all $x \in M$; then $[\nabla_X, \phi] = \nabla_X(\phi) \in so(n)$. Since $so(n)$ is closed under brackets, we conclude from the Campbell-Baker-Hausdorff formula, (2.32), that $c_j(\nabla_X, \nabla_Y)$ is of order two as a ψDO for $j \geq 2$.

Remark. It will be important to notice that as a ψDO, the operators $c_j(X,Y)$ contain NO differentiation; they are purely Clifford variable operators. This follows from the explicit form of the CBH formula.

In what follows, associate with t and s weight -1 and associate with ξ and e weight 1. We have that $\sigma((t\nabla_X)^a \circ (s\nabla_Y)^b)$ is the coefficient of $t^a s^b$ in the expression $\sigma(e^{t\nabla_X} \circ e^{s\nabla_Y})$. We need to show that the weight of each term of the coefficient of $t^a s^b$ is less than or equal to zero and to determine the term of weight zero. Applying the CBH formula we have:

$$\begin{aligned}\sigma(e^{t\nabla_X} \circ e^{s\nabla_Y}) &\equiv \sigma(e^{t\nabla_X + s\nabla_Y + \frac{1}{2}st\Omega(X,Y) + C(X,Y)}) \equiv \\ &\sigma(e^{t\nabla_X + s\nabla_Y + \frac{1}{2}\Omega_{\mathbb{S}}(X,Y) + C(X,Y)}).\end{aligned} \tag{2.37}$$

Define:

(1) $A = t\nabla_X + s\nabla_Y + \frac{1}{2}st\Omega_{\mathbb{S}}(X,Y)$ — this is weight 0;
(2) $B = C(X,Y)$— this is weight strictly less than 0.

We have:

$$e^{A+B} = e^{A+B}e^{-A}\,e^{A} = (e^{B+\{A,B\}})\,e^{A} \quad \text{by the CBH formula.} \tag{2.38}$$

We see from CBH that if the weight of B is less than zero, then the weight of the term$\{A,B\}$ is also less than zero; also the operator $e^{B+\{A,B\}}$ is purely a Clifford operator. Since this term is a Clifford operator, we can use equation 1.15 again to conclude that:

$$\sigma(e^{B+\{A,B\}}\,e^{A}) = \sigma(e^{B+\{A,B\}}) \wedge \sigma(e^{A}) \quad \text{mod lower order terms.} \tag{2.39}$$

Now $\sigma(e^{B+\{A,B\}})$ is of the form $1+f$, where f has weight strictly less than zero. Thus to finish the theorem, we need to show that $\sigma(e^A)$ has the desired form and compute the term of weight 0. To do this, we use the CBH formula again.

$$\begin{aligned}\sigma(e^{t\nabla_X + s\nabla_Y + \frac{1}{2}st\Omega_{\mathbb{S}}(X,Y)}) &= \sigma(e^{\frac{1}{2}ts\Omega_{\mathbb{S}}(X,Y)}e^{t\nabla_X + s\nabla_Y}) \\ &\text{mod Clifford terms of negitive weight .}\end{aligned} \tag{2.40}$$

Finally, using lemma 2.1, we have equation 2.40 becomes:

$$\begin{aligned}\sigma(e^{t\nabla_X + s\nabla_Y + \frac{1}{2}st\Omega_{\mathbb{S}}(X,Y)}) &= e^{\frac{1}{2}ts\Omega_{\mathbb{S}}(X,Y)}e^{it\langle X,\xi\rangle}e^{is\langle Y,\xi\rangle} \\ &\text{mod terms of negative weight .}\end{aligned} \tag{2.41}$$

Replacing X by iX, Y by iY and equating the powers of t and s, we obtain the conclusion of the theorem. ▮

3. Asymptotic Pseudodifferential Operators and their Calculus.

Widom developed a theory of asymptotics for pseudodifferential operators. If $p(x,\xi) \in \Sigma^m(\mathbb{E})$ is an ordinary symbol, Widom considered the family of pseudodifferential operators defined by $P_t = \theta(p(x,t\xi))$. When one considers this family instead of just the single operator P, one finds that the asymptotic expansions normally associated with heat kernels (as well as other expansions) follow naturally from the calculus that he developed. In our setting, we want to extend Widom's ideas to the class of s-symbols. The s stands for supersymmetric, and in supersymmetry Clifford multiplication is viewed as a differential operator.

Let $p \in S\Sigma^m(\mathbb{E})$, so that $p(x,\xi) = \sum_{i=0}^m p_i(x,\xi) \otimes \omega_i$, with $p_i \in \Sigma^{m-i}(\mathbb{E})$ and $\omega_i \in \Omega^i(M)$. We have the following definition, which was introduced in [**G**].

DEFINITION 3.1. *If $p(x,\xi) \in S\Sigma^m(\mathbb{E})$ then:*

$$p_t(x,\xi) = \sum_{i=0}^{m} p_i(x,t\xi) \otimes t^i\omega_i. \tag{3.1}$$

Next we need to recall the classical notion of an asymptotic expansion.

DEFINITION 3.2. *Let $p(x,\xi,t) \in S\Sigma^m(\mathbb{E})$ be a family of symbols; then $p(x,\xi,t)$ is called an asymptotic family if there exist symbols $p_n \in S\Sigma^{m-n}(\mathbb{E})$ such that the following asymptotic expansion holds:*

$$p(x,\xi,t) \sim \sum_{n=0}^{\infty} t^n p_n(x,\xi). \tag{3.2}$$

In other words, given $N > 0$, we have (recalling that the symbol spaces are Fréchet spaces):

$$\lim_{t\to 0} t^{-N}\left(p(x,\xi,t) - \sum_{i=0}^{N} t^i p_i(x,\xi)\right) = 0 \quad in S\Sigma^{m-n}(\mathbb{E}). \tag{3.3}$$

We will call the first term, $p_0(x,\xi)$, the leading symbol of p (not to be confused with the principal symbol).

As usual, the notion of an asymptotic expansion gives us an equivalence relation on the space of asymptotic families of symbols. This leads to the following definition:

DEFINITION 3.3. *Given an asymptotic family of symbols, $p(x,\xi,t)$, we call the equivalence class that p_t determines an asymptotic symbol. We will call a representative of the equivalence class an asymptotic symbol if no confusion is likely to arise.*

Let $H^s(\mathbb{E})$ be the s-th Sobolev space formed from the sections of $\mathbb{E} \otimes \mathbb{S}$, so $H^s(\mathbb{E})$ is the completion of $\Gamma(\mathbb{E} \otimes \mathbb{S})$ (see [**Gi**]). If P is an infinitely smoothing operator, then $P : H^s(\mathbb{E}) \to H^0(\mathbb{E})$, and we denote by $\|P\|_s$ the corresponding norm. This turns the space of smoothing operators into a Fréchet space.

Definition 3.4. *Let P_t be a family of smoothing operators; then P_t is called asymptotically zero if $P_t \sim 0$ in the Fréchet space topology. Thus, given $N > 0$, we have for all s:*

$$\lim_{t \to 0} t^{-N} \|P_t\|_s = 0. \tag{3.4}$$

Definition 3.5. *Given two families of operators, P_t and Q_t, we will say that P_t is equivalent to Q_t if the difference $P_t - Q_t$ is asymptotically zero. If P_t is equivalent to $\theta(p(x,\xi,t)_t)$ for p an asymptotic symbol, then we will call the equivalence class that P_t determines an asymptotic pseudodifferential operator, $A\psi DO$. As before, we will call a representative of this class an $A\psi DO$ if no confusion is likely to arise.*

Remark. Since the map θ depends on the choice of a cut-off function α, we should write $A\psi DO^\alpha$ for the corresponding collection of operators. However, the next lemma shows that this collection is independent of the choice of α .

Lemma 3.6. *Let $\phi \in C^\infty(M \times M)$ which is identically zero in a neighborhood of the diagonal $\Delta \subset M \times M$ and whose support is contained in the neighborhood of Δ where the exponential map is a diffeomorphism. Given any asymptotic symbol, $p(x,\xi,t)$, define $\bar{P}_t$ by:*

$$\bar{P}_t s(m) = \iint e^{i\langle X_m,\xi_m\rangle} \bar{\theta}(p(m,\xi,t)_t)\phi(m, Exp_m(X))\hat{s}_m(X_m)\, dX_m\, d\xi_m. \tag{3.5}$$

Then $\bar{P}_t$ is an asymptotically zero operator.

Proof: This is similar to lemma 1.2.6 of [**Gi**]. We can assume that $p(x,\xi,t) \in \Sigma^k(\mathbb{E})$, since p will be a finite sum of elements of this form times a form. Define a Laplican by $\Delta_\xi = \sum_j (i\frac{\partial}{\partial \xi_j})^2$; then $\Delta_\xi e^{i\langle X_m,\xi_m\rangle} = |X_m|^2 e^{i\langle X_m,\xi_m\rangle}$. Since $|x_m|^{-2l}\phi(m, X_m) \in C^\infty(T(M))$ for all l, we can integrate by parts to get:

$$\bar{P}_t s(m) = \iint e^{i\langle X_m,\xi\rangle} |X_m|^{-2l} \phi(m, X_m)\bar{\theta}(\Delta_\xi^l p(m, t\xi, t))\hat{s}_m(X_m)\, dX_m\, d\xi_m. \tag{3.6}$$

Now $\Delta_\xi^l p(x, t\xi, t) = t^{2l}(\Delta_\xi^l p)(x, t\xi, t) \in \Sigma^{k-2l}(\mathbb{E})$. Note that we get powers of t coming out from the repeated integration by parts. This is sufficient to prove the lemma. ∎

Corollary 3.7. *Let $p(x,\xi,t)$ be an asymptotic symbol; then the asymptotic operator $P_t = \theta(p(x,\xi,t)_t)$ is independent of the choice of the cut-off function α that was chosen to define θ.*

Proof: This follows from lemma 3.6 . ∎

Lemma 3.8. *Let $r(m, \xi_m, y, t) \in C^\infty(M, S\Sigma^d(\mathbb{E}))$ such that for all α, β, γ we have :*

$$\|\partial_m^\alpha \partial_\xi^\beta \partial_y^\gamma r(m, \xi_m, y, t)\| \leq C_{\alpha,\beta,\gamma}(1 + |\xi_m|)^{d-|\beta|}$$

where $C_{\alpha,\beta,\gamma}$ is independent of t. We denote the space of such functions by $\mathcal{S}$, which is a Fréchet space with the semi-norms:

$$\rho_{\alpha,\beta,\gamma}(r) = inf\{C_{\alpha,\beta,\gamma} | \quad \| \partial_m^\alpha \partial_\xi^\beta \partial_y^\gamma r(m, \xi_m, y, t)\| \leq C_{\alpha,\beta,\gamma}(1 + |\xi_m|)^{d-|\beta|} \}.$$

Furthermore, assume that $r(m, \xi, y, t)$ has an asymptotic expansion:

$$r(m, \xi, y, t) \sim \sum_{n=0}^{\infty} t^n r_n(m, \xi, y).$$

If we define:

$$R_t(s)(m) = \int e^{-i\langle X_m, \xi_m \rangle} r(m, \xi_m, exp_m(X_m), t)_t \hat{s}(X_m)\, dX_m\, d\xi_m$$

then R_t is an $A\psi DO$, and if $R_t \sim \theta(p_t)$ we have:

$$p(m, \xi_m, t) \sim \sum_{\alpha \geq 0} \frac{1}{\alpha!} \partial_{X_m}^\alpha \partial_{\xi_m}^\alpha r(m, \xi_m, exp_m(X_m), t)|_{X_m=0}.$$

PROOF: We will write $r(m, \xi_m, X_m, t)$ for $r(m, \xi_m, exp_m(X_m), t)$. In what follows we will assume that r has no form component. Note that since $\hat{s}(X_m)$ vanishes off a neighborhood of $\Delta \subset M \times M$, R_t depends only on $r(m, \xi_m, X_m, t)$ for X_m in a neighborhood of zero. Thus we will suppose WLG that $r(m, \xi_m, X_m, t)$ has compact support in the X_m variable. We will write

$$\tilde{s}(\xi_m) = \int e^{-i\langle X_m, \xi_m \rangle} \hat{s}(X_m)\, dX_m \quad \text{and}$$
$$q(m, \xi_m, \eta_m, t) = \int e^{-i\langle X_m, \eta_m \rangle} \hat{s}(X_m)\, dX_m.$$

Using the fact that the Fourier transform of the product is the convolution of the Fourier transforms, we get:

$$(3.7) \qquad \begin{aligned} (R_t s)(m) = \int e^{-i\langle X_m, \xi_m \rangle} r(m, t\xi_m, X_m, t) \hat{s}(X_m)\, dX_m\, d\xi_m = \\ \int q(m, t\xi_m, \xi_m - \eta_m, t) \tilde{s}(\eta_m)\, d\eta_m\, d\xi_m. \end{aligned}$$

We estimate $|q(m, t\xi_m, \xi_m - \eta_m, t)|$ in the usual way. The estimates on r and the fact that $q(m, \xi, \eta, t)$ is rapidly decreasing in the eta variable yields:

$$(3.8) \qquad |q(m, t\xi_m, \xi_m - \eta_m, t)| \leq C_k(1 + |t\xi|)^d (1 + |\xi - \eta|)^{-k}.$$

Now Peetre's inequality says: $(1 + |x + y|)^s \leq (1 + |x|)^s (1 + |y|)^{|s|}$. Using this inequality and the fact that for all l $|\tilde{s}(\eta)| \leq C_l(1 + |\eta|)^{-l}$, we get that

$|q(m, t\xi_m, \xi_m - \eta_m, t)\tilde{s}(\eta_m)|$ is integrable; thus we can interchange the order of integration . Define p by:

$$p(m, t\eta_m, t) = \int q(m, t\xi_m, \xi_m - \eta_m, t)\, d\xi_m = \int q(m, t\xi_m + t\eta_m, \xi_m, t)\, d\xi_m. \tag{3.9}$$

Then we have :

$$R_t(s)(m) = \int e^{-i\langle X_m, \xi_m\rangle} p(m, t\eta_m, t)\hat{s}(X_m)\, dX_m\, d\xi)_m. \tag{3.10}$$

Thus we only need to check that $p(m, \eta_m, t)$ is a symbol of order d and that it has the requisite asymptotic expansion.

Define a linear map $T : \mathcal{S} \to S\Sigma(\mathsf{E})$ by $T(r) = p$, where p is defined as above. We have that T is continuous from the Fréchet space $\mathcal{S}$ to the Fréchet space $S\Sigma(\mathsf{E})$. We have:

$$\begin{aligned}
&|\ \partial_m^\alpha \partial_{\eta_m}^\beta q(m, t\xi_m + \eta_m, \xi_m, t)\ | = \\
&|\int e^{-i\langle X_m, \xi_m\rangle} \partial_m^\alpha \partial_{\eta_m}^\beta r(m, t\xi_m + \eta_m, X_m, t)\, dX_m\ | = \\
&|\int e^{-i\langle X_m, \xi_m\rangle} \xi_m^{-\gamma} \partial_m^\alpha \partial_{\eta_m}^\beta \partial_{X_m}^\gamma r(m, t\xi_m + \eta_m, \xi_m, t)\, dX_m \leq \\
&KC_{\alpha,\beta,\gamma}(1 + |\xi_m|)^{-|\gamma|}(1 + |t\xi_m + \eta_m|)^{d-|\beta|} \leq \\
&KC_{\alpha,\beta,\gamma}(1 + |\xi_m|)^{-|\gamma|}(1 + |t\xi_m|)^{d-|\beta|}(1 + |\eta_m|)^{d-|\beta|}.
\end{aligned} \tag{3.11}$$

The constant K is independent of r and bounds the volume of the compact set that $r(m, \xi, X_m, t)$ has for its X_m support. If $|\gamma|$ sufficiently large we can integrate the above inequality to get:

$$\rho_{\alpha,\beta}(T(r) \leq K\rho_{\alpha,\beta,\gamma}(r)). \tag{3.12}$$

This shows that $T(r)$ is a symbol and that if $r(m, \xi, X_m, t)$ has an asymptotic expansion, then, $T(r)$ also has an asymptotic expansion.

Using Taylor's formula we have:

$$q(m, t\xi_m + \eta_m, t) \sim \sum_{\alpha \geq 0} \frac{1}{\alpha!} \frac{1}{t^{|\alpha|}} \partial_{\xi_m}^\alpha q(m, \eta_m, t)(t\xi_m)^\alpha. \tag{3.13}$$

Thus we have:

$$\begin{aligned}
p(m, \eta_m, t) &\sim \sum_{\alpha \geq 0} \frac{1}{\alpha!} \iint e^{-i\langle X_m, \xi_m\rangle} \xi_m^\alpha \partial_{\xi_m}^\alpha r(m, \eta_m, X_m, t)\, dX_m\, d\xi_m \sim \\
&\sum_{\alpha \geq 0} \frac{1}{\alpha!} \partial_{X_m}^\alpha \partial_{\xi_m}^\alpha r(m, \eta_m, X_m, t)|_{X_m = 0}.
\end{aligned} \tag{3.14}$$

∎

LEMMA 3.9. *If $P_t \sim \theta(p_t)$ is an $A\psi DO$, then the adjoint, P_t^*, is also an $A\psi DO$.*

PROOF: Recall that $\hat{s}_m(y)$ is parallel translation of $s(y)$ from $(\mathbb{E}\otimes\mathbb{S})_y$ to $(\mathbb{E}\otimes\mathbb{S})_m$times $\phi(m,y)$ where $\phi(m,y)$ is zero off of a neighborhood of $\Delta\subset M\times M$, where the exponential map is a diffeomorphism. Let s,h be sections and set $y=exp_m(X_m)$; then we have :

(3.15)

$$\langle P_t s,h\rangle = \int e^{-i\langle X_m,\xi_m\rangle}\langle p(m,t\xi_m,t)\hat{s}(X_m),h(m)\rangle\, dX_m\, d\xi_m\, dvol(m) =$$

$$\int e^{-i\langle X_m,\xi_m\rangle}\langle \phi(m,y)\tau(m,y)s(y),p(m,t\xi_m,t)^*h(m)\rangle\, dX_m\, d\xi_m\, dvol(m).$$

Recall that parallel translation is an isometry. Thus we have:

(3.16)

$$\langle \tau(m,y)s_m(y),p(m,t\xi_m,t)^*h(m)\rangle|_m = \langle s(y),\tau(y,m)\,(p(m,t\xi_m,t)^*h(m))\rangle|_y =$$

$$\langle s(y),(\tau(y,m)p(m,t\xi_m,t)^*)\,(\tau(y,m)h(m))\rangle|_y =$$

$$\langle s(y),(\tau(y,m)p(m,t\xi_m,t)^*)\,\hat{s}(X_y)\rangle|_y.$$

Let $dX_m = J(m,y)dvol(y)$, and $dvol(m) = J^{-1}(y,m)dX_y$; then we have $dX_m\, dvol(m) = a(m.y)dvol(y)\, dX_y$, where the function $a(m,y)$ has all of its derivatives bounded on the support of $\phi(m,y)$.

Now $exp_m(sX_m)=\gamma(s)$ is the unique geodesic with $\gamma(0)=m$ and $\gamma(1)=y$, thus $\gamma(1-s)$ is the unique geodesic from y to m. Since $\gamma\prime(0) = -\tau(y,m)(X_m)$, we see that $exp_y^{-1}(m) = -\tau(y,m)(X_m)$. Now $\langle X_m,\xi_m\rangle = \langle \tau(y,m)X_m,\tau(y,m)\xi_m\rangle$, so 3.15 equals:

$$\int e^{-i\langle X_y,\xi_y\rangle}\phi(m,y)a(m,y) \tag{3.17}$$

$$\langle s(y),\tau(y,m)p(m,\tau(m,y)t\xi_y,t)^*\hat{h}_y(X_y)\rangle\, dX_y\, d\xi_y\, dvol(y).$$

If we set $r(y,\xi_y,X_y,t) = \phi(m,y)a(m,y)\tau(y,m)p(m,\tau(m,y)t\xi_y,t)^* = \phi(m,y)a(m,y)\tau(y,exp_y(X_y))p(exp_y(X_y),\tau(m,y)t\xi_y,t)^*$, then we see that :

$$(P_t^*h)(y) = \int e^{-i\langle X_y,\xi_y\rangle}r(y,t\xi_y,X_y,t)\hat{h}(X_y)dX_y\, d\xi_y. \tag{3.18}$$

Once we show that r satisfies the conditions of lemma 3.8 we are done. ∎

LEMMA 3.9. *Suppose that $r(m,\xi_m,t)$ is a family of symbols in $S\Sigma^n(\mathbb{E})$ such that the symbol estimates hold uniformly in t. If $R_t=\theta(r_t)$, then $\sigma(R_t)_{t^{-1}}(m,\xi_m)\sim r(m,\xi_m,t)$.*

(It should be noted that we are NOT assuming that r is an asymptotic symbol.)

PROOF: Without loss of generality, we can assume that r has no form component. Now we compute, If $y = Exp_m(X_m)$, then we have:

$$\begin{aligned}
&\sigma(R_t)(\xi_m/t, m)s(m) = R_t\left(e^{-i\langle Exp_m^{-1}(y),\xi_m/t\rangle}\bar{s}(y)\right)|_{y=m} = \\
&\left(\int e^{i\langle X_m,\eta_m\rangle}e^{-i\langle X_m,\xi_m/t\rangle}r(m,t\eta_m,t)\phi(m,X_m)\,dX_m\,d\eta_m\right)s(m) = \\
&\left(\int e^{i\langle X_m,\eta_m\rangle}r(m,t\eta_m+\xi_m,t)\phi(m,X_m)\,dX_m\,d\xi_m\right)s(m).
\end{aligned} \tag{3.19}$$

Now we use Taylor's theorem on $r(m, t\eta_m + \xi_m, t)$ to get:

$$\begin{aligned}
&r(m,t\eta_m+\xi_m,t) = \sum_{|\alpha|<m}\frac{t^{|\alpha|}}{\alpha!}\partial_\eta^\alpha r(m,\xi_m,t) + E \\
&\text{where } E = \sum_{|\alpha|=m}\frac{m}{\alpha!}(t\alpha)^m\int_0^1 u^{m-1}\partial_\xi^\alpha r(m,\xi+(1-u)t\eta)\,du.
\end{aligned} \tag{3.20}$$

Recall that the function $\phi(m, X_m)$ is identically one in a neighborhood of zero; so we have:

$$\begin{aligned}
&\frac{t^{|\alpha|}}{\alpha!}\int e^{i\langle X_m,\eta_m\rangle}\eta_m^\alpha\partial_\eta^\alpha r(m,\xi_m,t)\,dX_m\,d\eta_m = \\
&\frac{t^{|\alpha|}}{\alpha!}\partial_\eta^\alpha r(m,\xi_m,t)\partial_{X_m}\phi(m,X_m)|_{X_m=0} = \begin{cases} r(m,\xi_m,t) \text{ if } \alpha=0 \\ 0 \text{ if } \alpha\neq 0\end{cases}.
\end{aligned} \tag{3.21}$$

To finish, we have to estimate the remainder term. First we note:

$$\begin{aligned}
&\|\ \partial_\xi^\alpha r(m,\xi+(1-u)t\eta,t)\ \| \le C(1+|\xi+(1-u)t\eta|)^{n-m} \le \\
&C(1+|\xi|)^{n-m}(1+|(1-u)t\eta|)^{|n-m|} \le C(1+|\xi|)^{n-m}(1+|\eta|)^{|n-m|}.
\end{aligned} \tag{3.22}$$

Therefore, integrating this estimate and remembering that $\int e^{i\langle X_m,\eta_m\rangle}\phi(m,X_m)\,dX_m$ is rapidly decreasing in η_m yields:

(3.23)

$$\begin{aligned}
&\|\ \int e^{i\langle X_m,\eta_m\rangle}E(m,\xi_m,\eta_m,t)\phi(m,X_m)\,dX_m\,d\eta_m\ \| \le \\
&Ct^m(1+|\xi_m|)^{n-m}\int\left|\int e^{i\langle X_m,\eta_m\rangle}\phi(m,X_m)\,dX_m\right|(1+|\eta_m|)^{|n-m|}\,d\eta_m \le \\
&Ct^m(1+|\xi_m|)^{n-m}
\end{aligned}$$

One can similarly estimate the derivatives of the error term. We conclude that $\forall m$:

$$\|\ r(m,\xi_m,t) - \sigma(R_t)_{t^{-1}}(m,\xi_m)\ \| \le Ct^m(1+|\xi_m|)^{n-m}. \tag{3.24}$$

∎

LEMMA 3.10. *Let $p(m,\xi,t)$ and $q(m,\xi,t)$ be two asymptotic symbols and let $P_t = \theta(p_t)$ and $Q_t = \theta(q_t)$ be the respective pseudodifferential operators; then $P_t \circ Q_t = \theta(r_t)$, where r is also an asymptotic symbol. Furthermore, the leading symbol of $P_t \circ Q_t$ is given by $a_0(p_0, q_0)$, where p_0 and q_0 are the leading symbols of p and q respectively.*

PROOF: We recall how $\theta(p_t)$ was defined. Let $\alpha(m,y)$ be a smooth function defined in a neighborhood of the diagonal $\Delta \subset M \times M$, where the exponential map is a diffeomorphism. Let $\tau(m,y)$ be parallel translation along the unique geodesic from m to y for $(m,y) \in supp(\alpha)$. For a section s we defined $\hat{s}_m(y) = \alpha(m,y)\tau(m,y)s(y)$. If $y = exp_m(X_m)$, then we wrote $\hat{s}(X_m)$ for $\hat{s}_m(y)$. The point to remember is that $\hat{s}(X_m)$ vanishes outside a neighborhood of zero.

In what follows we will omit the cut-off function α from most of the calculations to make the notation clearer. Now we compute. Let $y = exp_m(X_m)$ and $z = exp_y(X_y)$; then :

(3.25)

$P_t \circ Q_t(s)(m) =$

$$\int_{T^*_m}\int_{T_m} e^{i\langle X_m,\xi_m\rangle} p(m,t\xi_m)\tau^y_m \left(\int_{T^*_y}\int_{T_y} e^{i\langle X_y,\eta_y\rangle} q(y,t\eta_y)\tau^z_y s(z)\,dX_y\,d\eta_y\right) dX_m\,d\xi_m.$$

Let Y_m be the parallel translate along the geodesic from y to m of X_y and ζ_m the parallel translate of η_y along the geodesic from y to m. Since parallel translation is an isometry, we have that $\langle X_y, \eta_y\rangle_y = \langle Y_m, \zeta_m\rangle_m$, and the measures are also the same. Thus (3.21) becomes:

(3.26)

$$\int e^{i\langle Y_m,\zeta_m\rangle}\left(\int e^{i\langle X_m.\xi_m\rangle} p(m,t\xi_m)\tau^y_m q(y,t\tau^m_y(\zeta_m))\tau^y_m\tau^z_y\tau^m_z\,dX_m\,d\xi_m\right)$$
$$\tau^z_m s(exp_m(\phi(Y_m)))\,dY_m\,d\zeta_m$$

In the above equation we have written, $\phi(Y_m) = exp_m^{-1}(exp_y(\tau^y_m(Y_m)))$; note that $\phi(0) = X_m$.

Changing variables, with J the Jacobian of the transformation, yields:

(3.27)

$$\int e^{i\langle\phi^{-1}(Y_m),\zeta_m\rangle}\left(\int e^{i\langle X_m.\xi_m\rangle} p(m,t\xi_m)\tau^y_m q(y,t\tau^m_y(\zeta_m))\tau^y_m\tau^z_y\tau^m_z\,dX_m\,d\xi_m\right)$$
$$\tau^z_m s(exp_m(Y_m))J(Y_m)\,dY_m\,d\zeta_m.$$

Next note that $\phi^{-1}(0) = -X_m$ so by a result in **[Gi]** (bottom of the pg. 25); $\phi^{-1}(Y_m) = -X_m + A(Y_m)\cdot Y_m$, where $A(Y_m)$ is a linear map that is invertable in a neighborhood of zero. Thus equation (3.23) becomes:

(3.28)

$$\int e^{i\langle -X_m+A(Y_m)\cdot Y_m,\zeta_m\rangle}\int e^{i\langle X_m.\xi_m\rangle} p(m,t\xi_m)\tau^y_m q(y,t\tau^m_y(\zeta_m))\tau^y_m\tau^z_y\tau^m_z\,dX_m\,d\xi_m$$
$$\tau^z_m s(exp_m(\phi(Y_m)))J(Y_m)\,dY_m\,d\zeta_m.$$

Make another change of variables to get:

$$\int e^{\langle Y_m,\zeta_m\rangle} r(m,Y_m,t\zeta_m)\hat{s}_m(Y_m)\,dY_m\,d\zeta_m \qquad \text{where } r(m,t\zeta_m,Y_m)=$$

$$\int e^{i\langle X_m,\xi_m - A(Y_m)^t(\zeta_m)\rangle} p(m,t\xi_m)\tau_m^y q(y,t\tau_y^m(A(Y_m)^{-1t}(\zeta_m))) \tag{3.29}$$

$$\tau_m^y\tau_y^z\tau_z^m K(Y_m)\,dX_m\,d\xi_m.$$

Finally, we have:

$$r(m,t^{-1}\zeta_m,Y_m)_t =$$

$$\int e^{i\langle X_m,\xi_m\rangle} p(m,t\xi_m + B(Y_m)\zeta_m)\tau_m^y(q(y,\tau_y^m(C(Y_m))\zeta_m)S(m,y,z) \tag{3.30}$$

$$K(Y_m)dX_m\,d\xi_m.$$

Now standard estimates show that r is a symbol (see [**T**],vol. 1, pp. 23-29 theorem 3.2 and 3.3).

Next we neeed to show that the symbol of $P_t \circ Q_t$ has the requisite asymptotic expansion.

To do this, we appeal to Widom's calculation of the symbol of the composition of two pseudodifferential operators. Recall that in definition 2.4, given a pseudodifferential operator P, we defined $\mu(P) \in End(\mathbb{E}) \otimes End(\mathbb{S})$, which would be the symbol that is defined by Widiom. Widiom then shows that:

(3.31)

$$\mu(P\circ Q) \sim \sum_{m_1\ldots m_k\geq 2} \frac{i^{k-\sum_0^m p_i - \sum_1^k m_i}}{k!p_0!p_1!\ldots!p_k!m_1!\ldots!m_k!}\nabla^{p_1+m_1}l\ldots\nabla^{p_k+m_k}l\nabla^{p_0}\times$$

$$D^{\sum m_i}\mu(P)D^{\sum p_i}\mu(Q).$$

We can rephrase this in terms of theorem 2.1. Let $\mathbb{L}$ be the bundle over $T^*(M)$ defined in equation 2.7; then there exist differential operators $\bar{a}_n : \Gamma(\mathbb{L}\otimes\mathbb{L}) \to \Gamma(\mathbb{L})$, such that if $a_n = \sigma \circ \bar{a}_n$:

$$\sigma(\theta(p_t)\circ\theta(q_t)) \sim \sum t^n a_n(p,q)_t. \tag{3.32}$$

The degree of $a_n(p,q)$ is determined by two factors:

(1) how many times p and q are differentiated in the ξ variable; and
(2) the degree of the coefficents of a_n (since the coefficients of a_n will be in $End(\mathbb{E}) \otimes End(\mathbb{S}) \equiv End(\mathbb{E}) \otimes Cliff(T(M))$).

The a_n are determined by what they do on symbols that are polynomial in ξ (that is on differential operators). Suppose that $p = \omega\cdot\bar{p}$ with $\omega \in \Omega^s(M)$and $\bar{p}$ a homogeneous polynomial of degree $k-s$ in ξ. Then p is a symbol of degree k and $p_t = t^k p$. We make the same assumption about q, so that $q_t = t^l q$. Theorem 2.1 says:

$$\sigma(\theta(p_t)\circ\theta(q_t))_{t^{-1}} = t^{l+k}\sum_{n=0}^{\infty} a_n(p,q)_{t^{-1}}. \tag{3.33}$$

Now $a_n(p,q) \in S\Sigma^{l+k-n}(\mathbb{E})$, thus $a_n(p,q)_{t^{-1}} = t^{-(l+k-n)}a_n(p,q)$. If we put this into equation 3.33 we have:

$$\sigma(\theta(p_t) \circ \theta(q_t))_{t^{-1}} = \sum_{n=0}^{\infty} t^n a_n(p,q). \tag{3.34}$$

This shows that we have the requisite asymptotic expansion and identifies the first term of the expansion as $a_0(p,q)$. ∎

LEMMA 3.11. *Let $p_n \in S\Sigma^{l-n}(\mathbb{E})$, n=$n = 1, 2, \ldots$, then there exists an asymptotic symbol $p(m,\xi,t)$ such that $p(m,\xi,t) \sim \sum_{n=0}^{\infty} t^n p_n(m,\xi)$.*

PROOF: This is a standard argument (see [**W1**], lemma 4.2). One chooses a C^∞ function ϕ such that $\phi(x) = 0$ if $x \leq 1$ and $\phi(x) = 1$ if $x \geq 2$. Then choose ϵ_n, positive numbers tending to zero as $n \to \infty$. Set $p(m,\xi,t) = \sum_{n=0}^{\infty} t^n \phi(\epsilon_n(\frac{1}{t^2} + |\xi|^2))p_n(m,\xi)$. Then it is easy to show that $p(m,\xi,t) \sim \sum_{n=0}^{\infty} t^n p_n(m,\xi)$. ∎

DEFINITION 3.1. *Let P_t be an $A\psi DO$ with symbol $p(m,\xi,t) \sim \sum t^n p_n(m,\xi)$; then P_t is called asymptotically elliptic if the map $q \to a_0(p_0,q)$ is invertible.*

Remark. The notion of ellipticity for $A\psi DO's$ is slightly more subtle than it is for ordinary pseudodifferential operators. For example, let e_i be a local frame for $T^*(M)$ in a neighborhood U, and consider the operator $C(e_i)$, Clifford multiplication by e_i. Then $C(e_i)$ is an invertible operator in the neighborhood U. The asymptotic operator $tC(e_i)$, which is of order one, is, however, not invertible as an asymptotic operator, since it's inverse would be $t^{-1}C(e_i)$, which is not an $A\psi DO$. It is also clear that the map $p \to a_0(p,e_i)$ is not invertible, since $e_i \wedge e_i = 0$.

THEOREM 3.1. *If P_t is an asymptotically elliptic operator, then there exist Q_t, an asymptotic operator, such that $P_t \circ Q_t \sim 1$.*

PROOF: Let $\sigma(P_t)_{t^{-1}} \sim \sum_{n=0}^{\infty} t^n p_n$; then we find a symbol, $q \sim \sum_{n=0}^{\infty} t^n q_n$, by solving for the q_i's recursively. Using:

$$\sigma(\theta((p_1))_t \circ \theta((p_2)_t))_{t^{-1}} \sim \sum_{n=0}^{\infty} t^n a_n(p_1,p_2) \tag{3.35}$$

we have:

$$\sum_{n=0}^{\infty} t^n \left(\sum_{l+s+k=n} a_l(p_s,q_k) \right) - 1 = 0. \tag{3.36}$$

This says that $a_0(p_0, q_0) = 1$, or $q_0 = \{a_0(p_0, \cdot)\}^{-1}(1)$. For $n \geq 0$ we have :

$$\sum_{l+s+k=n} a_l(p_s, q_k) = 0 \text{ so we have:}$$

$$a_0(p_0, q_n) + \sum_{\substack{l+k+s=n \\ k<n}} a_l(p_s, q_k) \text{ or:}$$

(3.37)

$$q_n = \{a_0(p_0, \cdot)\}^{-1} \left\{ - \sum_{\substack{l+k+s=n \\ k<l}} a_l(p_s, q_k) \right\}.$$

Lemma 3.11 now says that there is an asymptotic symbol with the desired expansion. If we set $Q_t = \theta(q_t)$, then Q_t will be the desired inverse. ∎

THEOREM 3.2. *If P_t is an $A\psi DO$ of order less than zero, then $Tr_s(P_t)$ exist and:*

$$Tr_s(P_t) = (2\pi)^{-n}(\frac{2}{i})^{\frac{n}{2}} \int_{T^*(M)} tr^{\mathsf{E}}(\sigma(P_t))_{t^{-1}}\, d\xi. \tag{3.38}$$

PROOF: For the most part, the proof is standard and is obtained from computing the trace by integrating the kernel of an integral operator along the diagonal. For some of the details one can consult [**G**], (theorem 3.7). However, some explanation of notation is called for at this point.

Recall that we had an intermediate map, $\bar{\theta} : S\Sigma(\mathbb{E}) \to \Sigma(\mathbb{E} \otimes \mathbb{S})$, where $\Sigma(\mathbb{F})$ is the space of (ordinary) symbols for the bundle $\mathbb{F}$. The map $\bar{\theta}$ sent a form-valued symbol of $\mathbb{E}$ to a regular symbol of $\mathbb{E} \otimes \mathbb{S}$, where $\mathbb{S}$ is the bundle of spinors, by sending the form part to the corresponding operator in $End(\mathbb{S}) \cong Cliff(T^*(M))$.

Suppose that $P_t = \theta(p_t)$. If we denote by tr_s the supertrace on the bundle $\mathbb{E} \otimes \mathbb{S}$, then it is standard ([**W1**],pp. 46) that:

$$Tr_s(P_t) = (2\pi)^{-n} \int_{T^*(M)} tr_s(\bar{\theta}(p_t))\, dx\, d\xi. \tag{3.39}$$

Let $tr^{\mathsf{E}}(T)$ denote the trace of the endomorphism T on the bundle $\mathbb{E}$; using equation 1.22 we have that $tr_s(\bar{\theta}(p_t)) = (\frac{2}{i})^{\frac{n}{2}} top[tr^{\mathsf{E}}(p_t)]$. Here, $top[\omega]$ means the top dimensional piece of the form ω. Since only the top dimensional piece of a form contributes to the integral over the manifold M, equation 3.39 becomes:

$$Tr_s(P_t) = (2\pi)^{-n}(\frac{2}{i})^{\frac{n}{2}} \int_{T^*(M)} tr^{\mathsf{E}} \sigma(P_t)\, d\xi. \tag{3.40}$$

The symbol has not yet been inverse scaled. If we inverse scale the top piece, we introduce a factor of t^{-n} in front of the integral. However, the cotangent fiber also gets scaled by a factor of t^{-1} which introduces a factor of t^n in front of the integral, and the two factors cancel. Thus, we get exactly the formula 3.38 for the supertrace of P_t. ∎

COROLLARY 3.3. *If P_t is an $A\psi DO$ with leading symbol p_o of degree less than zero, then $Tr_s(P_t)$ admits an asymptotic expansion:*

$$Tr(P_t) \sim \sum_{m=0}^{\infty} b_m t^m, \quad with \quad b_0 = (2\pi)^{-n}(\frac{2}{i})^{\frac{n}{2}} \int tr^{\mathbb{E}}(p_o)\, d\xi. \tag{3.41}$$

PROOF: By assumption we have $\sigma(P_t)_{t^{-1}} \sim \sum_{m=0}^{\infty} t^m p_m$. Recalling that this expansion holds in the topology of the symbol space and using theorem 3.2 yields the desired expansion. ∎

Now we want to appy the above results to the Dirac operator on a compact spin manifold. Let $\mathbb{E}$ be a vector bundle over the compact spin manifold M and let $\nabla_{\mathbb{E}}$ be a connection on $\mathbb{E}$. We will denote by ∇ the tensor product connection on $\mathbb{E}\otimes\mathbb{S}$ (or $\mathbb{E}\otimes\mathbb{S}^{\pm}$). Let $\alpha : \mathbb{E}\otimes\mathbb{S}\otimes T^*(M) \to \mathbb{E}\otimes\mathbb{S}\otimes T(M)$ be defined by the inner product on $T(M)$. Let $C(X)$ denote Clifford multiplication by X, and let $c : \mathbb{E}\otimes\mathbb{S}\otimes T(M) \to \mathbb{E}\otimes\mathbb{S}$ be defined as follows:

$$c(e\otimes s\otimes X) = e\otimes c(X)\cdot s. \tag{3.42.}$$

The Dirac operator is defined by:

$$D(s) = (c\circ\alpha\circ\nabla)(s). \tag{3.43}$$

If $e_1, \cdots, e_n$ is a local orthonormal frame for the tangent bundle, then locally we have:

$$D(s) = \sum_{i=0}^{n} c(e_i)\cdot(\nabla_{e_i} s). \tag{3.44}$$

Remark. Note that the order of D is two, so tD is not an asymptotic operator; however, t^2D is an $A\psi DO$.

Next we need to know D^2. This is a standard computation, so we will only outline it. Fix a point $m \in M$ and choose a synchronous frame in a neighborhood of x ($\nabla_{e_i}(e_j)(x) = 0$, or $\Gamma^l_{ij}(x) = 0$). At the point x one computes:

$$D^2 = \sum_{i=1}^{n}(\nabla_{e_i})^2 + \frac{1}{4}\sum_{ijkl} R^i_{jkl} e_i e_j e_k e_l + \frac{1}{2}\sum_{ij} F_{ij} e_i e_j \tag{3.45}$$

where F_{ij} is the curvature of the bundle $\mathbb{E}$. By using the Bianchi identities and with K the scalar curvature of M, this reduces to :

$$D^2 = \sum_{i=1}^{n}(\nabla_{e_i})^2 + \frac{1}{2}\sum_{ij} F_{ij} e_i e_j + \frac{1}{4}K. \tag{3.46}$$

LEMMA 3.12. *Let $D_t = tD$, and let $\phi \in C^\infty(M)$; then:*

(1) *D_t^2 is an $A\psi DO$ with leading symbol $(-|\xi|^2 + \frac{1}{2}\sum F_{ij} f_i \wedge f_j)$; and*
(2) *$[D_t, \phi]$ is an $A\psi DO$ with leading symbol $d\phi$.*

PROOF: We have :

$$\sigma(t^2D^2) = -t^2|\xi|^2 + \frac{1}{2}t^2 \sum F_{ij} f_i \wedge f_j + \frac{1}{4}t^2 K. \tag{3.47}$$

Inverse scale to get:

$$\sigma(t^2D^2)_{t^{-1}} = (-|\xi|^2 + \frac{1}{2} \sum F_{ij} f_i \wedge f_j) + \frac{1}{4}t^2 K. \tag{3.48}$$

For the other part we have:

$$[D_t, \phi] = t \sum_{i=0}^{n} d\phi(e_i) \cdot e_i. \tag{3.49}$$

Consequently , $\sigma([D_t, \phi]) = td\phi$; so

$$\sigma([D_t, \phi])_{t^{-1}} = d\phi. \tag{3.50}$$

∎

Remark. We noted that D_t is not an $A\psi DO$; consequently, D_t^{-1} is not an $A\psi DO$ either. It is precisely this fact that makes Connes's original computation of the Chern character so difficult. One should note that in the letter to Quillen, Connes isolates the expression tD_t^{-1}, the point being that tD_t^{-1} is an $A\psi DO$. Specifically,

$$\sigma(tD_t^{-1}) = \sum t f_i \frac{1}{t\xi_i}. \tag{3.51}$$

Thus if we inverse scale we get:

$$\sigma(tD_t^{-1})_{t^{-1}} = \sum \frac{1}{\xi_i} f_i. \tag{3.52}$$

The crucial point here is that $c(e_i)^{-1} = c(e_i)$.

Given a connection and a "linear "function $l \in C^\infty(T^*(M) \times M)$, Widom defines a complete symbol for a pseudodifferential operator on M. It was shown in **[R]** that the function $\langle exp_m^{-1}(x), \xi_m \rangle = l(\xi_m, x)$ is such a linear function. If P_t is an $A\psi DO$ then we see that P_t is a family of pseudodifferential operators in the sense of Widom (the treatment of the Clifford variables as differential operators only adds a dependence on t that is uniform for $t \in [0,1]$, so Widoms theory works for our asymptotic operators.) We see that if we choose for the linear function $\langle exp_m^{-1}(x), \xi_m \rangle$, then Widom's symbol is exactly:

$$\sigma(P_t)(m, t^{-1}\xi). \tag{3.53}$$

Lemma 4.10 of **[W1]** can be stated as follows:

LEMMA 3.13. *The operator $e^{t^2D^2}$ is a $A\psi DO$.*

PROOF: Applying the function calculus to $e^{t^2D^2}$ gives:

$$e^{t^2D^2} = \frac{1}{2\pi}\int_\Gamma (t^2D^2-\lambda)^{-1}e^\lambda\, d\lambda. \tag{3.54}$$

Next , Widom shows that :

$$\sigma(e^{t^2D^2}) = \frac{1}{2\pi}\int_\Gamma \sigma((t^2D^2-\lambda)^{-1})e^\lambda\, d\lambda. \tag{3.55}$$

The delicate point is that the following expansion is uniform in λ:

$$\sigma((t^2D^2-\lambda)^{-1}) \sim \sum_{n=0}^{\infty} t^n(p_n(\lambda))_t. \tag{3.56}$$

Thus we can integrate the expansion term by term to obtain the desired asymptotic expansion for $\sigma(e^{t^2D^2})$. ∎

LEMMA 3.14. *Let p be the leading symbol of t^2D^2; then the leading symbol of $e^{t^2D^2}$ is $e^{a_0(p,\cdot)}(1)$.*

PROOF: Note that $(t^2D^2-\lambda)^{-1}$ has the leading symbol $a_0(p-\lambda,\cdot)^{-1}(1) = (a_0(p.\cdot)-\lambda)^{-1}(1)$. Thus $e^{t^2D^2}$ has the leading symbol:

$$\left(\frac{1}{2\pi}\int_\Gamma (a_0(p,\cdot)-\lambda)^{-1}e^\lambda\, d\lambda\right)(1) = e^{a_0(p,\cdot)}(1). \tag{3.57}$$

∎

COROLLARY 3.15. *With notation as above we have:*

$$\begin{aligned} Tr_s(e^{t^2D^2}) &= (2\pi)^{-n}(\frac{2}{i})^{\frac{n}{2}}\int_{T^*(M)} e^{a_0(p,\cdot)}(1)\, d\xi + O(t), \qquad \textit{hence} \\ ind(D) = \lim_{t\to 0} Tr_s(e^{t^2D^2}) &= (2\pi)^{-n}(\frac{2}{i})^{\frac{n}{2}}\int_{T^*(M)} e^{a_0(p,\cdot)}(1)(x,\xi)\, d\xi. \end{aligned} \tag{3.58}$$

Next we want to compute the integral in equation 3.58. Recall that R was the curvature of $T(M)$. Thus if X, Y are vector fields, $R(X,Y) \in End(T(M))$ and $R(X,Y)$ is an $so(n)$-valued endomorphism. Choose a local orthonormal frame, $e_1, \cdots, e_n$, for the tangent bundle and let $f_1, \cdots, f_n$ be the corresponding dual frame for the cotangent bundle. If the matrix of $R(X,Y)$ with respect to this frame is $(R(X,Y)_{ij})$, then the curvature of the spin bundle is given by:

$$\bar{R}(X,Y) = \frac{1}{2}\sum_{ij} R(X,Y)_{ij}e_ie_j \in Cliff(T(M)). \tag{3.59}$$

Thus we have:

$$\sigma(\bar{R}(X,Y)) = \frac{1}{2}\sum_{ij} R(X,Y)_{ij} f_i \wedge f_j. \tag{3.60}$$

The frame $f_1, \cdots, f_n$ determines local coordinates $(\xi_1, \cdots, \xi_n)$ on $T^*(M)$, and with respect to these coordinates we have the operator:

$$R(\frac{\partial}{\partial \xi}, \frac{\partial}{\partial \eta}) = 2\sum_{ij} \sigma(\bar{R}(e_i, e_j))\frac{\partial}{\partial \xi_i}\frac{\partial}{\partial \eta_j}. \tag{3.61}$$

In what follows we will simply write: $\sum_{ij} R(e_i, e_j)\frac{\partial}{\partial \xi_i}\frac{\partial}{\partial \eta_j}$, where $R(e_i, e_j)$ is the two-form $\sum_{kl} R^k_{lij} f_k \wedge f_l$. The formula for $a_0(p,q)$ is :

$$a_0(p,q) = e^{-\frac{1}{4}R(\frac{\partial}{\partial \xi}, \frac{\partial}{\partial \eta})} p(x,\xi) \wedge q(x,\eta)|_{\xi=\eta}. \tag{3.62}$$

Recall that F is the curvature of the bundle $\mathbb{E}$, so $\sigma(F) = \sum F_{ij} f_i \wedge f_j$. In what follows we will write $-|\xi|^2 + \frac{1}{2}F$ for $-|\xi|^2 + \frac{1}{2}\sigma(F)$. Given an arbitrary symbol q we have:

$$\begin{aligned} a_0(p,q) &= (-|\xi|^2 + \frac{1}{2}F) \wedge q(x,\xi) - \\ &\frac{1}{4}R(\frac{\partial}{\partial \xi}\frac{\partial}{\partial \eta})(-|\xi|^2 + \frac{1}{2}F) \wedge q(x,\eta)|_{\eta=\xi} + \\ &\frac{1}{2}\frac{1}{16}R(\frac{\partial}{\partial \xi}, \frac{\partial}{\partial \eta})\left(R(\frac{\partial}{\partial \xi}, \frac{\partial}{\partial \eta})\right)(-|\xi|^2 + \frac{1}{2}F) \wedge q(x,\eta)|_{\eta=\xi}. \end{aligned} \tag{3.63}$$

All other terms are zero. First we consider the last two terms. We have:

$$\begin{aligned} &R(\frac{\partial}{\partial \xi}, \frac{\partial}{\partial \eta})(-|\xi|^2 + \frac{1}{2}F) \wedge q(x,\eta)|_{\eta=\xi} = \\ &-2\sum_{ij} R(e_i, e_j)\xi_i \frac{\partial}{\partial \xi_i} q(x,\xi). \end{aligned} \tag{3.64}$$

We will denote the operator $\sum_{ij} R(e_i, e_j)\xi_i \frac{\partial}{\partial \xi_i}$ by $R(\xi, \frac{\partial}{\partial \xi})$.
For the next term we have:

$$\begin{aligned} &R(\frac{\partial}{\partial \xi}, \frac{\partial}{\partial \eta})R(\frac{\partial}{\partial \xi}, \frac{\partial}{\partial \eta})(-|\xi|^2\frac{1}{2}F) \wedge q(x,\eta) = \\ &\sum_{i,j,k,l} R(e_i, e_j)R(e_k, e_l)\frac{\partial}{\partial \xi_i}\frac{\partial}{\partial \eta_j}\frac{\partial}{\partial \xi_k}\frac{\partial}{\partial \eta_l}(-|\xi|^2 + \frac{1}{2}F) \wedge q(x,\eta). \end{aligned} \tag{3.65}$$

Now $-|\xi|^2 = -(\xi_1^2 + \cdots + \xi_n^2)$ so $\frac{\partial}{\partial \xi_i}\frac{\partial}{\partial \xi_k}(-|\xi|^2) = 0$ unless $i = k$ in which case it equals -2. Also $R(e_k, e_l) = -R(e_l, e_k)$; hence equation 3.65 becomes:

$$2\sum_{ijk} R(e_i, e_k)R(e_k, e_j)\frac{\partial}{\partial \xi_i}\frac{\partial}{\partial \xi_j} q(x,\xi). \tag{3.66}$$

We will write this as:

$$2\,(R\wedge R)\,(\frac{\partial}{\partial\xi},\frac{\partial}{\partial\xi})(q(x,\xi)). \tag{3.67}$$

Thus equation 3.67 becomes:

$$a_0(p,q)=\left(-|\xi|^2\frac{1}{2}F+\frac{1}{2}R(\xi,\frac{\partial}{\partial\xi})+\frac{1}{16}R\wedge R(\frac{\partial}{\partial\xi},\frac{\partial}{\partial\xi})\right)(q(x,\xi)). \tag{3.68}$$

Recall that we are writing 1 for the constant symbol $1(x,\xi)=1$, and we need to compute $e^{a_0(p,\cdot)}(1)$. Note that $\frac{1}{2}R(\xi,\frac{\partial}{\partial\xi})(-|\xi|^2)=-|\xi|^2$, so we have:

$$\begin{aligned} &e^{a_0(p,\cdot)}(1)=e^{\left(-|\xi|^2\frac{1}{2}F+\frac{1}{2}R(\xi,\frac{\partial}{\partial\xi})+\frac{1}{16}R\wedge R(\frac{\partial}{\partial\xi},\frac{\partial}{\partial\xi})\right)}(1)=\\ &e^{\frac{1}{2}F}e^{(-|\xi|^2+\frac{1}{16}R\wedge R(\frac{\partial}{\partial\xi},\frac{\partial}{\partial\xi}))}(1).\end{aligned} \tag{3.69}$$

Remark. Of course, 3.69 is only true for the constant symbol 1.

Next, we appeal to Mehler's formula,[**GJ**]. The operator $e^{-\frac{1}{2}(\xi^2-a^2(\frac{d}{d\xi})^2)}$ is an integral operator with kernel given by:

$$\begin{aligned} &k_a(\xi,\eta)=\\ &(2\pi a\sinh(a))^{-\frac{1}{2}}exp\left(-\frac{1}{2a\sinh(a)}(\cosh(a)(\xi^2+\eta^2)-2\xi\eta)\right).\end{aligned} \tag{3.70}$$

If A is a skew-symmetric matrix with eigenvalues $\pm ia_j$, then we can form the corresponding operator, $H=(|\xi|^2-A^2(\frac{\partial}{\partial\xi},\frac{\partial}{\partial\xi}))$. The kernel of e^{-H} satisfies:

$$\int k_H(\xi,\eta)\,d\xi\,d\eta=\prod_{j=1}^{n}\frac{2\pi a_j}{\sinh(a_j)}=\left(\det\frac{-2\pi iA}{\sinh(-iA)}\right)^{\frac{1}{2}}. \tag{3.71}$$

If we apply this to equation 3.58 and 3.69 we have:

$$\begin{aligned} &Ind(D)=\lim_{t\to 0}Tr_s(e^{\frac{1}{2}t^2D^2})=\\ &(2\pi)^{-n}(\frac{2}{i})^{\frac{n}{2}}\int_M e^{F/4}\left(\det\frac{-2\pi iR/4}{\sinh(-iR/4)}\right)^{\frac{1}{2}}.\end{aligned} \tag{3.72}$$

Given a matrix, B, recall that, ([**G**] pp. 99):

$$\hat{A}(B)=\left(\det\frac{B/4\pi}{\sinh(B/2\pi)}\right)^{\frac{1}{2}}. \tag{3.73}$$

Therefore, we have:

$$\begin{aligned} &\left(\det\frac{-2\pi i(R/4)}{\sinh(-i(R/4))}\right)^{\frac{1}{2}}=(2\pi)^{\frac{n}{2}}2^{\frac{n}{2}}\left(\det\frac{-i(R/4)/2}{\sinh(-i(R/4))}\right)^{\frac{1}{2}}=\\ &(2\pi)^{\frac{n}{2}}2^{\frac{n}{2}}\hat{A}(-2\pi i(R/4)).\end{aligned} \tag{3.74}$$

Next, recall that the Chern character, $Ch(B)$, is equal to $tr(e^{-B/2\pi i})$, so we have $tr(e^{F/4}) = Ch(-2\pi i(F/4))$. Thus:

$$(3.75)\quad Ind(D) = (2\pi)^{-n}(\frac{2}{i})^{\frac{n}{2}}(2\pi)^{\frac{n}{2}}2^{\frac{n}{2}}\int_M Ch(-2\pi i(F/4)) \wedge \hat{A}(-2\pi i(R/4)).$$

Recalling that the top degree piece of the intergrand is homogeneous of degree $\frac{n}{2}$ we have:

$$(3.76)\quad \begin{aligned} Ind(D) =&(2\pi)^{-n}(\frac{2}{i})^{\frac{n}{2}}(2\pi)^{\frac{n}{2}}2^{\frac{n}{2}}\left(\frac{-\pi i}{2}\right)^{\frac{n}{2}}\int_M Ch(F) \wedge \hat{A}(R) = \\ &(-1)^{\frac{n}{2}}\int_M Ch(F) \wedge \hat{A}(R). \end{aligned}$$

4. Calculation of the Cocycle for the Dirac Operator.

In this section we will apply the asymptotic calculus to the calculation of the cyclic cocycle corresponding to the Dirac operator on a compact spin manifold M. See [**C3**] and [**JLO**] for details on theta-summable fredholm modules. Let $H = H^+ \oplus H^-$ denote the Hilbert space of square-integrable spinors over M, and let ϵ be the corresponding grading operator, so $\epsilon|_{H^+} = 1$ and $\epsilon|_{H^-} = -1$. Let D be the Dirac operator on the space of spinors; thus:

$$(4.1)\quad D = \begin{pmatrix} 0 & D^- \\ D^+ & 0 \end{pmatrix}.$$

Here $D^+ : H^+ \to H^-$ and $D^- : H^- \to H^+$. Let $\mathfrak{A} = C^1(M)$; then $\mathfrak{A}$ is a Banach algebra (it is the completion of $C^\infty(M)$ with respect to the norm $||f|| = ||f||_\infty + ||df||_\infty$). Let $\bar{\mathfrak{A}} = \mathfrak{A}/\mathbb{C}$ and define $C_n(\mathfrak{A}) = \mathfrak{A} \otimes \bar{\mathfrak{A}}^{\otimes_\pi n}$ where $\otimes_\pi$ denotes the projective tensor product. The bar complex is $\sum_{n=0}^\infty C_N(\mathfrak{A})$ with a suitable topology. The cobar complex is the topological dual and the Chern character of a theta-summable Fredholm module is a cochain in the cobar complex, closed with respect to the coboundary operator $\partial = b + B$. Thus , for appropriate $B_i \in C_i(\mathfrak{A})^*$:

$$(4.2)\quad Ch((D, H)) = \sum_{i=0}^{\infty} B_i.$$

The Chern character of the module (H, D) as an entire cyclic cocycle is given by:

$$(4.3)\quad \begin{aligned} &(Ch((D, H)), (f_0, \ldots, f_{2k})) = B_{2k}(f_0, \ldots, f_{2k}) = \\ &\int_{\Delta_n} Tr_s\left(f_0 e^{-t_1 D^2}[D, f_1]e^{-(t_2-t_1)D^2} \cdots [D, f_{2k}]e^{-(1-t_{2k})D^2}\right) dt_1 \cdots dt_{2k}. \end{aligned}$$

Let $D_\alpha = \alpha D$. As noted before, D_α is not an asymptotic operator; however, $D_\alpha^2 = \alpha^2 D^2$ and $[D_\alpha, f] = \alpha C(df)$ are asymptotic operators. (Recall that $C(df)$ is Clifford multiplication by df.)

THEOREM 4.1.

(1) $(Ch((D_\alpha, H)), (f_0, \ldots, f_{2k}))$ *is convergent for* $\alpha \to 0$ *and for* $f_i \in \mathfrak{A}$.

(2) *The limit* $\tau_k(f_0, \ldots, f_{2k}) = \lim_{\alpha\to 0}(Ch((D_\alpha, H)), (f_0, \ldots, f_{2k})$ *gives a cocyle (considering all of the* τ_k*'s together)in the cobar complex, and:*

$$(4.4)\quad \tau_k(f_0, \ldots, f_{2k}) = vol(\Delta_{2k})(2\pi)^{-n}(\frac{2}{i})^{\frac{n}{2}} \int_M \hat{A}(-2\pi R/4) f_0\, df_1 \wedge \cdots \wedge df_{2k}.$$

PROOF: That the collection of τ_k's form a cyclic cocycle follows from the existence of the limit and the cooresponding fact for $Ch((D,H))$, which can be found in [**GS**]. Consider the operator:

$$(4.5)\qquad P_\alpha = f_0 e^{-t_1 D_\alpha^2}[D_\alpha, f_1] e^{-(t_2-t_1)D_\alpha^2} \cdots [D_\alpha, f_{2k}] e^{-(1-t_{2k})D_\alpha^2}.$$

By the above remarks, each of the individual terms is an asymptotic operator; hence, so is their product. To compute $\lim_{\alpha\to 0} Tr_s(P_\alpha)$, it suffices to compute the leading symbol. Recall that the leading symbols of $e^{\frac{-t}{2}D_\alpha^2}$ and $[D_\alpha, f]$ are:

$$(4.6)\qquad \begin{array}{ll} (a) & e^{-t(|\xi|^2 - \frac{1}{16}R\wedge R(\partial_\xi, \partial_\xi))}(1) \\ (b) & df. \end{array}$$

By the formula for the composition of symbols, 4.6a and 4.6b commute, since df contains no differentiation and 4.6a is a form of even order. Thus we have:

(4.7)

$$\lim_{\alpha\to 0} Tr_s(P_\alpha) = \lim_{\alpha\to 0}(2\pi)^{-n}(\frac{2}{i})^{\frac{n}{2}} \int (\sigma(P_\alpha))_{\alpha^{-1}} d\xi =$$

$$(2\pi)^{-n}(\frac{2}{i})^{\frac{n}{2}} \int \sigma(f_0 e^{-t_1 D_\alpha^2}[D_\alpha, f_1] e^{-(t_2-t_1)D_\alpha^2} \cdots [D_\alpha, f_{2k}] e^{-(1-t_{2k})D_\alpha^2})_{\alpha^{-1}} =$$

$$(2\pi)^{-n}(\frac{2}{i})^{\frac{n}{2}} \int f^0 df_1 \wedge \cdots \wedge df_{2k} \wedge e^{-t_1-(t_2-t_1)-\cdots-(1-t_{2k})(|\xi|^2 - \frac{1}{16}R\wedge R(\partial_\xi, \partial_\xi))} =$$

$$(2\pi)^{-n}(\frac{2}{i})^{\frac{n}{2}} \int_M f_0 df_1 \wedge \cdots \wedge df_{2k} e^{a(p,\cdot)}(1) =$$

$$(2\pi)^{-n}(\frac{2}{i})^{\frac{n}{2}} \int_M \hat{A}(-2\pi i R/4) f_0 df_1 \wedge \cdots \wedge df_{2k}.$$

∎

REFERENCES

[**A**] l. Alvarez-Gaumé, *Supersymmetry and the Atiyah-Singer Index Theorem*, Commun. Math. Physics **90** (1983), 161-173.

[**ABP**] M. Atiyah, R. Bott, S. Patodi, *On the Heat Equation and the Index Theorem*, Invent. Math. **19** (1973), 279–330.

[**ABS**] M. Atiyah, R. Bott, A. Shapiro, *Clifford Modules*, Topology **3** (1964), 1-38.

[**B**] J. Bokobza-Haggiag, *Opérateurs Psaeudodifférentiel sue une Variété Différentiable*, Ann.Inst.Fourier (Grenoble) **19** #**1** (1969), 125-177.

[C1] A. Connes, *Non-Commutative Differential Geometry*, Pub. Math. IHES **62** (1985), 41-144.

[C2] A. Connes, *Letter to Quillen.*

[C3] A. Connes, *Entire Cyclic Cohomology of Banach Algebras and Characters of Theta-Summable Fredholm Modules*, K-Theory **1** (1988), 519-548.

[G] E. Getzler, *Pseudodifferential Operators on Supermanifolds and the Atiyah-Singer Index Theorem*, Commun. Math. Physics **92** (1983), 163-178.

[Gi] P. Gilkey, "Invariance Theory, the Heat Equation, and the Atiyah-Singer Index Theorem," Publish or Perish Inc., Wilmington, Delaware, 1984.

[GJ] J. Glimm, A. Jaffe, "Quantum Physics," Springer-Verlag, Ney Yourk,NY, 1981.

[GS] E. Getzler, A. Szenes, *On the Chern Character of a Theta-Summable Fredholm Module*, preprint.

[JLO] A. Jaffe, A. Lesniewski, K. Osterwalder, *Ouantum K-Theory*, preprint.

[MF] V. Maslov, M.Fedoriuk, "Semi-Classical Approximation in Quantum Mechanics," D. Reidel, Boston, MA, 1981.

[R] J. Rempala, *A Semi-Global Taylor Formula for Manifolds*, Annales Polonici Mathematici **XLVIII** (1988).

[T] F. Treves, "Introduction to Pseudodifferential and Fouier Intergal Operators," Plenumn Press, New York, NY, 1980.

[W1] H. Widom, *A Complete Symbol Calculus for Pseudodifferential Operators*, Bull, Sc. Math., 2nd Series **104** (1980), 19-63.

[W2] H. Widom, *Families of Pseudodifferential Operators*, "Topics in Functional Analysis," Academic Press, New York, NY, 1978, pp. 345-395.

Department of Mathematics, M.I.T. , Cambridge, Mass. 02139
Department of Mathematics-Campus Box 426, University of Colorado, Boulder CO 80309

Contemporary Mathematics
Volume **105**, 1990

A Lefschetz Theorem on Open Manifolds

JAMES L. HEITSCH AND CONNOR LAZAROV†

Abstract. A Leftschetz theorem is proven for a class of endomorphisms of a Dirac complex of bounded geometry over a complete Riemannian manifold of bounded geometry. The Dirac complex must satisfy an auxilliary condition which insures that, as t goes to infinity, the kernel of the heat operator converges uniformly to the kernel of the projection onto the harmonic sections.

1. Introduction. In this note we show how the techniques of [HL] can be adapted to give a Lefschetz theorem for certain endomorphisms of a Dirac complex over a complete Riemannian manifold of bounded geometry provided that the manifold satisfies certain conditions (see Theorem 4). This paper should be viewed as an adjunct to the papers of Roe [RI,RII] on an index theorem for open manifolds. We assume that the reader is familiar with these papers as well as [HL].

Our proof employs the heat equation method. Given an endomorphism T of a Dirac complex (E,d) over the manifold M, we consider $T^s(Te^{-t\Delta})$ where T^s is a super trace determined by a regular exhaustion of M and Δ is the Laplacian associated to (E,d). We show that $T^s(Te^{-t\Delta})$ is independent of t and that as $t \to 0$ it is asymptotic to an expression which depends only on objects restricted to the fixed point set of the quasi isometry of M over which T sits. This local Lefschetz number is the analogue of the local expression for the Lefschetz number in the classical Lefschetz Theorem. As $t \to \infty$, one would like to prove that $T^s(Te^{-t\Delta})$ converges to the super trace of T acting on the reduced L^2 cohomology of (E,d), i.e. the global Lefschetz number of T. In general this is false. In Theorem 3, we give several conditions, each of which insures that this convergence takes place.

The essential idea of this paper is to replace the condition of compactness of the ambient manifold by conditions which permit unifrom estimates to be made. This allows us to commute the taking of limits with the application of the super trace.

We would like to thank Leon Karp and John Roe for helpful conversations.

1980 *Mathematics Subject Classification* (1985 *Revision*). Primary 58C30.

+First author partially supported by NSF Grant DMS 8503029, second author by a PSC-CUNY Grant

0271-4132/90 $1.00 + $.25 per page

2. Dirac complexes and geometric endomorphisms. Let M be an m dimensional, oriented, complete Riemannian manifold of bounded geometry [CG], [CGT], [D], [RI]. If N is a submanifold of M, we denote its tangent bundle by TN. The distance function on M is written $\rho(x, y)$. If K is any subset of M and $\epsilon > 0$, we set

$$\begin{aligned}\text{Pen}^+(K, \epsilon) &= \{x \in M | \rho(K, x) \leq \epsilon\} \\ \text{Pen}^-(K, \epsilon) &= \{x \in M | \rho(M - K, x) \geq \epsilon\}.\end{aligned}$$

If E is a vector bundle, $C^\infty(E)$ is the space of smooth sections of E and $L^2(E)$ is the space of L^2 sections. The fiber of E over a point x is written E_x and if $A : E \to F$ is a vector bundle map, the restriction of A to E_x is written A_x. Finally, $E|K$ denotes the restriction of E to K.

Let (E, d) be a Dirac complex of bounded geometry over M, [GL], [RI], [HL]. That is

1) $E = (E_0, E_1, \ldots, E_k)$ is a family of smooth finite dimension Hermitian vector bundles over M with connections, ∇^i on E_i.
2) $d = (d_0, \ldots, d_{k-1})$ is a family of differential operators where $d_i : C^\infty(E_i) \to C^\infty(E_{i+1})$ and $d_{i+1} \circ d_i = 0$.
3) Let $d^*_{i-1} : C^\infty(E_i) \to C^\infty(E_{i-1})$ be the adjoint of d_{i-1} and let D be the operator $D = \oplus(d_i + d^*_{i-1}) : C^\infty(E) \to C^\infty(E)$, where $E = \oplus E_i$. Give E the Hermitian metric coming from the E_i. E is required to be a Clifford module over the Clifford bundle of M, $\nabla = \oplus \nabla^i$ is required to be a connection compatible with the Riemannian connection on the Clifford bundle of M, and D is required to be the associated Dirac operator.
4) The curvature tensor of ∇ is bounded on M as are all of its covariant derivatives.

Example. If M is a complete Riemannian manifold of bounded geometry, then each of the classical complexes (deRham, Signature, Spin, Dolbeault) which can be defined over M is a Dirac complex of bounded geometry.

We now describe the endomorphisms of the Dirac complex (E, d) which we will be considering. This is a generalization of the geometric endomorphisms appearing in [ABI,II]. Let $f : M \to M$ be a quasi-isometry and let $A_i : f^*E_i \to E_i$ be a C^∞ bundle map. Define $T_i : C^\infty(E_i) \to C^\infty(E_i)$ by

$$(T_i\, s)(x) = A_{i,x}(s(f(x)))$$

for $s \in C^\infty(E_i)$. As

$$A_{i,x} : (f^*E_i)_x = E_{i,f(x)} \to E_{i,x}$$

we have that $(T_i s)(x) \in E_{i,x}$. We assume that the A_i are chosen so that

$$d_i \circ T_i = T_{i+1} \circ d_i,$$

and we call $T = (T_0, \ldots, T_k)$ the geometric endomorphism determined by f and $A = (A_0, \ldots, A_k)$. We assume that

i) $\rho(x, f(x))$ is bounded on M
ii) A is bounded on M.

The natural example of such an endomorphism is f^*, the map induced by f on the deRham complex, provided f satisfies i) above. Such an f is given by the time t flow of a vector field on M which is bounded. Of course f induces a similar endomorphism, also denoted f^*, for the other classical complexes provided that f preserves the defining structures.

In addition to assuming that f is a quasi-isometry, we must make some assumptions about the fixed point set N of f, and about the behaviour of f with respect to N. In particular, we assume

1) $N = \bigcup_j N_j$ where each N_j is a closed oriented submanifold of M and that the N_j are ϵ disjoint, i.e. there is $\epsilon > 0$ so that $\rho(N_i, N_j) > \epsilon$ for $i \neq j$.
2) f is uniform off N. That is, given $\epsilon > 0$, there is $\delta > 0$ so that for all x with $\rho(x, N) > \epsilon$, $\rho(x, f(x)) > \delta$.
3) f is uniformly non-degenerate along N. Let $\nu = (TM|N)/TN$ be the normal bundle of N in M. The differential df of f induces a bundle map df_ν of ν covering the identity map of N. To say f is uniformly non-degenerate along N means that there is $\epsilon > 0$ so that for all $x \in N$, $|\det(I - df_\nu)_x| > \epsilon$, where I is the identity map of ν.
4) For $x \in N$, put df_x in Jordan normal form, and denote the subspace consisting of the generalized eigenvectors for eigen values different from 1 by Λ_x. 3) above implies that Λ_x and TN_x are complementary subspaces of TM_x. Let a_x be the angle between Λ_x and TN_x (If $\Lambda_x = TM_x$ or $TN_x = TM_x$, we set $a_x = \pi/2$). We require that there is $\epsilon > 0$ so that for all $x \in N$, $a_x > \epsilon$.

3. Traces and the basic theorem. Recall that a regular exhaustion of M is an increasing sequence $\{M_i\}$ of compact subsets of M whose union is M satisfying for $r \geq 0$

$$\lim_{i\to\infty} \operatorname{vol}(\operatorname{Pen}^+(M_i, r))/\operatorname{vol}(\operatorname{Pen}^-(M_i, r)) = 1.$$

We assume that M admits a regular exhaustion. Recall the following facts and notation from [RI], sections 5 and 6. Denote by $U(M, E)$ the space of uniform operators on E, and by $U_{-\infty}(M, E)$ the ideal of uniform operators of order $-\infty$.

1) If $f \in \mathcal{S}(\mathbf{R})$, the Schwartz space of $\mathbf{R}$, then the operator $f(D)$ defined by the spectral theorem is in $U_{-\infty}(M, E)$.

Let $\mathcal{S}(\mathbf{R}^+)$ be those functions on $\mathbf{R}$ whose restriction to $[0, \infty)$ coincides with the restriction of an element of $\mathcal{S}(\mathbf{R})$ to $[0, \infty)$. If we set $\Delta = D^2$ and $f \in \mathcal{S}(\mathbf{R}^+)$, then $f(\Delta) \in U_{-\infty}(M, E)$. (This follows from the fact that the spectrum of Δ is contained in $[0, \infty)$).

2) Any $B \in U_{-\infty}(M, E)$ is represented by a bounded smoothing kernel $k(x, y)$, i.e. for all $s \in C^\infty(E)$, $(Bs)(x) = \int_M k(x, y)s(y)d\operatorname{vol}(y)$.
3) Each regular exhaustion $\{M_i\}$ of M defines at least one trace on $U_{-\infty}(M, E)$, denoted $\mathcal{T}$.

$\mathcal{T}$ is defined as follows. Given any bounded m form a on M, consider the sequence $(1/\mathrm{vol}\, M_i) \int_{M_i} a$. This sequence does not necessarily converge. However, it is always possible to choose a subsequence M_{i_m} so that for *all* such a $(1/\mathrm{vol})\, M_{i_m} \int_{M_{i_m}} a$ does converge. This limit defines a linear functional on the bounded m forms on M. For any operator B on $C^\infty(E)$ with smooth bounded Schwartz kernel $k(x,y)$, we set

$$\mathcal{T}(B) = \lim_{m\to\infty} (1/\mathrm{vol}\, M_{i_m}) \int_{M_{i_m}} tr\ k(x,x) d\mathrm{vol}(x)$$

and $\mathcal{T}(B)$ defines a trace on $U_{-\infty}(M,E)$. We assume that the regular exhaustion has been chosen so that

$$\mathcal{T}(B) = \lim_{i\to\infty} (1/\mathrm{vol}\, M_i) \int_{M_i} tr\ k(x,x) d\mathrm{vol}(x).$$

In order to state our main theorem we shall need a variant $\mathcal{T}^s$ of $\mathcal{T}$ called the supertrace associated to $\{M_i\}$. Let $E_e = \oplus E_{2i}$ and $E_o = \oplus E_{2i+1}$. As $C^\infty(E) = C^\infty(E_e) \oplus C^\infty(E_o)$, we may represent any linear operator B on $C^\infty(E)$ as $B = \begin{bmatrix} B_{11} & B_{12} \\ B_{21} & B_{22} \end{bmatrix}$, where $B_{11} : C^\infty(E_e) \to C^\infty(E_e)$, $B_{12} : C^\infty(E_o) \to C^\infty(E_e)$, etc. Denote the Schwartz kernel of B_{ij} by k_{ij}. If B has smooth bounded Schwartz kernel, we set

$$\mathcal{T}^s(B) = \lim_{i\to\infty} (1/\mathrm{vol}\, M_i) \int_{M_i} tr\ k_{11}(x,x) - tr\ k_{22}(x,x) d\mathrm{vol}(x).$$

We shall henceforth write $tr^s k(x,x)$ for $tr\ k_{11}(x,x) - tr\ k_{22}(x,x)$. An operator B is called even if $B_{12} = B_{21} = 0$ and odd if $B_{11} = B_{22} = 0$. Note that if B and C are both even elements of $U_{-\infty}(M,E)$, then $\mathcal{T}^s(CB) = \mathcal{T}^s(BC)$, while if B and C are both odd, $\mathcal{T}^s(CB) = -\mathcal{T}^s(BC)$.

LEMMA. *Suppose B and C are in $U_{-\infty}(M,E)$ and that T is as above, i.e. $\rho(x, f(x))$ and A are bounded on M. Then $\mathcal{T}^s(TBC)$ and $\mathcal{T}^s(CTB)$ both exist. If both B and C are even, $\mathcal{T}^s(TBC) = \mathcal{T}^s(CTB)$. If both are odd $\mathcal{T}^s(TBC) = -\mathcal{T}^s(CTB)$.*

PROOF. $BC \in U_{-\infty}(M,E)$ and so has bounded smooth Schwartz kernel $k(x,y)$. The kernel of TBC is then $A_x k(f(x),y)$ which is bounded and smooth. Thus $\mathcal{T}^s(TBC)$ exists.

Denote the kth Sobolev space of E by $W_k(E)$. As A_x is bounded and f is a quasi-isometry with $\rho(x, f(x))$ bounded, T defines a quasi-local continuous map from $L^2(E) = W_0(E)$ to $W_0(E)$. Given any k and r, B maps W_k to W_0 continuously and quasi-locally. Similarly for C from W_0 to W_{k-r}. Thus CTB maps W_k to W_{k-r} continuously and quasi-locally so $CTB \in U_{-\infty}(M,E)$ and $\mathcal{T}^s(CTB)$ is defined.

To finish the proof, we refer the reader to the proof of Theorem 6.7 of [RII]. (Note that TB satisfies the conclusions of Proposition 5.4 of [RII]).

Now for $t > 0$, the function $e^{-tx} \in \mathcal{S}(\mathbf{R}^+)$ so $e^{-t\Delta} \in U_{-\infty}(M,E)$. Our basic theorem is:

THEOREM 1. *$\mathcal{T}^s(Te^{-t\Delta})$ is independent of t.*

PROOF. Set $d = \oplus d_i$ and $d^* = \oplus d_i^*$. Then both d and d^* are odd operators on E and bounded geometry implies that both are uniform operators. Note that $\Delta = D^2 = dd^* + d^*d$, that T is an even operator, and by assumption $Td = dT$. As $d\Delta = \Delta d$, we have that for any bounded measurable function f, $df(\Delta) = f(\Delta)d$ (see [HL]) and as Δ is even, so is $f(\Delta)$. Now given $s, t > 0$ set

$$\varphi(x) = e^{-tx} - e^{-sx}$$

$$\Psi(x) = \frac{e^{-tx} - e^{-sx}}{x}$$

$$\Psi_1(x) = \frac{e^{-tx/2} - e^{-sx/2}}{x}$$

$$\Psi_2(x) = e^{-tx/2} + e^{-sx/2}.$$

Each of these functions is in $\mathcal{S}(\mathbf{R}^+)$ and $\varphi(x) = x\Psi(x)$, $\Psi(x) = \Psi_1(x)\Psi_2(x)$. We have

$$\begin{aligned}
&\mathcal{T}^s(Te^{-t\Delta}) - \mathcal{T}^s(Te^{-s\Delta})\\
=&\mathcal{T}^s(T\varphi(\Delta)\\
=&\mathcal{T}^s(T\Delta\Psi(\Delta))\\
=&\mathcal{T}^s(Tdd^*\Psi(\Delta) + Td^*d\Psi(\Delta))\\
=&\mathcal{T}^s(Tdd^*\Psi(\Delta)) + \mathcal{T}^s(Td^*d\Psi(\Delta)).
\end{aligned}$$

Now

$$\begin{aligned}
&\mathcal{T}^s(Tdd^*\Psi(\Delta))\\
=&\mathcal{T}^s(Tdd^*\Psi_1(\Delta)\Psi_2(\Delta))\\
=&\mathcal{T}^s(\Psi_2(\Delta)Tdd^*\Psi_1(\Delta))\\
=&\mathcal{T}^s(\Psi_2(\Delta)dTd^*\Psi_1(\Delta))\\
=&-\mathcal{T}^s(Td^*\Psi_1(\Delta)\Psi_2(\Delta)d)\\
=&-\mathcal{T}^s(Td^*\Psi(\Delta)d)\\
=&-\mathcal{T}^s(Td^*d\Psi(\Delta)),
\end{aligned}$$

and we are done. Of course, to be completely rigorous, we must worry about the domains of the various operators we are manipulating. For this see the Appendix of [HL].

In the next two sections, we examine the behavior of $\mathcal{T}^s(Te^{-t\Delta})$ as $t \to 0$ and as $t \to \infty$.

4. The limit as $t \to 0$. Recall that $N = \cup_j N_j$ where the N_j are closed oriented submanifolds which are ϵ disjoint. In this section we will use some of the results of [HL] to show

THEOREM 2. *To each N_j we can associate a smooth dim N_j form a_j which depends only on f, A, the symbol of Δ, the metrics, and their derivatives to a finite order only on N_j so that*

$$\mathcal{T}^s(Te^{-t\Delta}) = \lim_{i\to\infty} \frac{1}{vol\, M_i} \sum_i \int_{M_i \cap N_j} a_j.$$

For the classical complexes, we can identify the a_j explicitly. Suppose (E, d) is a classical complex (deRham, signature, spin, Dolbeault) and suppose $T = f^*$. Then a_j is the usual local integrand given by the Atiyah-Singer G index theorem. If f is the identity map, we recover Roe's index theorem on open manifolds [RI,II]. If (E, d) is an arbitrary Dirac complex and N_j is a point, say x, then

$$a_j(x) = \frac{\sum_{i=1}^{k} (-1)^i tr\, A_{i,x}}{|\det(I - df)_x|}.$$

For the classical complexes with $T = f^*$ and N_j a point, a_j may be further identified and is given by the classical formulae. See [ABII]. In particular, for the deRham complex, $a_j = \text{sign } \det(I - df)_x$.

Proof of Theorem 2. Choose $\epsilon > 0$ so small that the N_i are ϵ disjoint and for each i, $\text{Pen}^+(N_i, \epsilon)$ is an embedded normal disc bundle for N_i in M. That we can choose ϵ so that the second requirement is satisfied follows from the fact that M has bounded geometry. We shall now denote $\text{Pen}^+(K, \epsilon)$ by $\eta(K)$ for any subset K of M. Let $k_t^T(x, y)$ be the Schwartz kernel of $Te^{-t\Delta}$. Since f is uniform off N, there is $\delta > 0$ so that for all $x \in M - \eta(N)$, $\rho(x, f(x)) > \delta$. This fact combined with the proof of Theroem 5.5 of [HL] gives that $k_t^T(x, x)$ is uniformly asymptotic to 0 for all $x \in M - \eta(N)$. Thus given a there is a constant C_a depending only on a so that

$$|tr^s k_t^T(x, x)| < C_a t^a$$

for t near 0 and $x \in M - \eta(N)$. Now we have

$$(1/\text{vol}\, M_i)\left|\int_{M_i - \eta(N)} tr^s k_t^T(x, x) d\text{vol}(x)\right| \leq (1/\text{vol}\, M_i) \int_{M_i - \eta(N)} C_a t^a d\text{vol}(x)$$
$$\leq C_a t^a,$$

so

$$\begin{aligned}\mathcal{T}^s(Te^{-t\Delta}) &= \lim_{t\to 0} \mathcal{T}^s(Te^{-t\Delta}) \\ &= \lim_{t\to 0}\lim_{i\to\infty} (1/\text{vol}\, M_i) \int_{M_i} tr^s k_t^T(x, x) d\text{vol}(x) \\ &= \lim_{t\to 0}\lim_{i\to\infty} (1/\text{vol}\, M_i) \int_{M_i \cap \eta(N)} tr^s k_t^T(x, x) d\,\text{vol}(x).\end{aligned}$$

LEMMA. $\lim_{i\to\infty} (1/vol\, M_i) \int_{M_i \cap \eta(N)} tr^s k_t^T(x, x) d\, vol(x)$

$= \lim_{i\to\infty} (1/vol\, M_i) \int_{\eta(N)|M_i \cap N} tr^s k_t^T(x, x) d\, vol(x)$

where $\eta(N)$ is thought of as an embedded normal disc bundle of N.

PROOF. As $\{M_i\}$ is a regular exhaustion of M,

$$\lim_{i\to\infty} \mathrm{vol}(\mathrm{Pen}^+(M_i,\epsilon))/\mathrm{vol}\, M_i = 1$$

and

$$\lim_{i\to\infty} \mathrm{vol}(\mathrm{Pen}^-(M_i,\epsilon))/\mathrm{vol}\, M_i = 1.$$

Now $|tr^s k_t^T(x,x)|$ is bounded, say by C, on M and the symmetric difference of $\eta(N)|M_i \cap N$ and $M_i \cap \eta(N)$ is a subset of $\mathrm{Pen}^+(M_i,\epsilon) - \mathrm{Pen}^-(M_i,\epsilon)$. Thus

$$\lim_{i\to\infty} (1/\mathrm{vol}\, M_i)\Big| \int_{\eta(N)|M_i\cap N} tr^s k_t^T(x,x)d\,\mathrm{vol}(x) - \int_{M_i\cap\eta(N)} tr^s k_t^T(x,x)d\,\mathrm{vol}(x)\Big|$$

$$\leq \lim_{i\to\infty} (1/\mathrm{vol}\, M_i) \int_{\mathrm{Pen}^+(M_i,\epsilon)-\mathrm{Pen}^-(M_i,\epsilon)} C\ d\,\mathrm{vol}(x)$$

$$= \lim_{i\to\infty} \frac{C\cdot(\mathrm{vol\ Pen}^+(M_i,\epsilon) - \mathrm{vol\, Pen}^-(M_i,\epsilon))}{\mathrm{vol}\, M_i} = 0.$$

Thus we have

$$\mathcal{T}^s(Te^{-t\Delta}) = \lim_{t\to 0}\lim_{i\to\infty} (1/\mathrm{vol}\, M_i) \int_{\eta(N)|M_i\cap N} tr^s k_t^T(x,x)d\,\mathrm{vol}(x).$$

$$= \lim_{t\to 0}\lim_{i\to\infty} (1/\mathrm{vol}\, M_i) \sum_j \int_{\eta(N_j)|M_i\cap N_j} tr^s k_t^T(x,x)d\,\mathrm{vol}(x).$$

If we denote integration over the fiber of $\eta(N_j)$ by $\displaystyle{\fint_{\eta(N_j)}}$ then

$$\int_{\eta(N_j)|M_i\cap N_j} tr^s k_t^T(x,x)d\,\mathrm{vol}(x) = \int_{M_i\cap N_j} \fint_{\eta(N_j)} tr^s k_t^T(x,x)d\,\mathrm{vol}(x).$$

Let k be a positive integer. Proposition 5.8 of [HL] combined with the bounded geometry of M and the fact that there is $\epsilon > 0$ so that for all $x \in N$, $|\det(I - df_\nu)_x| > \epsilon$ and $a_x > \epsilon$, gives us that for each j, $\fint_{\eta(N_j)} tr^s k_t^T(x,x)d\,\mathrm{vol}(x)$ is uniformly (on *all* of N) asymptotic to

$$(*) \qquad \sum_{s\geq -n_j}^{k} d_{j,s} t^{s/2}$$

That is, there is a constanct C so that for all t near 0 and all j,

$$\left| \fint_{\eta(N_j)} tr^s k_t^T(x,x)d\,\mathrm{vol}(x) - \sum_{s=-n_j}^{k} d_{j,s} t^{s/2} \right| < Ct^{k+1/2}.$$

In (*), $n_j = \dim N_j$ and $d_{j,s}$ is a smooth n_j form on N_j which depends only on f, A, the symbol of Δ, the metrics, and their derivatives to a finite order only on N_j. [Note that in adapting the proof of 5.8 of [HL] to use here, the fact that $a_x > \epsilon$ implies that as $t \to 0$ dw ranges over the fiber of η_j^L uniformly for all $x \in N$ (notation is that of [HL]), and the fact that $|\det(I - df_\nu)_x| > \epsilon$ implies that the estimates made on $a_m(x, w, \xi, t)$ are uniform on N].

For $-m \leq s < -n_j$ set $d_{j,s} = 0$ where $m = \dim M$. Since the asymptotic convergence is uniform, we have

$$\lim_{i\to\infty} (1/\mathrm{vol}\, M_i) \sum_j \int_{M_i \cap N_j} \fint_{\eta(N_j)} tr^s k_t^T(x,x) d\,\mathrm{vol}(x)$$

is asymptotic as $t \to 0$ to $\lim\limits_{i\to\infty} (1/\mathrm{vol}\, M_i) \sum\limits_j \int_{M_i \cap N_j} \sum\limits_{s\geq -m}^{k} d_{j,s} t^{s/2}$. We claim that this last expression equals

$$(**) \qquad \sum_{s\geq -m}^{k} \left[\lim_{i\to\infty} (1/\mathrm{vol}\, M_i) \sum_j \int_{M_i \cap N_j} d_{j.s} \right] t^{s/2}.$$

To prove this we need only show that for all s, $\lim\limits_{i\to\infty} (1/\mathrm{vol}\, M_i) \sum\limits_j \int_{M_i \cap N_j} d_{j,s}$ exists. Now $d_{j,s}$ is an n_j form on N_j constructed (using a universal algorithm) out of objects all of which are bounded on M. It follows that $d_{j,s}$ is bounded independently of j. Let ϕ_j be a bounded $m - n_j$ form on M whose support is contained in $\eta(N_j)$ satisfying $\int_{\eta(N_j)} \phi_j = 1$. Denote by $\phi_j \wedge d_{j,s}$ the bounded m form on M $\phi_j \wedge \pi_j^* d_{j,s}$ where $\pi_j : \eta(N_j) \to N_j$ is the projection. Then $\phi \wedge d_s = \sum\limits_j \phi_j \wedge d_{j,s}$ is a bounded m form on M and we have from [RI], Proposition 6.4 that $\lim\limits_{i\to\infty} (1/\mathrm{vol}\, M_i) \int_{M_i} \phi \wedge d_s$ exists. As above we have

$$\lim_{i\to\infty} (1/\mathrm{vol}\, M_i) \int_{M_i} \phi \wedge d_s =$$

$$\lim_{i\to\infty} (1/\mathrm{vol}\, M_i) \sum_j \int_{M_i \cap \eta(N_j)} \phi_j \wedge d_{j,s} =$$

$$\lim_{i\to\infty} (1/\mathrm{vol}\, M_i) \sum_j \int_{\eta(M_i \cap N_j)} \phi_j \wedge d_{j,s} =$$

$$\lim_{i\to\infty} (1/\mathrm{vol}\, M_i) \sum_j \int_{M_i \cap N_j} \fint_{\eta(N_j)} \phi_j \wedge d_{j,s} =$$

$$\lim_{i\to\infty} (1/\mathrm{vol}\, M_i) \sum_j \int_{M_i \cap N_j} d_{j,s}.$$

Now we are finished, for (**) gives an asymptotic expansion as $t \to 0$ for $\mathcal{T}^s(Te^{-t\Delta})$, which is independent of t. Thus all the coefficients in (**) must be zero except for the coefficient of t^0, namely

$$\lim_{i\to\infty} (1/\mathrm{vol}\, M_i) \sum_j \int_{M_i \cap N_j} d_{j,0},$$

which must equal $\mathcal{T}^s(Te^{-t\Delta})$. Setting $a_j = d_{j,0}$, we are done. The identification of the a_j in the classical cases is then carried out in the usual way. See [ABII], [G], [ABP].

Let $M, \{M_i\}$ and $\mathcal{T}^s$ be as above, and let (E, d) be the deRham complex on M. We define the average Euler character of M to be

$$X_{\mathcal{T}}(M) = \lim_{i\to\infty} (1/\text{vol}\, M_i) \int_{M_i} X(\Omega)$$

where $X(\Omega)$ is the Pfaffian applied to the curvature of the Levi-Civita connection on E. We say that a vector field X on M with isolated zeros is uniformly nondegenerate provided that there is $\epsilon > 0$ so that $|\det X_x| > \epsilon$ for all $x \in M$ with $X(x) = 0$. Here X_x denotes the linear map which the Lie derivative $\mathcal{L}_X$ defines on TM_x.

THEOREM 3. *Let X be a vector field on M which is bounded and assume that the zero set N of X consists of isolated points which are ϵ disjoint and that X is uniformly nondegenerate. Assume that there is an $\epsilon > 0$ so that for $0 < s < \epsilon$, the time s flow of X, f_s, has fixed point set N and that f_s is uniform off N. Then*

$$(***) \qquad X_{\mathcal{T}}(M) = \lim_{i\to\infty} (1/\mathit{vol}\, M_i) \sum_{x\in M_i\cap N} \deg(X, x)$$

where $\deg(X, x)$ *is the usual degree of the zero of X at x.*

PROOF. For $0 < s < \epsilon$, f_s satisfies the conditions we need to apply Theorems 1 and 2. (The fact that X is uniformly nondegenerate implies that each f_s is uniformly nondegenerate along N). In addition, $\deg(X, x) =$ sign $\det(I - df_{s,x})$ for each $x \in N$, and $0 < s < \epsilon$. Thus the right hand side of (***) is just $\mathcal{T}^s(f_s^* e^{-t\Delta})$. The left hand side is $\mathcal{T}^s(I^* e^{-t\Delta}) = \mathcal{T}^*(e^{-t\Delta})$, where $I = f_0$ is the identity map of M. Thus to prove the theorem we need only show that the Schwartz kernel of $f_s^* e^{-t\Delta}$ converges uniformly to $k_t(x, y)$, the Schwartz kernel of $e^{-t\Delta}$, as $s \to 0$.

Denote by A_x^s the map induced by f_s from $\Lambda^* T^* M_{f(x)}$ to $\Lambda^* T^* M_x$. The Schwartz kernel of $f_s^* e^{-t\Delta}$ is $A_x^s k_t(f_s(x), y)$. Now $k_t(x, y)$ is uniformly continuous on $M \times M$ (see [RI]), and as X is bounded, $f_s(x)$ converges uniformly to x as $s \to 0$. Thus $k_t(f_s(x), y)$ converges uniformly to $k_t(x, y)$ as $s \to 0$. Similarly, A_x^s converges uniformly to the identity. The theorem follows.

5. The limit as $t \to \infty$. The question of what happens to $\mathcal{T}^s(Te^{-t\Delta})$ as $t \to \infty$ is a bit more problematic than the limit as $t \to 0$. The sort of answer one wants is the following. As A is bounded and f is a quasi isometry, T induces a map $T: L^2(E) \to L^2(E)$. Now $Td = dT$ so T induces a map T^* on the reduced L^2 cohomology of (E, d), and we would like to compute the $\mathcal{T}$ super trace of this operator. To do so we note that the reduced L^2 cohomology of (E, d) is naturally isomorophic to the kernel $\Delta \subset L^2(E)$. See [HL], the appendix. If $P : L^2(E) \to (\text{kernel}\ \Delta)$ is the projection, with smooth bounded Schwartz kernel

$k^P(x,y)$, then TP is an even smoothing operator on E with smooth bounded Schwartz kernel $k^{TP}(x,y) = A_x k^P(f(x),y)$. We define

$$L_T(T^*) = \lim_{i\to\infty} (1/\mathrm{vol}\, M_i) \int_{M_i} tr^s k^{TP}(x,x) d\,\mathrm{vol}(x).$$

Then we would like to prove

$$\lim_{t\to\infty} T^s(Te^{-t\Delta}) = L_T(T^*).$$

In general this is false as the examples in [RII] show. However, we have the following theorem.

THEOREM 4. *If any of the conditions below holds then*

$$\lim_{t\to\infty} T^s(Te^{-t\Delta}) = L_T(T^*),$$

and so

$$L_T(T^*) = \lim_{i\to\infty} (1/\mathit{vol}\, M_i) \sum_j \int_{M_i \cap N_j} a_j.$$

That is we have an analogue of the classical Lefschetz Theorem.

1. *Given* $\delta > 0$, *then for all sufficiently large* t, $\sup\{|e^{-t\Delta}s(x) - Ps(x)| : x \in M\} \leq \delta \|s\|$ *for all* $s \in L^2(E)$. *($\|s\|$ is the* L^2 *norm of* s*).*

2. *There is* $\delta > 0$ *so that* $\langle \Delta s, s\rangle \geq \delta \|s\|^2$ *for all* $s \in L^2(E_e)$.

3. *The curvature operator* R *appearing in the Weitzenbock formula for* Δ *is a nonnegative operator.*

4. *For sufficiently large* k, $\lim_{t\to\infty} \|e^{-t\Delta} - P\|_{-k,k} = 0$, *where* $\|\ \|_{-k,l}$ *is the norm of an operator from* $W_{-k}(E)$ *to* $W_l(E)$.

PROOF. Each condition implies that the Schwartz kernel $k_t(x,y)$ of $e^{-t\Delta}$ converges uniformly to the Schwartz kernel $k^P(x,y)$ of P. In case 3. we also have that $P \equiv 0$. For the proof that conditions 1,2, and 3 imply uniform convergence see [RII]. Condition 1 comes from Lemma 1.5 (replace $\epsilon_{x,v}$ by $\epsilon_{y,v}$). Condition 2 comes from Prop. 1.3 and condition 3 from Prop. 1.11.

For 4., let δ_x be the distributional section of E given by $\langle \delta_x, s\rangle = s(x)$ where $\langle\ ,\ \rangle$ is the natural inner product on sections of E. Bounded geometry implies that there is a $k > 0$ so that $\delta_x \in W_{-k}(E)$ for all $x \in M$ and the Sobolev norm $\|\delta_x\|_{-k}$ is bounded independently of x. Now

$$\begin{aligned} |k_t(x,y) - k^P(x,y)| &= |\langle (e^{-t\Delta} - P)\delta_y, \delta_x\rangle| \\ &\leq \|e^{-t\Delta} - P\|_{-k,k} \|\delta_y\|_{-k} \|\delta_x\|_{-k}. \end{aligned}$$

and we have the result. Note that condition 4 is fulfilled provided that there is an interval of the form $(0,a)$ so that $(0,a) \cap$ spectrum $\Delta = \phi$. This follows immediately from estimates given by the spectral mapping theorem.

Now the Schwartz kernel of $Te^{-t\Delta}$ is given by $A_x k_t(f(x),y)$ which converges uniformly to $A_x k^P(f(x),y)$ the Schwartz kernel of TP. The theorem follows immediately.

THEOREM 5. *Suppose one of the conditions of Theorem 4 is satisfied. Then $\mathcal{T}^s(PTP)$ exists and equals $L_{\mathcal{T}}(T^*)$.*

Given a quasi isometry f of M, we write $L_{\mathcal{T}}(f)$ for $L_{\mathcal{T}}(f^*)$. As f is a quasi isometry, the map $A_x : \Lambda^* T^* M_{f(x)} \to \Lambda^* T^* M_x$ is bounded so $L_{\mathcal{T}}(f)$ exists.

COROLLARY. *Let (E, d) be the deRham or signature complex of M. Suppose that one of the conditions of Theorem 4 is satisfied and that f_s is a smooth one parameter family of quasi isometries of M. The $L_{\mathcal{T}}(f_s)$ is independent of s.*

Proof of Corollary. Lemma 6.3 of [HL] implies that the map Pf_s^*P is independent of s.

Proof of Theorem 5.

$$\begin{aligned} L_\nu(T) &= \lim_{t\to\infty} \mathcal{T}^s(Te^{-t\Delta}) \\ &= \lim_{t\to\infty} \mathcal{T}^s(e^{-t\Delta/2}Te^{-t\Delta/2}) \\ &= \lim_{t\to\infty} \mathcal{T}^s(e^{-t\Delta}Te^{-t\Delta}). \end{aligned}$$

Thus we need only show that the Schwartz kernel of $e^{-t\Delta}Te^{-t\Delta}$ converges uniformly to the Schwartz kernel of PTP.

If A is a self adjoint smoothing operator and $s \in L^1(E) \cap L^2(E)$,

$$\begin{aligned} ||As||_{L^2}^2 = \langle As, As\rangle = \langle s, A^2 s\rangle &\le ||s||_{L^1} ||A^2 s||_{L^\infty} \\ &\le ||s||_{L^1} ||A||_{L^2, L^\infty} ||As||_{L^2}. \end{aligned}$$

Thus $||As||_{L^2} \le ||s||_{L^1} ||A||_{L^2,L^\infty}$ where $||A||_{L^2,L^\infty}$ is the norm of A as an operator from $L^2(E)$ to $L^\infty(E)$. Now each of the first three conditions of Theorem 4 implies that for all $s \in L^2(E)$,

$$||(e^{-t\Delta} - P)s||_{L^\infty} \le c(t)||s||_{L^2}$$

where $c(t) \to 0$ as $t \to \infty$, i.e. $||e^{-t\Delta} - P||_{L^2,L^\infty} \le c(t)$. $e^{-t\Delta} - P$ is self adjoint and smoothing so for any $s \in L^1(E) \cap L^2(E)$

$$||(e^{-t\Delta} - P)s||_{L^2} \le c(t)||s||_{L^1}.$$

Now let $s \in L^1(E) \cap L^2(E)$, let $t \ge 1$, and note that $P = Pe^{-\Delta}$ and $||P||_{L^2} \le 1$. Then since

$$e^{-t\Delta}Te^{-t\Delta}s - PTPs = (e^{-t\Delta} - P)TPs + e^{-t\Delta}T(e^{t\Delta} - P)s,$$

we have

$$\begin{aligned} &||(e^{-t\Delta}Te^{-t\Delta} - TPT)s||_{L^\infty} \le \\ &||(e^{-t\Delta} - P)TPs||_{L^\infty} + ||e^{-t\Delta}T(e^{-t\Delta} - P)s||_{L^\infty} \le \\ &c(t)||TPe^{-\Delta}s||_{L^2} + A||T||_{L^2}||(e^{-t\Delta} - P)s||_{L^2} \le \\ &c(t)||T||_{L^2}||e^{-\Delta}s||_{L^2} + A||T||_{L^2}c(t)||s||_{L^1} \le \\ &c(t)||T||_{L^2}A||s||_{L^1} + A||T||_{L^2}c(t)||s||_{L^1} = \\ &2Ac(t)||T||_{L^2}||s||_{L^1} \end{aligned}$$

where A is a bound (say for $t \geq 1$) for $||e^{-t\Delta}||_{L^2,L^\infty}$ which exists by the usual Sobolev embedding argument.

We have shown that the norm of $e^{-t\Delta}Te^{-t\Delta} - PTP$, as an operator from $L^1(E)$ to $L^\infty(E)$, goes to 0 as $t \to \infty$. This implies uniform convergence of the Schwartz kernels. (This proof was kindly provided by John Roe.)

If condition 4 is satisfied, we have the following. The Sobolev embedding theorem implies that for sufficiently large k, there is a constant C_k so that $||e^{-t\Delta} - P||_{L^2,L^\infty} \leq C_k||e^{-t\Delta} - P||_{0,k}$. The diagram

$$\begin{array}{ccccc} W_0(E) & \xrightarrow{\ i\ } & W_{-k}(E) & \xrightarrow{e^{-t\Delta} - P} & W_k(E) \\ \big\lfloor & & \xrightarrow{\qquad e^{-t\Delta} - P \qquad} & & \big\uparrow \end{array}$$

commutes, so $||e^{-t\Delta}-P||_{0,k} \leq ||i||_{0,-k}||e^{-t\Delta}-P||_{-k,k}$ and $\lim\limits_{t\to\infty} ||e^{-t\Delta}-P||_{L^2,L^\infty} = 0$. To finish we repeat the argument above.

References

[ABI] M.F. Atiyah and R. Bott, *A Leftschetz fixed point formula for elliptic complexes I*, Annals of Math. **86** (1967), 374–407.

[ABII] M.F. Atiyah and R. Bott, *A Leftschetz fixed point formula for elliptic complexes II*, Applications, Annals of Math. **88** (1968), 451–491.

[APB] M.F. Atiyah, R. Bott, and V.K. Patodi, *On the heat equation and the index theorem*, Invent. Math. **19** (1973), 279–330. Errata ibid **28** (1975), 277–280.

[CG] J. Cheeger and M. Gromov, *Bounds on the von Neumann dimension of L^2-cohomology and the Gauss-Bonnet theorem for open manifolds*, J. Differential Geometry **21** (1985), 1–34.

[CGT] J. Cheeger, M. Gromov, and M. Taylor, *Finite propagation speed, Kernel estimates for functions of the Laplace operator, and the geometry of complete Riemannian manifolds*, J. Differential Geometry **17** (1982), 15–54.

[D] J. Dodziuk, *L^2 harmonic forms on complete manifolds*, Annals of Math. Studies, No. **102**, Princeton University Press, (1982), 291–302.

[G] P.B. Gilkey, *Invariance Theory, The Heat Equation, and the Atiyah-Singer Index Theorem*, Publish or Perish, Wilmington, 1984.

[GL] M. Gromov and H.B. Lawson, Jr., *Positive scalar curvature and the Dirac operator*, Pub. Math. IHES **58** (1983), 83–196.

[HL] J.L. Heitsch and C. Lazarov, *A Leftschetz theorem for foliated manifolds*, to appear, Topology.

[RI] J. Roe, *An index theorem on open manifolds I*, J. of Differential Geometry **27** (1988), 87–113.

[RII] J.Roe, *An index theorem on open manifolds II*, J. of Differential Geometry **27** (1988), 115–136.

University of Illinois at Chicago
Lehmann College, CUNY

Contemporary Mathematics
Volume **105**, 1990

Eta Invariants and the Odd Index Theorem for Coverings

STEVEN HURDER*

1. Introduction. In this paper we develop the "odd" analog of the Atiyah-Singer Index Theorem for coverings [1, 50]. The odd index theorem for a compact manifold gives a topological formula for the integer index of a Toeplitz operator constructed from the compression of a unitary multiplier to the positive space of an elliptic self-adjoint psuedo-differential operator [8, 40]. Equivalently, this integer is a spectral flow invariant [6, 52]. When this construction is performed on an infinite covering the elliptic operator is lifted from the base, but there is a wider range of choices for the multiplier than exist on the compact base. The Toeplitz index or spectral flow invariant is calculated using a trace on an appropriate von Neumann algebra associated to the multiplier. If the multiplier is lifted from the base, then this index is an integer and we have the obvious direct generalization of the usual index theorem for coverings. However, if the multiplier is not a lift, then real-valued indices occur. We give several applications of the odd coverings index theorem, including the realization of the relative eta-invariant [3] as a type II-index, as announced in [28] and first proved in detail in [31].

Let M be a compact, oriented manifold without boundary of dimension m, with fundamental group $\pi_1(M)$. Consider a Galois covering

$$\Gamma \to \widetilde{M} \xrightarrow{\pi} M \tag{1.1}$$

with a surjection $\pi_1(M) \to \Gamma$ such that composing with the *right* action of Γ on $\widetilde{M}$ yields the "deck" action of $\pi_1(M)$ on $\widetilde{M}$. A subset M_0 of $\widetilde{M}$ is a *fundamental domain* if

(1.2a) M_0 is compact and path connected with connected interior $\overset{\circ}{M}_0$.

(1.2b) The boundary $\partial M_0 \equiv M_0 - \overset{\circ}{M}_0$ has Lebesgue measure zero.

(1.2c) M_0 maps onto M, and $\overset{\circ}{M}_0$ maps one-to-one into M.

1980 *Mathematics Subject Classification* (1985 *Revision*). Primary 58G10, 46L80, 57R20.

*Supported in part by NSF Grant DMS 86-01976 and a Sloan Foundation Fellowship.

We choose and fix a fundamental domain M_0. For each $\gamma \in \Gamma$, let M_γ denote the right translate $M_0 \cdot \gamma$.

Consider an elliptic, first-order differential operator,

$$D : C^\infty(M, E_0) \to C^\infty(M, E_1) \tag{1.3}$$

acting on the smooth sections of Hermitian vector bundles E_0 and E_1. Let $\mathcal{H}_0$ and $\mathcal{H}_1$ denote the corresponding Hilbert space completions of these smooth section spaces. Let D also denote the unique closure of D to a densely defined unbounded operator from $\mathcal{H}_0$ to $\mathcal{H}_1$, and D^* will denote its adjoint. When $E_0 = E_1$ then D is *(essentially) self-adjoint* if $D = D^*$. The results of this paper are all formulated for D as above. They can also be proven for D pseudo-differential with distributional kernel supported in an ϵ-tube around the diagonal in $M \times M$, where ϵ is less than the injectively radius of M, but the necessary techniques are clumsy and so for clarity will be avoided. (The necessary methods are developed in sections 3 and 5 of [31].)

The operator D determines an even K-homology class, $[D] \in K_0^a(M)$. There is a pairing between K-homology and ordinary K-theory,

$$\mathrm{Ind} : K^0(M) \boxtimes K_0^a(M) \to \mathbf{Z}. \tag{1.4}$$

Given a complex vector bundle, $\xi \to M$, we equip it with an Hermitian connection ∇_ξ, then the pairing is defined as the analytic index of the extended operator

$$D \otimes \nabla^\xi : C^\infty(M, E_0 \otimes \xi) \to C^\infty(M, E_1 \otimes \xi) \tag{1.5}$$

denoted by $\mathrm{Ind}^a(D \otimes \nabla^\xi)$. This, of course, is the motivation for, and simplest example of, the external product in Kasparov's KK-theory (cf. [36], [40]).

Let us return to the problem of index theorems for coverings. We set $\widetilde{E}_i = \pi^! E_i$, $i = 0, 1$ and then lift the operator D to a differential operator $\widetilde{D}$ acting on *compactly supported* smooth sections

$$\widetilde{D} : C_c^\infty(\widetilde{M}, \widetilde{E}_0) \to C_c^\infty(\widetilde{M}, \widetilde{E}_1), \tag{1.6}$$

with formal adjoint $\widetilde{D}^*$. A standard, basic fact is that $\widetilde{D}$ is uniquely closable on the corresponding Hilbert space closures of these section spaces (cf. [1]), and we let $\widetilde{D}, \widetilde{D}^*$ also denote the closures. The key observation of Atiyah and Singer was that the projection operators onto $\ker(\widetilde{D})$ and $\ker(\widetilde{D}^*)$ are represented by *smooth* kernels $\tilde{p}_0(y_0, y_1)$ and $\tilde{p}_1(y_0, y_1)$ respectively on $\widetilde{M} \times \widetilde{M}$. Moreover, these kernels are invariant under the deck action

$$\tilde{p}_i(y_0 \cdot \gamma, y_1 \cdot \gamma) = \tilde{p}_i(y_0, y_1); i = 0, 1; \gamma \in \Gamma \tag{1.7}$$

and *their restrictions to the diagonal satisfy*

$$\tilde{p}_i(y, y) = p_i(\pi(y), \pi(y)) \tag{1.8}$$

where p_i is the smooth kernel on $M \times M$ representing the projection onto $\ker(D)$ and $\ker(D^*)$ as $i = 0$ or 1. Thus, if Tr denotes the fiberwise trace on $\mathrm{End}(\widetilde{E}_i)$ and we set

$$\mathrm{Ind}_\Gamma(\widetilde{D}) = \int_{M_0} \mathrm{Tr}(\tilde{p}_0(y, y)) - \mathrm{Tr}(\tilde{p}_1(y, y)) dy \tag{1.9}$$

then we have obtained part of the index theorem for coverings:

THEOREM (ATIYAH [1], SINGER [50]). *With respect to the* Γ*-module structure on* $\widetilde{\mathcal{H}}_0$, $\widetilde{\mathcal{H}}_1$ *the operator* $\widetilde{D}$ *is Breuer Fredholm, and its* Γ*-index is given by* $Ind_\Gamma(\widetilde{D})$. *Moreover,*

$$(1.10) \qquad Ind_\Gamma(\widetilde{D}) = Ind^a(D)$$

$$(1.11) \qquad = (-1)^m \int_M \psi^{-1}(ch(\sigma_D)) \cup Td(M)$$

where ψ *is the Thom isomorphism.* □

The justification for the claim that $\mathrm{Ind}_\Gamma(\widetilde{D})$ is the Breuer index requires some delicate analysis. A recent proof of this via the methods that are the basis for this work is given in (Chapter 13, [47]). Of course, the second line (1.11) is the usual Atiyah-Singer Index Theorem.

The primary application of the index theorem for coverings was to construct non-trivial, locally finite subspaces of $\widetilde{\mathcal{H}}_0$ or $\widetilde{\mathcal{H}}_1$ on which Γ represents. For Γ a compact lattice and D a generalized Dirac operator, Atiyah and Schmid [4] realized the discrete series representations of Γ in this way on spaces of harmonic spinors. A far-reaching generalization of the coverings index theorem (again starting from the observations (1.7) and (1.8)) to pairings with cyclic cocycles over the group algebra of Γ has been given by Connes and Moscovici, which has deep topological applications to the Novikov Conjecture [22, 25, 26].

We conclude this introduction to the "even" index theorem for coverings with a formulation of it via Kasparov's bivariant KK-theory. The principal symbol of the lifted operator $\widetilde{D}$ is elliptic and Γ-invariant, so determines a class denoted

$$(1.12) \qquad [D]_\Gamma \in KK(C(M)),\ C_r^*(\Gamma))$$

where $C_r^*(\Gamma)$ is the reduced C^*-algebra of Γ (cf. [7, 32]). The map to a point, $M \to \mathrm{pt}$, induces a class $i^* \in KK(\mathbf{C}, C(M))$ and the exterior product with i^* defines a map, μ, for which there is a commutative diagram.

$$(1.13) \qquad \begin{array}{ccc} KK(C(M), C_r^*(\Gamma)) & \leftarrow & KK(C(M), \mathbf{C}) \\ \downarrow \mu & \swarrow \mu_\Gamma & \downarrow \mu \\ KK(\mathbf{C}, C_r^*(\Gamma)) & \leftarrow & KK(\mathbf{C}, \mathbf{C}) \\ \downarrow \mathrm{tr}_\Gamma & & \cong\downarrow \\ \mathbf{R} & \supset & \mathbf{Z} \end{array}$$

The composition $\mu_\Gamma([D])$ is the C^*-index of the Γ-equivariant operator $\widetilde{D}$ and lies in the domain of the natural trace, tr_Γ, on $C_r^*(\Gamma)$. The coverings index theorem implies that

$$(1.14) \qquad \mathrm{Ind}_\Gamma(\widetilde{D}) = \mathrm{tr}_\Gamma(\mu_\Gamma([D]))$$

and commutativity of (1.13) implies that this is an integer. Let us now see how this formulation of the coverings index theorem motivates the "odd" version of the theorem.

Let $E = E_0 = E_1$ and D be self-adjoint. Then the analytic index of D vanishes. The fundamental observation of Baum-Douglas [8,9] and Kasaparov [40] is that in spite of this, D still carries index information in the form of a pairing corresponding to (1.4) where now $[D] \in K_1^a(M)$:

$$\text{Ind} : K^1(M) \boxtimes K_1^a(M) \to \mathbf{Z}. \tag{1.15}$$

The Toeplitz index interpretation of (1.15) associates to $[D]$ the projections $P^{\pm}$ onto the positive, respectively non-negative, spectral subspaces of D. Then for $u \in K^1(M)$ represented by a unitary $\varphi : M \to U(N) \subset GL(N.\mathbf{C})$ we form the multiplier operator on $\mathcal{H}^N$, again denoted by φ. The operator

$$T_D(\varphi) = P^+ \circ \varphi \circ P^+ - P^- \tag{1.16}$$

is Fredholm on $\mathcal{H}^N$ and (1.15) assigns to u and $[D]$ the integer $\text{Ind}^a(T_D(\varphi))$. This is the approach developed by Baum and Douglas (cf. §20, [8] and [9].) The spectral flow interpretation of (1.15) observes that the involutions

$$\mathcal{E}_0 = P^+ - P^- \text{ and } \mathcal{E}_1 = \varphi \circ (P^+ - P^-) \circ \varphi^*$$

differ by a compact operator (as P^+ is pseudo-differential with symbol of order 0 on M that commutes with the multiplier, φ). Thus, there is an integer invariant, the essential codimension, $EC(\mathcal{E}_1, \mathcal{E}_0)$, which measures the spectral flow (cf. [6]) of a family of operators $\mathcal{E}_t$ interpolating between $\mathcal{E}_0$ and $\mathcal{E}_1$. Wojciechowski [52] proves that

$$EC(\mathcal{E}_1, \mathcal{E}_0) = \text{Ind}^a(T_D(\varphi)) \tag{1.17}$$

so that $EC(\mathcal{E}_1, \mathcal{E}_0)$ also evaluates the pairing (1.15).

With the odd index theorem for coverings, the pairing (1.15) is the crucial aspect to be understood. Let us consider first the simplest case of how to generalize this pairing. Just as the elliptic operator D from $\mathcal{H}_0$ to $\mathcal{H}_1$ determined an even class $[D]_\Gamma$ in (1.12), a self-adjoint operator determines an *odd* class

$$[D]_\Gamma \in KK^1(C(M),\ C_r^*(\Gamma)). \tag{1.18}$$

A unitary $u \in K^1(M)$ determines a bivariant class $[u] \in KK^1(\mathbf{C}, C(M))$, and we can replace the index map $\mu \equiv i^* \boxtimes \_\_$ with the pairing map $\mu_* \equiv [u] \boxtimes \_\_$. The latter changes the parity of the KK-group, so that upon composing we obtain a map

$$[u]_\Gamma : KK^1(C(M), \mathbf{C}) \to KK(\mathbf{C}, C_\Gamma^*)). \tag{1.19}$$

A straightforward calculation of the exterior product with $[u]$ then yields our first line of the following extension of the coverings index theorem to self-adjoint operators:

THEOREM 1. *For D self-adjoint geometric operator on M,*

$$\operatorname{tr}_\Gamma([u]_\Gamma([D])) = \operatorname{tr}_\Gamma(\mu_\Gamma([T_D(\varphi)])) \tag{1.20}$$
$$= \operatorname{Ind}^a(T_D(\varphi)). \tag{1.21}$$

The content of Theorem 1 is that it calculates the analytic index of the Toeplitz operator $T_{\widetilde{D}}(\varphi)$ on $\widetilde{M}$, formed by lifting φ to a multiplier on $\widetilde{M}$ then compressing to the positive eigenspace of $\widetilde{D}$, in terms of data on M. The spectrum of $\widetilde{D}$ need not be discrete (in fact, it could be all of $\mathbf{R}$), so that we obtain from $\operatorname{Ind}(T_D(\varphi)) \neq 0$ the non-triviality of a continuous spectral flow for $\widetilde{D}$ acting on L^2-sections of $\widetilde{E} \to \widetilde{M}$. This is ellaborated on in §2 below. The proof of Theorem 1 is given in §5.

Our interest in the odd index theorem lies deeper, though. We will formulate in section 2 the pairing (1.15) directly on $\widetilde{M}$ in terms of the spectrum of $\widetilde{D}$ and unitary multipliers on $\widetilde{M}$. The classes $u \in K^1(M)$ correspond to the Γ-periodic multipliers obtained by lifting from M to $\widetilde{M}$. It is of great interest to study other classes of unitaries on $\widetilde{M}$ for which the generalized Toeplitz compressions have an "index", with the constraint that it is possible to construct a continuous dimension function that measures the real-valued index of this compression. We discuss in detail this extention for the Γ-almost periodic multipliers which arise from finite-dimensional unitary representations, as described in sections 4 and 5. The more general construction of section 5 corresponds to "compactifications" of Γ, and allows consideration of multipliers with uniform recurrence under the Γ-action (denoted Γ-uniform in section 5), and multipliers which are Γ-random, corresponding to parameter spaces X which are Borel measure spaces. These latter cases are only briefly discussed.

The basic theme is that associated to each unitary multiplier, φ, is a *hull completion* X_φ with continuous Γ-action. From this we obtain a C^*-algebra, A_φ, which is constructed from an associated foliated space, $(V, \mathcal{F}_\varphi)$. The algebra A_φ is always stably isomorphic to the cross-product $C(X_\varphi) \rtimes \Gamma$. If φ is Γ-*amenable*, then there is a trace, $\operatorname{tr}_{A_\varphi}$, on A_φ which defines a continuous dimension function on $KK(\mathbf{C}, A_\varphi)$. The multiplier φ extends to Φ on the foliated space V, which induces a map in KK-theory

$$[\Phi] : KK^1(C(V), A_\varphi) \to KK(\mathbf{C}, A_\varphi) \tag{1.22}$$

that is the "hull-extension" of $[u]_\Gamma$. Composing $[\Phi]$ with $\operatorname{tr}_{A_\varphi}$ yields the generalized index map

$$\operatorname{tr}_{A_\varphi} \circ [\Phi] : KK^1(C(V), A_\varphi) \to \mathbf{R}. \tag{1.23}$$

The operator D lifts to a leafwise operator on $\mathcal{F}_\varphi$, and in anology to the construction of $[D]_\Gamma$, there is a class $[D_\mathcal{F}] \in KK^1(C(V), A_\varphi)$. The real number $\operatorname{tr}_{A_\varphi}([\Phi]([D_\mathcal{F}]))$ then represents the generalized von Neumann index of the

Toeplitz operator $T_{\widetilde{D}}(\varphi)$. (In the case where we allow Φ to be a "random multiplier", we forego the above KK-formalism and directly construct the random Toeplitz operators $T_\epsilon(\varphi_x)$ for $x \in X$ some measure space. The index class is a difference of projections in a von Neumann algebra, W_φ, with the index obtained by applying a trace on this algebra (cf. [17]).

In either the C^*-completion or W^*-completion case, the abstract index $\mathrm{tr}_{A_\varphi}([\Phi]([D_{\mathcal{F}}]))$ has a topological formula derived from the measured foliation index theorem of Connes [18, 19].

Our passing from the Γ-index for coverings to a foliation index is parallel to the use of the foliation index theorem by Connes and Moscovici in their announcement [24], in place of the coverings index theorem used by Atiyah and Schmid [4]. On a technical level, for X_φ compact there is an inclusion $C_r^*(\Gamma) \to A_\varphi$, and the trace tr_{A_φ}, on A_φ restricts to the usual trace on $C_r^*(\Gamma)$. The coverings index theorem applies to operators which are in the commutant of the representatiom of $C_r^*(\Gamma)$ on $\widetilde{\mathcal{H}}$ (cf. [50]). This allows for precisely the Γ-invariant Toeplitz operators. By extending the trace, and representing on a field of Hilbert spaces associated to the foliation, we enlarge the commutant to include our class of "amenable" multipliers and their compressions.

The abstract odd index theory developed in sections 2 through 5 below is applied in section 6 to obtain a relation between eta invariants and a continuous spectral flow on $\widetilde{M}$. Fix the data: D is a self-adjoint first-order elliptic operator on M, and $\alpha : \Gamma \to U(N)$ is an injective representation. We form the associated Hermitian flat $\mathbf{C}^N$-bundle, $E_\alpha \to M$, with Hermitian connection ∇^α. Assume also the existence of a trivialization $\Theta : E_\alpha \cong M \times \mathbf{C}^N$; then the product bundle $M \times \mathbf{C}^N$ has both a horizontal flat connection, ∇^0, and an Hermitian flat connection $\nabla^{\overline{\alpha}} = \Theta_*(\nabla^\alpha)$. Define symmetric extensions $D_0 = D \otimes \nabla^0$ and $D_1 = D \otimes \nabla^{\overline{\alpha}}$ acting on smooth sections of $E^N = E \oplus \ldots \oplus E$. The family

$$\{D_t = (1-t)D_0 + tD_1 \mid 0 \le t \le 1\}$$

will consist of elliptic self-adjoint operators to which we can associate a 1-parameter family of eta-onvariants, and following [3] we define the relative eta invariant

$$\langle [D], [\overline{\alpha}] \rangle_\eta \equiv \eta(D, \alpha, \Theta) \tag{1.24}$$

$$= \int_0^1 \dot{\eta}(D_t)dt. \tag{1.25}$$

The equality (1.24) is to indicate the heuristic viewpoint put forward by Atiyah, Patodi and Singer that $\eta(D, \alpha, \Theta)$ is a real-valued index obtained by pairing D with the "real" K-theory class of $[\overline{\alpha}] = (\alpha, \Theta) \in K^1(M) \otimes \mathbf{R}$ (cf. §7, [3]). In fact, in this geometric context it is possible to replace the right-side column of (1.13) with the above "eta-cup product" (1.24) with $[\overline{\alpha}]$ to obtain a map

$$\begin{array}{c} K^1(C(M), \mathbf{C}) \\ \downarrow [\overline{\alpha}]_\eta \\ \mathbf{R} \end{array} \tag{1.26}$$

We will show that there exists a corresponding φ and algebra A_φ as discussed previously so that the new diagram commutes:

THEOREM 2. *Let M be an odd-dimensional closed manifold and D a self-adjoint geometric operator on M. Then*

$$\operatorname{tr}_{A(\varphi)}([\Phi]([D_{\mathcal{F}}])) = \langle [D], [\overline{\alpha}] \rangle_{\eta}. \tag{1.27}$$

The content of Theorem 2 is to interpret the eta-invariant coupling of Atiyah, Patodi and Singer as a Breuer index in a naturally associated von Neumann context. The left-side of (1.27) was denoted by $\langle [D], [\overline{\alpha}] \rangle_{\Gamma}$ in the announcement [28]. We will give a new proof of (1.27) using methods distinct from those of [29, 31] where the first proofs were given.

The results of this paper grew out of the author's attempt to understand the main theorem of [28] in a context independent of cyclic cohomology. It is a pleasure to thank R. Douglas and J. Kaminker for many conversations which have had relevance to this work, and also M. Ramachandran for his helpful comments. The contents of this paper are an expanded version of the author's talk at the Joint Summer Research Conference of the NSF on "Geometric and Topological Invariants of Elliptic Operators, Bowdoin, July 1988." The author is grateful to the organizers for the opportunity to present this work.

§2. Toeplitz Operators for Coverings.

In this section we introduce the most general class of smooth Toeplitz operators on the covering $\widetilde{M}$ associated to a Γ-periodic elliptic first-order differential operator $\widetilde{D}$. The "index" of these operators will be defined via a boundary map in algebaric K-theory. Our development of the odd index theory on $\widetilde{M}$ exactly parallels J. Roe's discussion of even index theory on open manifolds in (§4, [44]). We postpone to later sections the discussion of two fundamental problems: first, to describe the possible closures of these Toeplitz algebras; and second, the related question of when there exists an appropriate trace with which to assign a real-valued dimension to the abstract indices. The first problem is the odd analog of the theme of Roe [45], with some simplification possible due to the periodicity of D. In section 3 below, we give one answer to the second problem using invariant means on Γ, which parallels (§6, [44]) in our context. However, a principal theme of this paper is that other solutions are possible by considering the "trace" to be define only on a subalgebra of the Toeplitz algebra generated by operators whose symbols come from some "geometric completion" of Γ.

The theory of (smooth) Toeplitz operators on the real line with symbols in the classes we consider here have been thoroughly studied, first by Curto, Muhly and Xia [27], Ji and Xia [39] and Ji and Kaminker [38]. A technical theme developed in Douglas, Hurder and Kaminker [30] and employed essentially in this section, is that Fourier transform on the line is replaced in higher dimensions by the use of the wave operator associated to $\widetilde{D}$. The use of this technique in index theory was pioneered by Roe in [44, 45,46].

We fix a covering $\widetilde{M} \to M$ with Galois group Γ and the lift $\widetilde{D}$ acting on $C_c^{\infty}(\widetilde{M}, \widetilde{E})$. A choice of Riemannian metric on TM lifts to a metric of bounded geometry on $\widetilde{M}$ (cf. §2, [44]). Introduce the Frechet-uniform algebra of smooth complex-valued functions,

$$\mathcal{A} \equiv C^{\infty}_{\text{unif}}(\mathcal{M})$$

which is characterized by the property that given $f \in \mathcal{A}$, for any contractable open set $U \subset M$ and single covering sheet $\widetilde{U} \subset \widetilde{M}$, the restriction $f \mid \widetilde{U}$ defines a function on U with uniform C^{∞}-estimates for the Frechet norm on M. Let $M(N, \mathcal{A})$ denote the algebra of $N \times N$ matrices with entries from $\mathcal{A}$.

Let P^+ denote the projection onto the positive eigenspaces of the closure $\widetilde{D}$ acting in $\widetilde{\mathcal{H}} = L^2(\widetilde{M}, \widetilde{E})$. As $\widetilde{D}$ need not have (any) discrete spectrum, the more precise definition of P^+ is to first introduce the characteristic function

$$h_0(x) = \chi_{(0,\infty)}(x) = \begin{cases} 1 & x > 0 \\ 0 & x \leq 0 \end{cases} \tag{2.1}$$

and use the functional calculus to define $P^+ = h_0(\widetilde{D})$. Then set $P^- = \text{Id} - P^+$. We also need approximating functions

$$h_\epsilon(x) = \begin{cases} 1 & x \geq \epsilon \\ 0 & x \leq \epsilon \end{cases} \tag{2.2}$$

defined by $h_\epsilon(x) = h_1(x/\epsilon)$ where we fix h_1 smooth with $h_1' \geq 0$ and $h_1'(0) = 0$. Set $P_\epsilon = h_\epsilon(\widetilde{D})$ for $\epsilon > 0$, and $P_0 = P^+$. For each $N \geq 1$, by abuse of notation we let P_ϵ also denote the diagonal extension to the direct sum

$$\widetilde{\mathcal{H}}^N = \widetilde{\mathcal{H}} \oplus \ldots \widehat{\oplus} \widetilde{\mathcal{H}} \equiv L^2(\widetilde{M}, \widetilde{E} \otimes \mathbf{C}^N). \tag{2.3}$$

The last ingredient we need for the construction of the Toeplitz extension is the theory of uniform operators on $\widetilde{M}$ from Roe (§5, [44]). For $k \in \mathbf{Z}$, an operator

$$Q : C_c^{\infty}(\widetilde{M}, \widetilde{E} \otimes \mathbf{C}^N) \to C^{\infty}(\widetilde{M}, \widetilde{E} \otimes \mathbf{C}^N) \tag{2.4}$$

is *uniform of order* $\leq k$ if for each $r \in \mathbf{R}$, it has a continuous extension to a quasi-local operator on the Sobolev closures,

$$Q_r : W^r(\widetilde{E} \otimes \mathbf{C}^N) \to W^{r-k}(\widetilde{E} \otimes \mathbf{C}^N). \tag{2.5}$$

By abuse of notation, we let $\mathcal{U}_k$ denote the collection of uniform operators of order $\leq k$, which for $k \leq 0$ form an algebra. We recall the characteristic property of $Q \in \mathcal{U}_k$.

PROPOSITION (5.4, [44]). *Let $Q \in \mathcal{U}_k$, then Q is represented by a distributional kernel $k_Q(x,y)$ on $\widetilde{M} \times \widetilde{M}$:*

$$Qu(x) = \int_{\widetilde{M}} k_Q(x,y)u(y)dy. \tag{2.6}$$

Moreover, k_Q is smooth off the diagonal in $\widetilde{M} \times \widetilde{M}$ and for $k < -m$ there is a function $\nu = \nu(r)$ tending to zero as $r \to \infty$ such that for $r > 0$,

$$\int_{\widetilde{M} - B(x,r)} \{k_Q(x,y)^2 + k_Q(y,x)^2\} dy \leq \nu(r). \tag{2.7}$$ □

In the above, $B(x,r)$ is the metric ball of radius r in $\widetilde{M}$ about x. A key technical tool for index theory is Roe's formulation of kernel estimates due to Cheeger, Gromov and Taylor [54] for operators $f(\widetilde{D})$ defined by the functional calculus. Introduce the space $S^m(\mathbf{R})$ of *symbols* of order $\leq m$ on $\mathbf{R}$ defined as the subspace of $C^\infty(\mathbf{R})$ consisting of functions f with estimates

$$|f^{(\ell)}(x)| \leq C_\ell \cdot (1+|x|)^{m-\ell}. \tag{2.8}$$

The best constants, C_ℓ, for (2.8) define semi-norms on $S^m(\mathbf{R})$, so that this is a Frechet space.

THEOREM (5.5, [44]). *Let D be a geometric operator on M and $f \in S^m(\mathbf{R})$. Then the operator $f(\widetilde{D}) \in \mathcal{U}_m$.* □

We now construct the smooth Toeplitz extension of $\mathcal{A}$ by $\mathcal{U}_{-1}$. Fix $\epsilon > 0$ and let $\mathcal{T}_\epsilon^\infty$ be the algebra of finite sums and products of elements from $\mathcal{U}_{-1}$ and the space of operators

$$\{P_\epsilon \cdot \varphi \cdot P_\epsilon \mid \varphi \in \mathcal{A}\}.$$

There are also matrix versions of $\mathcal{T}_\epsilon^\infty$ for $\varphi \in M(N,\mathcal{A})$. For simplicity of exposition, we discuss the scalar case, and leave to the reader the elementary extensions to the matrix case.

LEMMA 2.1. *For $\varphi \in \mathcal{A}$, the commutator $[P_\epsilon, \varphi] = P_\epsilon\varphi - \varphi P_\epsilon \in \mathcal{U}_{-1}$.*

PROOF: P_ϵ is a pseudo-differential operator of order 0 on $\mathcal{M}$ whose principal symbol commutes with φ, so that $[P_\epsilon, \varphi]$ is a pseudo-differential operator of order -1 and induces maps $S^r(\widetilde{E}) \to S^{r-1}(\widetilde{E})$. We need to show that it is a uniform operator. Introduce the function $g_\epsilon(x) = h_\epsilon(x)/x$ and define $Q_\epsilon = g_\epsilon(\widetilde{D}) \in \mathcal{U}_{-1}$ by the above result of Roe. Calculate

$$\begin{aligned} P_\epsilon\varphi - \varphi P_\epsilon &= DQ_\epsilon\varphi - \varphi D Q_\epsilon \\ &= D[Q_\epsilon, \varphi] + [D,\varphi]Q_\epsilon. \end{aligned} \tag{2.9}$$

The assumption that φ is uniform is easily seen to imply $[Q_\epsilon, \varphi] \in \mathcal{U}_{-2}$ and $[D,\varphi] \in \mathcal{U}_{-1}$. By (Proposition 5.2 [44]) the products $D \cdot [Q_\epsilon, \varphi]$ and $[D,\varphi]Q_\epsilon$ lie in $\mathcal{U}_{-1}$. □

LEMMA 2.2. *The algebra $\mathcal{T}_\epsilon^\infty$ is independent of the choice of $\epsilon > 0$.*

PROOF: Let $\epsilon, \delta > 0$, then the difference $(h_\epsilon - h_\delta) \in S^{-\infty}$, so that

$$(P_\epsilon - P_\delta) \in \mathcal{U}_{-\infty} \subset \mathcal{U}_{-1}.$$ □

PROPOSITION 2.3. *There exists an exact sequence of uniform operators*

$$0 \to \mathcal{U}_{-1} \to \mathcal{T}^\infty \xrightarrow{\sigma} \mathcal{A} \to 0. \tag{2.10}$$

PROOF: Let $\mathcal{T}^\infty = \mathcal{T}_\epsilon^\infty$ for $\epsilon > 0$, and note that $h_\epsilon \in S^0$ so that $\mathcal{T}^\infty \subset \mathcal{U}_0$. The linear map

$$\begin{aligned} T_\epsilon &: \mathcal{A} \to \mathcal{T}^\infty \\ &: \varphi \to T_\epsilon(\varphi) = P_\epsilon \varphi P_\epsilon \end{aligned} \tag{2.11}$$

descends to an algebra map

$$\widehat{T} : \mathcal{A} \to \mathcal{T}^{\infty}/\mathcal{U}_{-1} \tag{2.12}$$

by Lemma 2.1. The map $\widehat{T}$ is clearly onto, and independent of ϵ as $\mathcal{U}_{-1}$ is stable under conjugation by elements of $\mathcal{A}$. So it suffices to show that $\widehat{T}$ is monic and set $\sigma = \widehat{T}^{-1}$. This is a corollary of the following, which we leave to the reader as an exercise. We say that an operator B on $\widetilde{\mathcal{H}}$ is *locally compact* if for all compactly supported functions $f_0, f_1 \in C_c(\widetilde{M})$ the composition $f_0 B f_1$ is a compact operator. Note that each element of $\mathcal{U}_{-1}$ is locally compact.

LEMMA 2.4. *For $\varphi \in \mathcal{A}$ non-zero, the smooth Toeplitz operator $T_\epsilon(\varphi)$ is not locally compact.* □

We call (2.10) the *smooth Toeplitz extension* on $\widetilde{M}$ determined by $\widetilde{D}$. Our notation corresponds to both the smoothness of the symbol algebra $\mathcal{A}$ and the property that the quasi-projector P_ϵ is a uniform operator.

There are several variations on the construction of Toeplitz extensions. First, the algebra $\mathcal{A}$ can be replaced by an algebra of uniformly continuous (matrix-valued) functions on $\widetilde{M}$. However, our Lemma 2.1 fails for φ only continuous, which neccesitates replacing $\mathcal{U}_{-1}$ with a larger algebra. For $\mathcal{M}$ compact, it is customary to replace $\mathcal{U}_{-1}$ with its $*$-closure the Banach algebra of compact operators. For $\widetilde{M}$ identified with the leaf of a foliation, by restricting the symbols to continuous functions which extend continuously to the ambient foliated (compact) manifold, then $\mathcal{U}_{-1}$ can be replaced with the C^*-algebra of the foliation. This is the approach taken in previous works on Toeplitz extensions on open manifolds (cf. [27,30,38,39].) One goal of our approach is to develop the Toeplitz index for the full $\mathcal{A}$, then to study its "completion" as a process associated with defining a numerical index of the Toeplitz operators.

The second alternative Toeplitz extension is formed by replacing P_ϵ with the projection P_0 in $\widetilde{\mathcal{H}}$. Define $\mathcal{T}_0^{\infty}$ to be the algebra generated by finite sums and products of elements of $\mathcal{U}_{-1}$ and the space of bounded operators

$$\{T_0(\varphi) = P_0 \varphi P_0 \mid \varphi \in \mathcal{A}\}.$$

Introduce the class S_0^{-1} of symbols with a jump discontinuity at the origin. More precisely, for a function f on $\mathbf{R}$ and constant $z \in \mathbf{C}$ set:

$$f_z(x) = \begin{cases} f(x) + ze^{-x^2}, & x > 0 \\ f(x), & x \leq 0 \end{cases} \quad \text{, and}$$

$$S_0^{-1} = \{f \mid f_z \in S^{-1} \text{ for some } z \in \mathbf{C}\}. \tag{2.13}$$

Let $\widehat{\mathcal{U}}_{-1}$ be the algebra generated by finite sums and products of elements of $\mathcal{U}_{-1}$ and the space $\{f(\widetilde{D}) \mid f \in S_0^{-1}\}$. Each operator $f(\widetilde{D})$ for $f \in S_0^{-1}$ is locally compact, and as $P_0 = P_\epsilon + f_\epsilon(\widetilde{D})$ where $f_\epsilon = h_0 - h_\epsilon \in S_0^{-1}$, the method of Proof of Proposition 2.4 also yields:

PROPOSITION 2.5. *There is an exact sequence*

$$0 \to \widehat{\mathcal{U}}_{-1} \to \mathcal{T}_0^{\infty} \to \mathcal{A} \to 0 \tag{2.14}$$ □

The algebras $\mathcal{U}_{-1}$ and $\widehat{\mathcal{U}}_{-1}$ are generally distinct, but this depends upon the spectral measure of $\widetilde{D}$ at 0. For example, $\widetilde{M} = \mathbf{R}$ and $\widetilde{D} = -i\frac{d}{dx}$ then the spectral measure is simply $d\xi$, Lebesgue measure on the line, and the Theorem 2.5 of Ji and Xia shows that $K_1(\mathcal{U}_{-1})$ and $K_1(\widehat{\mathcal{U}}_{-1})$ differ (cf. Proposition 3.1, [39]). The operators $f_z(\widetilde{D})$ define bounded maps from $W^r(\widetilde{E} \otimes \mathbf{C}^N)$ to itself by the spectral theorem, but the difficulty is that they need not be uniform operators. The function f_z is not absolutely continuous, hence its Fourier transform is not integrable so that the estimator constructed by Roe (Formula 5.6, [44]) will not tend to zero at infinity. It is possible to impose hypotheses on $\widetilde{D}$ which give $f_z(\widetilde{D})$ is uniform - for example, if the spectral measure of $\widetilde{D}$ is smooth at 0 with vanishing first derivative. Such hypotheses tend to be restrictive on the geometry of $\widetilde{M}$ and operator $\widetilde{D}$ (cf. Chapter 7, [37]).

We next consider the "index" invariants associated to the smooth Toeplitz extension. The algebras $\mathcal{A}$ and $\mathcal{U}_{-1}$ are not Banach closed so that their K-theory need not be periodic. Nonetheless, we investigate only the group $K_1(\mathcal{A})$, as the higher groups $K_n(\mathcal{A})$ require more sophisticated methods of cyclic cohomology (cf. [21,23]). The connecting map in the six-term exact sequence of algebraic K-theory,

$$\partial : K_1(\mathcal{A}) \to K_0(\mathcal{U}_{-1}) \tag{2.15}$$

is called the abstract index map. Following the model of (§4, [44]), we will make this map completely explicit.

Recall that each class in $K_1(\mathcal{A})$ is represented by some $\varphi \in M(N, \mathcal{A})$ such that the pointwise evaluations of φ are invertible and lie in a compact neighborhood of the identity in $GL(N, \mathbf{C})$. For scalar φ, this is the condition that there exists a constant $c > 0$ for which $c > |\varphi| > 1/c$, and the class of φ is equivalent to its unitary part $\varphi/|\varphi|$.

We must add a unit to $\mathcal{T}^{\infty}$, so let $\widetilde{\mathcal{T}}^{\infty}$ denote the algebra spanned by $\mathcal{T}^{\infty}$ and the identity operator, I, on $\widetilde{\mathcal{H}}^N$. Define an operator via the spectral theorem,

$$P_\epsilon^- = h_\epsilon(-\widetilde{D}).$$

Then $I = P_\epsilon + P_\epsilon^- + F_\epsilon$, where $F_\epsilon \in \mathcal{U}_{-\infty} \subset \mathcal{U}_{-1}$. We augment the algebra $\mathcal{U}_{-1}$ by adding P_ϵ^{-1} to obtain $\widetilde{\mathcal{U}}_{-1}$, then there is an exact sequence

$$0 \to \widetilde{\mathcal{U}}_{-1} \to \widetilde{\mathcal{T}}^{\infty} \to \mathcal{A} \to 0. \tag{2.16}$$

We will first construct the index map

$$\widetilde{\partial} : K_1(\mathcal{A}) \to K_0(\widetilde{\mathcal{U}}_{-1}).$$

Fix $\varphi \in GL(N, \mathcal{A})$ representing $[\varphi] \in K_1(\mathcal{A})$. Define operators on $\widetilde{\mathcal{H}}^N$:

$$\begin{aligned} P &= \widetilde{T}_\epsilon(\varphi) = P_\epsilon \varphi P_\epsilon - P_\epsilon^- \\ Q &= \widetilde{T}_\epsilon(\varphi^{-1}) = P_\epsilon \varphi^{-1} P_\epsilon - P_\epsilon^{-1} \\ S &= I - PQ \\ T &= I - QP \end{aligned}$$

We then observe the identities

$$(2.17)\qquad \begin{cases} P_\epsilon P_\epsilon^- &= 0 = P_\epsilon^- P_\epsilon \\ QS &= TQ \\ SP &= PT \end{cases}$$

LEMMA 2.6. $S, T \in \mathcal{U}_{-1}$.

PROOF:

$$(2.18)\qquad \begin{cases} I - PQ &= I - (P_\epsilon \varphi P_\epsilon - P_\epsilon^-)(P_\epsilon \varphi^{-1} P_\epsilon - P_\epsilon^-) \\ &= (I - (P_\epsilon^-)^2) - P_\epsilon \varphi P_\epsilon^2 \varphi^{-1} P_\epsilon \end{cases}$$

which lies in $\mathcal{U}_{-1}$ as the principal symbols of φ and P_ϵ commute, so that both expressions in the last line of (2.18) are congruent to P_ϵ modulo $\mathcal{U}_{-1}$. A similar calculation applies for $T = I - QP$. □

For each integer $\ell > 0$ define

$$(2.19)\qquad \begin{aligned} A_\ell &= (I + S + \ldots + S^{2\ell-1})P - I \\ &= P(I + T + \ldots + T^{2\ell-1}) - I. \end{aligned}$$

PROPOSITION 2.7. *For each integer $\ell > 0$, the class $\widetilde{\partial}[\varphi] \in K_0(\widetilde{\mathcal{U}}_{-1})$ is represented by the difference of projections in $M(2, \widetilde{\mathcal{U}}_{-1})$*

$$(2.20)\qquad \begin{bmatrix} (I + A_\ell)Q & (I + A_\ell)T^\ell \\ T^\ell Q & T^{2\ell} \end{bmatrix} - \begin{bmatrix} I & 0 \\ 0 & 0 \end{bmatrix}.$$

PROOF: For $\ell = 1$ the connecting homomorphism $\widetilde{\partial}$ of algebraic K-theory is explicitly evaluated by Roe (§4. [44]), so we indicate only the necessary changes. Let $R \in GL(2, \widetilde{\mathcal{T}}^\infty)$ be the matrix

$$(2.21)\qquad R = \begin{bmatrix} I + A_\ell & -S^\ell \\ T^\ell & Q \end{bmatrix}$$

with inverse

$$R^{-1} = \begin{bmatrix} Q & T^\ell \\ -S^\ell & I + A_\ell \end{bmatrix}.$$

Note that $R = \text{Ind}$ modulo $\mathcal{U}_{-1}$, then evaluating $\{R\pi R^{-1} - \pi\}$ yields (2.20). □

Note that the difference in (2.20) can be rewritten as

$$(2.22)\qquad \widetilde{\partial}[\varphi] \sim \begin{bmatrix} -S^\ell & * \\ * & T^\ell \end{bmatrix} \in M(2, \mathcal{U}_{-1})$$

using Lemma 2.6, the identities (2.17) and the fact that $P_\epsilon^- \in \mathcal{U}_0$ is a multiplier on $\mathcal{U}_{-1}$. We can therefore identify $\widetilde{\partial}[\varphi]$ with a class $\partial[\varphi] \in K_0(\mathcal{U}_{-1})$ in the kernel of the augmentation-induced map

$$(2.23)\qquad K_0(\mathcal{U}_{-1}) \to K_0(\widetilde{\mathcal{U}}_{-1}) \to K_0(\widetilde{\mathcal{U}}_{-1}/\mathcal{U}_{-1}).$$

A standard diagram chase shows that $\partial[\varphi]$ represents the boundary map of (2.10) applied to $[\varphi]$.

The formula (2.22) for $\partial[\varphi]$ is familiar as the standard expression to which applying a trace yields the index of the "elliptic" operator P (cf. Proposition 6, Appendix 2, [20].) However, the above derivation of (2.22) is purely algebraic without making assumptions on the spectrum of $\widetilde{D}$. We next impose the hypothesis on $\widetilde{D}$ that P_ϵ is a projection, so that $\widetilde{D}$ has a gap in the spectrum between 0 and ϵ. (We could also proceed with the weaker hypothesis that $P_0 \in \mathcal{U}_{-1}$ and $\lim_{\epsilon\to 0} P_\epsilon = P_0$ in the uniform topology; this case is left to the reader.) The formula (2.22) can then be simplified to resemble the more familiar expression involving Hankel operators that is associated with the index of Toeplitz operators.

Let $U(N, \mathcal{A}) \subset GL(N, \mathcal{A})$ be the subgroup of unitary matrices.

Let $\varphi \in U(N, \mathcal{A})$ be unitary-valued with the inverse denoted by φ^*. Set

$$\begin{aligned} T^+(\varphi) &= P_\epsilon \varphi P_\epsilon \\ T^-(\varphi) &= (I - P_\epsilon)\varphi(I - P_\epsilon) \\ H^+(\varphi) &= (I - P_\epsilon)\varphi P_\epsilon \\ H^-(\varphi) &= P_\epsilon\varphi(1 - P_\epsilon). \end{aligned}$$

The identities $\varphi\varphi^* = I = \varphi^*\varphi$ yield corresponding identities for the Toeplitz operators $T^\pm(\varphi)$ and the Hanekl operators $H^\pm(\varphi)$. We need only the following

$$T^+(\varphi)T^+(\varphi^*) + H^-(\varphi)H^+(\varphi^*) = P_\epsilon. \tag{2.24}$$

Proposition 2.8. *Let $\widetilde{D}$ have no spectrum in the interval $(0, \epsilon)$, and assume $[\varphi] \in K_1(\mathcal{A})$ is represented by a unitary-valued $\varphi \in GL(N, \mathcal{A})$. Then*

$$\partial[\varphi] \sim \begin{bmatrix} -(H^-(\varphi)H^+(\varphi^*))^\ell & * \\ * & (H^-(\varphi^*)H^+(\varphi))^\ell \end{bmatrix}. \tag{2.25}$$

Proof: In the proof of Proposition 2.7, we replace the operator P_ϵ^- with $(I - P_\epsilon)$ in the definition of P and Q. The remainder of the proof proceeds as before, as we need only the property

$$(I - P_\epsilon) \cdot P_\epsilon = P_\epsilon - P_\epsilon^2 = 0 \tag{2.26}$$

corresponding to the first identity (2.17). Then formula (2.18) simplifies so that from (2.24) we obtain

$$S = H^-(\varphi)H^+(\varphi^*).$$

The corresponding formula for T then yields (2.25). □

For P_ϵ a projection, the Hankel operator $H^+(\varphi)$ represents the spectral flow from the range of P_ϵ to the range of $(1-P_\epsilon)$ induced by multiplication by φ. This spectral flow is heuristically the kernel of $P_\epsilon\varphi P_\epsilon - (I - P_\epsilon)$. Similarly, $H^-(\varphi)$ represents the spectral flow from range P_ϵ to range $(1 - P_\epsilon)$ induced by φ^*, so represents the kernel of the adjoint $P_\epsilon\varphi^* P_\epsilon - (I - P_\epsilon)$. Our next task is to assign "dimensions" to these spectral flows.

§3. Amenable Covers.

In this section we assume that the Galois group Γ of the covering $\widetilde{M} \to M$ is amenable. Following (§6, [44]) we call $\widetilde{M}$ an *amenable covering*. This hypothesis yields a "trace", tr_Γ, defined on the algebra $\mathcal{U}_{-k}$ for $k > m$ so that we define the Γ-Toeplitz index

$$(3.1) \qquad \begin{aligned} \mathrm{Ind}_\Gamma : & K_1(\mathcal{A}) \to \mathbf{R} \\ & [\varphi] \to \mathrm{tr}_\Gamma(\partial[\varphi]). \end{aligned}$$

The dimension function defined by tr_Γ is a renormalized dimension $\widetilde{\mathcal{H}}^N$, so that the spectral flow induced by $[\varphi]$ is assigned a renormalized dimension by (3.1) via a process analogous to the procedure for measuring the index of almost periodic operators on $\mathbf{R}^m$ by dividing the index on domains by the volume of the domain. The Γ-trace is the natural generalization of this idea (cf. §6, [44]).

There are three common classes of amenable discrete groups, each more general than the previous, but as the generality increases the known results about their Γ-index theory decreases:

(QA) Quasi-abelian. There is a finite group G and an exact sequence

$$0 \to \mathbf{Z}^m \to \Gamma \to G \to 1.$$

(QN) Quasi-nilpotent. There is a finite group G, nilpotent group N and exact sequence

$$1 \to N \to \Gamma \to G \to 1.$$

(QS) Quasi-solvable. There is a finite group G, solvable group S and exact sequence

$$1 \to S \to \Gamma \to G \to 1.$$

A theme of this paper is that while the full algebra $\mathcal{A}$ of symbols may be very hard to explicitly describe, there are important subalgebras of symbol classes corresponding to the representation theory of the group Γ, and for symbols from these subalgebras the index map (3.1) can be reduced to a more familiar (topological) form. For the class (QA), there are the quasi-periodic functions in $\widetilde{M}$ with additional symmetry group G. For Γ quasi-nilpotent, the Kirillov theory parametrizing the unitary representations provides a large class of "Γ-quasi-periodic" symbols. Similarly, finite unitary representations of Γ-quasi-solvable lead to distinguished symbols in $\mathcal{A}$, but for this case both the representation theory and the Toeplitz index theory are less-well understood.

Recall that by Fölner's theorem, Γ is amenable if and only if Γ admits a Fölner sequence: there is an ascending sequence of finite subsets,

$$(3.2) \qquad \{1\} = \Gamma_0 \subset \Gamma_1 \subset \ldots \subset \Gamma; \quad \bigcup_{n=0}^{\infty} \Gamma_n = \Gamma$$

so that for each $\gamma \in \Gamma$ and $\epsilon > 0$, there is $N(\gamma, \epsilon)$ such that $n > N(\gamma, \epsilon)$ implies

$$(3.3) \qquad \mathrm{Card}(\Gamma_n \, \Delta \, \Gamma_n \cdot \gamma) \leq \epsilon \cdot \mathrm{Card}(\Gamma_n)$$

where $(D\Delta B) = (A\backslash B) \cup (B\backslash A)$. For example, when $\Gamma = \mathbf{Z}^m$ we take for Γ_n the cube $\{-n, \dots, n\}^m$. For the classes (QN) and (QS), an inductive procedure starting with the case $\mathbf{Z}^\ell$ yields explicit choices for the Γ_n. (cf. Feldman [33].) We will assume that a choice of $\{\Gamma_n\}$ for Γ has been made.

Define an averaging sequence for $\widetilde{M}$ via the Fölner sequence for Γ by setting

$$M_n = M_0 \cdot \Gamma_n = \bigcup_{\gamma \in \Gamma_n} M_\gamma. \tag{3.4}$$

Each closed set M_n is a union of translates of the fundamental domain whose overlaps have measure zero. Thus $\text{vol}(M_n) = \text{Card}(\Gamma_n) \cdot \text{vol}(M_0)$. For a closed set $A \subset \widetilde{M}$, introduce its r-boundary for $r > 0$:

$$\partial_r(A) = \{y \in \widetilde{M} \mid \text{dist}(y, A) < r, \text{dist}(y, \widetilde{M} - A) < r\}. \tag{3.5}$$

An averaging sequence is *Fölner* if for all $r > 0$,

$$\lim_{n \to \infty} \frac{\text{vol}(\partial_r(M_n))}{\text{vol}(M_n)} = 0. \tag{3.6}$$

PROPOSITION 3.1. *$\{\Gamma_n\}$ is a Fölner sequence for Γ if and only if $\{M_n\}$ is a Fölner averaging sequence for $\widetilde{M}$.*

PROOF: This is a direct consequence of ideas of Plante [42]. Details of the proof are given in Proposition 6.6 of Roe [44]. □

A key idea of Roe's work [44,45] is that a Fölner averaging sequence of $\widetilde{M}$ defines a trace, denoted tr_Γ, on the algebras $\mathcal{U}_{-\ell}$ for $\ell > m$. By a *trace* we mean a linear functional $\tau : \mathcal{U}_{-\ell} \to \mathbf{C}$ such that $\tau(AB) = \tau(BA)$ for all $A, B \in \mathcal{U}_{-\ell}$. A *normal trace* will mean that for suitable topology on $\mathcal{U}_{-\ell}$, if there is a monotone increasing sequence $\{A_1, A_2, \dots\}$ with limit A_∞ and $\lim \tau(A_n) = a_\infty$, then $\tau(A_\infty)$ exists and equals a_∞. The traces obtained from Fölner averaging sequences are generally not normal on all of $\mathcal{U}_{-\ell}$.

To define the linear functional, tr_Γ, first note that each operator $Q \in \mathcal{U}_{-\ell}$ for $\ell > m$ is represented by a continuous kernel, k_Q, on $\widetilde{M} \times \widetilde{M}$, so that its restriction to the diagonal is well-defined. (cf. Proposition 5.8, [47]). For each n, set

$$m_n(Q) = \text{vol}(M_n)^{-1} \cdot \int_{M_n} \text{Tr}(k_Q(y, y))dy. \tag{3.7}$$

The integral in (3.7) depends only on the C^0-topology on kernels, so we can use the weak-$*$ compactness of the bounded linear functionals on $C^0(\widetilde{M})$ to select a convergent net, then set

$$\text{tr}_\Gamma(Q) = \lim{}^* \{m_n(Q)\}. \tag{3.8}$$

THEOREM (THEOREM 6.7, [44]). *The linear functional tr_Γ defines a trace of $\mathcal{U}_{-\ell}$ for $\ell > m$.* □

We remark that a key technical idea behind the proof of this theorem is the *uniform* decay of the kernels in $\mathcal{U}_{-\ell}$, including the L^2-estimate (2.7).

Note that our notation tr_Γ is imprecise, as the weak-$*$ limit process can (and will for $\widetilde{M}$ not compact) produce an uncountable number of inequivalent means, with the axiom of choice used to select one of these. Each such choice yields a distinct "continuous dimension" function on projections in $K_0(\mathcal{U}_{-\ell})$. However, it is often the case that $\mathcal{U}_{-\ell}$ admits preferred subalgebras on which all choices for tr_Γ agree. For example, the kernels on $\mathbf{R}^m$ whose diagonal restriction is almost-periodic have this property. For more general groups Γ we can define Γ-almost-periodic functions on $\widetilde{M}$ with similar properties. Thus, we settle on the notation tr_Γ in recognition that for most practical examples, tr_Γ is determined by Γ.

The tracial property of tr_Γ implies that it defines a linear map $\text{tr}_\Gamma : K_0(\mathcal{U}_{-\ell}) \to \mathbf{R}$ for $\ell > m$. Combining this remark with Roe's Theorem we obtain

PROPOSITION 3.2. *For $\epsilon > 0$, $\ell > m$ and unitary $\varphi \in U(N, \mathcal{A})$, there is a well-defined index, independent of ϵ and ℓ,*

$$\text{Ind}_\Gamma(T_\epsilon(\varphi)) = \text{tr}_\Gamma(I - QP)^\ell - \text{tr}_\Gamma(I - PQ)^\ell. \tag{3.9}$$ □

The Toeplitz operator $T_0(\varphi) = P_0\varphi P_0$ is the strong operator limit of the operators $T_\epsilon(\varphi) = P_\epsilon \varphi P_\epsilon$, and it is natural to ask whether the value of (3.9) is equal to that for $\epsilon = 0$, corresponding to the classical case. However, if tr_Γ is not normal then we cannot show this for all symbols in $\mathcal{A}$. The difficulty is the same as that encountered in Roe [45]. In section 5 we discuss additional regularity hypotheses on the symbol φ which will ensure that $\epsilon = 0$ case yields the same Toeplitz index.

§4. $K_1(\mathcal{A})$ and Representation Theory.

In this section we consider the symbols $\varphi \in GL(N, \mathcal{A})$ arising from the finite-dimensional unitary representations of Γ. We introduce the concept of a *homotopy trivial* representation, then parametrize the resulting symbols by the *gauge group* of maps from M to $U(N)$.

Let $\alpha : \Gamma \to U(N)$ be a representation. A smooth function $\varphi : \widetilde{M} \to U(N)$ is *α-related* if

$$\varphi(x \cdot \gamma^{-1}) = \alpha(\gamma) \cdot \varphi(x); \; x \in \widetilde{M}, \; \gamma \in \Gamma. \tag{4.1}$$

Let $\mathcal{R}_\alpha$ denote the set of α-related maps. Γ acts via isometries on $\widetilde{M}$, so that $\mathcal{R}_\alpha \subset GL(N, \mathcal{A})$. Let us obtain an alternative characterization.

The representation α determines a flat $U(N)$-bundle over M,

$$P_\alpha = \widetilde{M} \times U(N)/(x, g) \sim (x\dot{\gamma}^{-1}, \alpha(\gamma) \cdot g). \tag{4.2}$$

Denote by $C^\infty(M, P_\alpha)$ the set of smooth sections of $\pi : P_\alpha \to M$.

PROPOSITION 4.1. $\mathcal{R}_\alpha \cong C^\infty(M, P_\alpha)$.

PROOF: A section $s : M \to P_\alpha$ is equivalent to a Γ-equivariant map

$$\tilde{s} : \widetilde{M} \to \widetilde{M} \times U(N)$$
$$\tilde{x} \to (\tilde{x}, \varphi(\tilde{x}))$$

where

$$\tilde{s}(\tilde{x}\gamma^{-1}) = (\tilde{x}\cdot\gamma^{-1}, \varphi(\tilde{x}\cdot\gamma^{-1}))$$
$$= (\tilde{x}\cdot\gamma^{-1}, \alpha(\gamma)\cdot\varphi(\tilde{x}))$$

yields that φ is α-related. Conversely, given $\varphi \in \mathcal{R}_\alpha$ we define a smooth section s by specifying $\tilde{s}(\tilde{x}) = (\tilde{x}, \varphi(\tilde{x}))$ then observe that (4.1) implies that $\tilde{s}$ is Γ-equivariant. □

We will say that α is *homotopically trivial* if the bundle $P_\alpha \to M$ admits a continuous section. As a continuous map always admits smooth approximations, this is equivalent to the set $\mathcal{R}_\alpha$ being non-empty. Moreover, we can use Proposition 4.1 to completely characterize the set $\mathcal{R}_\alpha$ when it is non-empty.

Given $s_0 \in C^\infty(M, P_\alpha)$ and a smooth function $g : M \to U(N)$, we use the right $U(N)$-action on P_α to obtain $s_0 \cdot g \in C^\infty(M, P_\alpha)$, where

$$\widetilde{s_0 \cdot g}(\tilde{x}) = (\tilde{x}, s_0(\tilde{x})\cdot g(\pi(\tilde{x}))). \tag{4.3}$$

Conversely, given two sections $s_0, s_1 \in C^\infty(M, P_\alpha)$ for each $x \in M$ there is a unique $g(x) \in U(N)$ such that $s_0(x)\cdot g(x) = s_1(x)$. The resulting function $g : M \to U(N)$ is smooth and satisfies $s_1 = s_0 \cdot g$. Define the gauge group

$$\mathcal{G}(N) = C^\infty(M, U(N)) \tag{4.4}$$

then we have shown

PROPOSITION 4.2. *For α homotopically trivial, choose $s_0 \in C^\infty(M, P_\alpha)$. Then there is topological equivalence*

$$\mathcal{R}_\alpha \cong s_0 \cdot \mathcal{G}(N) \equiv \{s_0 \cdot g \mid g \in \mathcal{G}(N)\}. \tag{4.5}$$ □

COROLLARY 4.3. *Each $\varphi \in \mathcal{R}_\alpha$ determines a homomorphism*

$$\varphi^\# \cdot K^1(M) \to K_1(\mathcal{A}). \tag{4.6}$$

PROOF: Identify $K^1(M) \cong K_1(C^\infty(M))$, then for $g : M \to U(N)$ let $\varphi^\#([g])$ be the class of $\varphi \cdot \tilde{g} \in U(N, \mathcal{A})$. This is well-defined since the inclusion $\pi^* : C^\infty(M) \to \mathcal{A}$ is continuous and the product $\mathcal{A} \times \mathcal{A} \to \mathcal{A}$ is continuous. □

For Γ amenable we can compose the map (4.6) with the Γ-index map, (3.9):

$$\mathrm{Ind}_\Gamma \circ \varphi^\# \cdot K^1(M) \to \mathbf{R}. \tag{4.7}$$

This index map generalizes the periodic symbol index map $\mathrm{tr}_\Gamma \circ [u]_\Gamma$ of (1.19) and (1.20). For α the trivial representation and $\varphi \equiv e$ the constant map onto the identity in $U(N)$, we have for $u = [\varphi]$

$$\mathrm{Ind}_\Gamma \circ e^\#(u) = \mathrm{tr}_\Gamma \circ [u]_\Gamma([1]). \tag{4.8}$$

Unlike the index formula for (1.21), the topological formula for (4.7) derived in section 5 will involve the characteristic classes of the flat bundle foliation on P_α. Equivalently, we must incorporate the Cheeger-Chern-Simons classes of P_α with section s_0 into the index formula.

Let us conclude this section with a discussion of the example $\Gamma = \mathbf{Z}^m$ and $\widetilde{M} = \mathbf{R}^m$, so that $M = \mathbf{T}^m$. A representation $\alpha : \Gamma \to U(N)$ is equivalent to specifying:

(4.9) a compact toral subgroup $\mathbf{T}^\ell \subset U(N)$.

(4.10) a choice of m points $\{\tilde{x}_1, \ldots, \tilde{x}_m\}$ in the cover $\mathbf{R}^\ell \to \mathbf{T}^\ell$ whose images $\{x_1, \ldots, x_m\}$ in $\mathbf{T}^\ell$ generate a dense subgroup.

From this characterization, it is clear that each representation α can be continuously deformed to the trivial representation. Thus, the bundle P_α is always topologically a product, so α is homotopically trivial. Given the data (4.9) and (4.10), there is a canonical (homotopy class of a) section s_0 of P_α, determining $\varphi_0 \in \mathcal{R}_\alpha$. Let α_t be the representation determined by the data (4.9) and the new points $\{t \cdot x_1, \ldots, t \cdot x_m\}$ in (4.10), so that $\alpha = \alpha_1$ and α_0 is trivial. The family of representations for $t \in I \equiv [0,1]$ defines a (U,N)-bundle $\pi : \widetilde{P}_\alpha \to M \times I$. Its restriction to $M \times \{0\}$ is explicitly a product $M \times U(N) \times \{0\}$, with section $(x,0) \to (x,e,0)$. Extend the flat partial connection on $\widetilde{P}_\alpha$ defined by the representations α_t to a full connection on $M \times I$, so that parallel transparent in $\widetilde{P}_\alpha$ along the I-coordinate is defined. Then we define $s_0(x)$ to be the transport from $t = 0$ to $t = 1$ of $(x,e,0) \in \pi^{-1}(x,0)$ to $s_0(x) \in \pi^{-1}(x,1)$. The resulting section s_0 is smooth, and depends up to homotopy only on the choice of points $\{x_0, \ldots, x_m\}$ in (4.10). When applied to the simplest case $\Gamma = \mathbf{Z}$ and $N = 1$, this procedure defines a product structure on $P_\alpha \cong \mathbf{T}^2$ for which the flat foliation is by lines of slope $= x_1$ (cf. examples of [29]).

The integral Chern character

$$\mathrm{ch}^* : K^1(\mathbf{T}^m) \to \bigoplus_{i \text{ odd}} H^i(\mathbf{T}^m; \mathbf{Z}) \tag{4.11}$$

is injective onto a subgroup of finite index. Thus, for $N > m/2$ and for each odd degree cohomology class z on $\mathbf{T}^m$ in the image of ch^*, there corresponds a map $g_z \in \mathcal{G}(N)$. The lifted map, $\varphi_0 \cdot \tilde{g}_x \in \mathcal{R}_\alpha$, then has the properties that it is α-related, and its restriction to each fundamental domain M_γ carries the cohomological structure of z.

The manifold $\mathbf{T}^m$ is Spin$_\mathbf{C}$, so that the Dirac operator D acting on the spinors over $\mathbf{T}^m$ is a geneator of the K-homology group $K_1^a(\mathbf{T}^m)$. This means that for index theory on $\mathbf{T}^m$, the operators $D_\xi \equiv D \otimes \nabla^\xi$, for $\xi \to \mathbf{T}^m$ an Hermitian vector bundle, form a complete set of representatives. Let $\widetilde{D}_\xi$ denote the lift to $\mathbf{R}^m$, and form the corresponding Toeplitz index map

$$\mathrm{Ind}_{\Gamma,\xi} : K_1(\mathcal{A}) \to \mathbf{R}. \tag{4.12}$$

This map depends only on the K-theory class of ξ, so composing with a morphism $\varphi^\#$ we obtain

PROPOSITION 4.4. *For $\Gamma = \mathbf{Z}^m$ and α specified by (4.9) and (4.10), there is a well-defined non-degenerate pairing*

$$\mathrm{Ind}_\Gamma \circ \varphi_0^\# : K^0(\mathbf{T}^m) \times K^1(\mathbf{T}^m) \to \mathbf{R}. \tag{4.13}$$ □

The non-degeneracy of (4.13) follows from the topological formula for index. Note that when α is trivial, (4.13) is the K-theory Poincaré Duality pairing of the $\mathrm{Spin}_{\mathbf{C}}$-oriented manifolds $\mathbf{T}^m$. (cf. [9]). Thus, (4.13) represents Poincaré Duality in "α-twisted K-theory."

The spectrum of the lifted operators D_ξ is of band type; that is, a countable union of closed intervals (cf. Chapter 7, [37]). When the origin lies in a gap in the spectrum, we can apply the analysis at the end of section 3 along with non-degeneracy of (4.13) to conclude there are Toeplitz operators $T_0(\varphi)$ on $\mathbf{R}^m$ with non-trivial renormalized spectral flows.

§5. Hull Closures and the Foliation Toeplitz Index Theorem.

In this section we discuss the most general class of unitary multipliers for which we can define a continuous dimension function on their corresponding Toeplitz index class. We begin with the hull-completion construction - this is analogous to the technique used in the study of almost periodic operators to construct the Bohr space on which the operator admits a continuous extension. We then consider a generalization of the hull-completion in which the Γ-orbits need not be dense. For Γ amenable, or for an appropriate hypothesis on the Γ-action on the hull otherwise, there exists a trace $\mathrm{tr}_{\mathfrak{m}}$ on the index classes arising from the symbols defined over the hull. These Toeplitz operators are part of the theory of foliation Toeplitz operators, which we relate to the Toeplitz extension defined in section 2. The measured foliation index theorem yields a topological formula for the Toeplitz indices with respect to $\mathrm{tr}_{\mathfrak{m}}$. To illustrate our final results, we consider two special cases. The first case of periodic operators yields Theorem 1 of section 1. The second case, for α-related multipliers, introduces Cheeger-Simons characters into the Toeplitz index.

The C^1-hull-closure, X_φ, of a symbol $\varphi \in U(N, \mathcal{A})$ is a topological space with a continuous left Γ-action. We first define the set of translates

$$\Gamma \cdot \varphi = \{(\gamma \cdot \varphi)(y) = \varphi(y \cdot \gamma) \mid \gamma \in \Gamma\}. \tag{5.1}$$

The uniform C^1-norm on $U(N.\mathcal{A})$ is given by

$$||\varphi||_1 = \sup_{y \in \widetilde{M}} ||\varphi(y)|| + \sup_{y \in \widetilde{M}} ||\nabla\varphi\,|_y\,||. \tag{5.2}$$

This induces a topology on $\Gamma \cdot \varphi$, and we let $X_\varphi = \overline{\Gamma \cdot \varphi}$ denote the sequential closure of $\Gamma \cdot \varphi$.

LEMMA 5.1. *There is a canonical continuous Γ-action on X_φ.*

PROOF: Each $\gamma \in \Gamma$ acts uniformly continuously on $\mathcal{A}$ in the uniform C^1-norm. Given $x = \{\gamma_n \cdot \varphi\} \in X_\varphi$, the sequence $\{\gamma \cdot \gamma_n \cdot \varphi\}$ will again be convergent. We set

$$\gamma \cdot \{\gamma_n \cdot \varphi\} = \{\gamma \cdot \gamma_n \cdot \varphi\}.$$

This action is unformly continuous on the dense set $\Gamma \cdot \varphi$, so its continuous extension to X_φ is continuous. □

For each $\gamma \cdot \varphi \in \Gamma \cdot \varphi$, let $\widehat{\gamma\varphi} \in X_\varphi$ denote the stationary sequence it determines. The diagonal action of Γ on $\widetilde{M} \times X_\varphi$ is defined by $\gamma \cdot (y, x) = (y \cdot \gamma^{-1}, \gamma \cdot x)$.

LEMMA 5.2. *There is a continuous map* $\widetilde{\Phi} : \widetilde{M} \times X_\varphi \to U(N)$ *such that*

a) $\widetilde{\Phi}(\gamma(y,x)) = \widetilde{\Phi}(y,x)$, *so descends to a continuous map*

$$\Phi : V \equiv \Gamma\backslash(\widetilde{M} \times X_\varphi) \to U(N).$$

b) *For each* $x \in X_\varphi$, *the restriction*

$$\varphi_x \equiv \widetilde{\Phi}_x : \widetilde{M} \times \{x\} \to U(N)$$

is uniformly C^1.

c) *For each* $\gamma \cdot \varphi \in \Gamma \cdot \varphi$, *the restriction*

$$\widetilde{\Phi} : \widetilde{M} \times \{\widehat{\gamma \cdot \varphi}\} \to U(N)$$

is equal to $\gamma \cdot \varphi$.

PROOF: Given $y \in \widetilde{M}$ and $x = \{\gamma_n \cdot \varphi\}$, set

$$\widetilde{\Phi}(y,x) = \lim_{n\to\infty} \varphi(y \cdot \gamma_n). \tag{5.3}$$

This converges C^1-uniformly, so that for x fixed, $\widetilde{\Phi}(y,x)$ is uniformly C^1 in y. Moreover, the uniform C^1-norm of φ_x depends continuously on x.

The stationary sequence $\gamma \cdot \varphi$ is defined by taking $\gamma_n = \gamma$ for all n, so c) follows from (5.3) directly.

Finally, to prove b) we calculate

$$\begin{aligned}\widetilde{\Phi}(\gamma \cdot (y,x)) &= \widetilde{\Phi}(y \cdot \gamma^{-1}, \gamma x) \\ &= \lim_{n\to\infty} \varphi((y \cdot \gamma^{-1}) \cdot (\gamma \cdot \gamma_n)) \\ &= \lim_{n\to\infty} \varphi(y \cdot \gamma_n) \\ &= \widetilde{\Phi}(y,x).\end{aligned}$$

□

We use the C^1-norm to define X_φ due to b) above. To obtain that the commutator $[P_\epsilon, \varphi] \in \mathcal{U}_{-1}$ in section 2 we needed φ to be uniformly C^1. Thus, b) guarantees a similar result for each multiplier φ_x, $x \in X_\varphi$. The definition of X_φ is obviously modeled on the hull-completion technique used in the theory of almost periodic operators cf. [10,51]. However, in that case, no commutator estimates are needed, so that the uniform C^0-topology for closing $\Gamma \cdot \varphi$ suffices. It is easy to construct multipliers φ on $\mathbf{R}^m$ for which the C^0 and C^1-closures differ, so this is an essential point.

In analogy with the almost periodic case on $\mathbf{R}^n$, we make the following definitions for $\varphi \in U(N, \mathcal{A})$:

φ is γ-*uniform* if X_φ admits a Borel probability measure, $\mathfrak{m}$, which is Γ-invariant.

φ is Γ-*normal* if φ is Γ-uniform and X_φ is compact.

φ is Γ-*almost periodic* if φ is Γ-normal, and $\mathfrak{m}$ is the *unique* Γ-invariant Borel probability measure on X_φ.

For example, if φ is $\mathbf{Z}^m$-normal then a standard lemma of Fourier series theory implies that φ is Γ-almost periodic, as X_φ is a compact abelian group with $\mathbf{Z}^m$ as a dense subgroup. If Γ is not abelian, then there are examples which are normal but not almost periodic.

For Γ amenable, every continuous action on a compact space admits an invariant probability measure $\mathfrak{m}$. Thus, if X_φ is compact then φ is Γ-normal.

In Lemma 5.2 we introduced the quotient space $V = \Gamma\backslash(\widetilde{M} \times X_\varphi)$. Note that V has a foliation, $\mathcal{F}_\varphi$, whose leaves are the images of the level sets $\{\widetilde{M} \times \{x\} \mid x \in X_\varphi\}$. In general, V is only a topological manifold, but the leaves have a natural C^∞-manifold structure. The multiplier φ on $\widetilde{M}$ induces a global function $\Phi : V \to U(N)$, whose restrictions to leaves in C^1, uniformly in V.

Let us consider a special case of this construction, for $\varphi \in \mathcal{R}_\alpha$ where $\alpha : \Gamma \to U(N)$. First, observe

$$\Gamma \cdot \varphi = \{\alpha(\gamma) \cdot \varphi \mid \gamma \in \Gamma\}.$$

Multiplication by $U(N)$ on $U(N, \mathcal{A})$ is C^1-uniformly continuous, so we can identify

$$X_\varphi = \overline{\Gamma \cdot \varphi} = \mathrm{Hull}(\alpha) \cdot \varphi \tag{5.4}$$

where $\mathrm{Hull}(\alpha)$ denotes the closure of the subgroup $\alpha(\Gamma) \subset U(N)$. Clearly, Γ acts on X_φ via multiplication on $\mathrm{Hull}(\alpha)$ on the left with dense orbits. By the compactness of $U(N)$, hence of $\mathrm{Hull}(\alpha)$, and uniqueness of Haar measure on compact Lie groups we conclude that φ is Γ-almost periodic. If α is faithful, then the Γ-action on X_φ will be fixed-point free. Finally, identify $X_\varphi = \mathrm{Hull}(\alpha) \subset U(N)$, then $\widetilde{\Phi}$ is given by

$$\widetilde{\Phi} : \widetilde{M} \times \mathrm{Hull}(\alpha) \to U(N)$$
$$\widetilde{\Phi}(y, g) = g \cdot \varphi(y).$$

The construction of the foliated space V above has an extension that yields the most general class of symbols for which a trace function will be defined. Let X denote a topological space and $\mathfrak{m}$ a Borel probability measure. Denote by $\mathrm{Homeo}(X, \mathfrak{m})$ the subsets of the homeomorphisms of X that preserve the measure $\mathfrak{m}$. Given a group homomorphism

$$\alpha : \Gamma \to \mathrm{Homeo}(X, \mathfrak{m}),$$

we associate a topological manifold

$$V = \widetilde{M} \times X/(y, x) \sim (y \cdot \gamma^{-1}, \alpha(\gamma) \cdot x)$$

with foliation $\mathcal{F}_\alpha$ whose leaves L_x are the image of level sets $\{\widetilde{M} \times \{x\} \mid x \in X\}$. Again, the leaves of $\mathcal{F}_\alpha$ have a natural smooth manifold structure. Each leaf of $\mathcal{F}_\alpha$ is a covering of M via the fibration map $\pi : V \to M = \Gamma\backslash\widetilde{M}$.

There are several variations on the above definition. We can restrict attention to X a compact smooth manifold with a volume form ω of total mass 1, then let $\alpha : \Gamma \to \mathrm{Diffeo}(X,\omega)$ represents into ω-preserving diffeomorphisms. The quotient V will be a smooth compact foliated manifold. At the other extreme, we can take X to be a Borel measure space with probability measure $\mathfrak{m}$, and let $\mathrm{Aut}(X, \mathfrak{m})$ be the group of Borel isomorphisms of X that preserve $\mathfrak{m}$. Then V is a foliated measure space in the broadest sense of [41].

Define $\mathcal{C}(V/\mathcal{F}_\alpha)$ to be the set of uniformly continuous maps $\varphi : V \to \mathbf{C}$ whose restrictions to leaves of $\mathcal{F}_\alpha$ are C^1 uniformly in V. Given φ, for each $x \in X$ we thus obtain a uniformly C^1-map

$$\varphi_x : \widetilde{M} = \widetilde{M} \times \{x\} \to \Gamma\backslash(\widetilde{M} \times X) \to \mathbf{C}. \tag{5.5}$$

(In the case X is a Borel measure space, we consider the functions $\mathcal{M}(V/\mathcal{F}_\alpha)$ which are Borel on V and whose restrictions to leaves are uniformly C^1, with Borel dependence on the parameter x.)

We fix the differential operator D on $C^\infty(M, E)$, then use the covering property of the leaves of $\mathcal{F}_\alpha$ to lift it to a leafwise operator $D_{\mathcal{F}}$ for $\mathcal{F}_\alpha$ acting on sections (over the holonomy covers of leaves) of $\widetilde{E} = \pi^* E$. (cf. §3, [31]). These operators are *leafwise* essentially self-adjoint, so that we can leafwise apply the spectral theorem to obtain a family of operators,

$$\begin{cases} P_\epsilon &= \{P_{\epsilon,x} : \widetilde{\mathcal{H}}_x^N \to \widetilde{\mathcal{H}}_x^N \mid x \in X\} \\ \widetilde{\mathcal{H}}_x^N &= L^2(\widetilde{M}_x, \widetilde{E}\mid_{\widetilde{M}_x} \otimes \mathbf{C}^N). \end{cases} \tag{5.6}$$

where $\widetilde{M}_x \to M$ is the covering of M that is also the holonomy cover of the leaf, L_x, through x.

For a symbol $\varphi \in U(N, \mathcal{C}(V/\mathcal{F}_\alpha))$ we construct leafwise Toeplitz operators

$$T_\epsilon(\varphi) = \{T_{\epsilon,x}(\varphi) = P_{\epsilon,x} \cdot \varphi \cdot P_{\epsilon,x} \mid x \in X\}. \tag{5.7}$$

The proof of Lemma 2.1 required only that φ be uniformly C^1, so that it applies leaf-by-leaf in the above context. Introduce the algebra $\mathcal{U}_{-1}(V/\mathcal{F}_\alpha)$ of leafwise operators on $\widetilde{\mathcal{H}}$, so that the representing kernels $k_{Q,x} \in \mathcal{U}_{-1}(\widetilde{M}_x)$ and depend uniformly continuously on x. We then obtain the foliated version of (2.10):

PROPOSITION 5.3. *There exists an exact sequence of uniform foliation operators*

$$0 \to \mathcal{U}_{-1}(V/\mathcal{F}_\alpha) \to \mathcal{T}(V/\mathcal{F}_\alpha) \to \mathcal{C}(V/\mathcal{F}_\alpha) \to 0. \tag{5.8}$$ □

By its construction, the sequence (5.8) maps naturally to the Toeplitz C^*-extension associated to $D_{\mathcal{F}}$. For each $x \in X$, there is a representation of the algebras $\mathcal{U}_{-1}(V/\mathcal{F}_\alpha)$ and $\mathcal{T}(V/\mathcal{F}_\alpha)$ on $\widetilde{\mathcal{H}}_x$ which defines a C^*-semi-norm. Let $C^*(V/\mathcal{F}_\alpha)$ and $\overline{\mathcal{T}}(V/\mathcal{F}_\alpha)$ denote their respective closures with respect to the supremum over X of these semi-norms.

PROPOSITION 5.4. *There is a commuting diagram of exact sequences*

$$\begin{array}{ccccccccc} 0 & \to & \mathcal{U}_{-1}(V/\mathcal{F}_\alpha) & \to & \mathcal{T}(V/\mathcal{F}_\alpha) & \to & \mathcal{C}(V/\mathcal{F}_\alpha) & \to & 0 \\ & & \downarrow & & \downarrow & & \downarrow & & \\ 0 & \to & C^*(V/\mathcal{F}_\alpha) & \to & \overline{\mathcal{T}}(V/\mathcal{F}_\alpha) & \xrightarrow{\sigma} & C^0(V) & \to & 0 \end{array} \tag{5.9}$$

PROOF: Exactness of the bottom row is proved in [30]. □

A Γ-invariant probability measure, $\mathfrak{m}$, on X induces a transverse invariant measure for $\mathcal{F}_\alpha$ (cf. [42]). This in turn, defines a densely defined trace $\mathrm{tr}_{\mathfrak{m}}$ on $C^*(V/\mathcal{F}_\alpha)$ (cf. Chapter 4, [41]), which is given (locally) by restricting a kernel to the leaf diagonals and integrating, then integrating transversally with $\mathfrak{m}$. In particular, the trace is defined on the image of $\mathcal{U}_{-\ell}(V/\mathcal{F}_\alpha)$ for $\ell > m$ which consists of operators represented by continuous kernels.

The K-theory connecting map for the bottom row of (5.9),

$$\partial_{\mathcal{F}} : K_1(C^0(V)) \to K_0(C^*(V/\mathcal{F}_\alpha)) \tag{5.10}$$

has range in the domain of $\mathrm{tr}_{\mathfrak{m}}$, so we obtain an odd analytic index map

$$\mathrm{Ind}_{\mathfrak{m}} = \mathrm{tr}_{\mathfrak{m}} \circ \partial_{\mathcal{F}} : K_1(C^0(V)) \to \mathbf{R}. \tag{5.11}$$

The Toeplitz operator analogue of Connes' measured foliation index theorem [18] yields a topological formula for this index. Let $C_{\mathfrak{m}}$ denote the Ruelle-Sullivan homology class on V associated to $\mathfrak{m}$. The following is proved in [30]:

THEOREM. *For $\varphi : V \to U(N)$,*

$$Ind_{\mathfrak{m}}[\varphi] = (-1)^m \langle \psi^{-1} ch^*(\sigma_D) \cup ch^*(\varphi) \cup Td(\mathcal{F}), C_{\mathfrak{m}} \rangle \tag{5.12}$$

where ψ is the Thom isomorphism for the leafwise unit cotangent bundle $S^1\mathcal{F} \to V$, $ch(\sigma_D)$ is the Chern character for the positive eigenbundle over $S^1\mathcal{F}$ determined by the fiberwise involution σ_D which is the principal symbol of $D_{\mathcal{F}}$, and $Td(\mathcal{F})$ is the leafwise Todd class. □

REMARK 5.5: The index formula (5.12) also holds for symbols $\varphi \in \mathcal{M}(V/\mathcal{F}_\alpha)$ which are essentially bounded on V. This uses the full force of the measured foliation index theorem, as it applies to foliated measure spaces with uniformly bounded leaf geometry. The current $C_{\mathfrak{m}}$ is transverse to $\mathcal{F}$, so all of the differential data in (5.12) is integrated leafwise. Thus, $\mathrm{ch}^*(\varphi)$ need only be a leafwise form for the right-hand-side of (5.12) to be defined. □

We let $\mathcal{C}(\widetilde{M})$ denote the closure of $\mathcal{A}$ in the uniform C^1-topology, and introduce $\mathcal{T}$, the algebra generated by $\mathcal{U}_{-1}$ and operators $P_\epsilon \varphi P_\epsilon$ for $\varphi \in \mathcal{C}(\widetilde{M})$. We again have an exact sequence

$$0 \to \mathcal{U}_{-1} \to \mathcal{T} \to \mathcal{C}(\widetilde{M}) \to 0, \tag{5.13}$$

containing (2.10) as a sub-exact-sequence. We want to define a map from the top row of (5.9) to (5.13). For each $x \in X$ the composition (5.5) defines a restriction map

$$r_x : \mathcal{C}(V/\mathcal{F}_\alpha) \to \mathcal{C}(\widetilde{M}). \tag{5.14}$$

However, to obtain maps between the first two terms of these exact sequences, we need an additional geometric hypothesis.

The action $\alpha : \Gamma \times X \to X$ is *essentially free at* $x_0 \in X$ if for each $e \neq \gamma \in \Gamma$, x_0 is *not* an interior point of the fixed-point set

$$X_\gamma = \{x \mid \alpha(\gamma)x = x\}.$$

We say α is *essentially free* if X_γ has no interior for all $\gamma \in \Gamma$ not the identity.

LEMMA 5.6. *Let X_φ be the hull-completion of φ. If $\hat{\varphi} \in X_\varphi$ is not a fixed-point for all non-identity $\gamma \in \Gamma$, then $\Gamma \times X_\varphi \to X_\varphi$ is essentially free.*

PROOF: Let $x \in U \subset X_\varphi$ and suppose that $U \subset X_\gamma$ is open. Then $x \in \overline{\Gamma \cdot \hat{\varphi}}$ implies some $\delta \cdot \hat{\varphi} \in U$. Thus $\gamma \cdot \delta\hat{\varphi} = \delta\hat{\varphi}$, so $\hat{\varphi}$ is a fixed-point for $\delta^{-1}\gamma\delta \in \Gamma$ which must be the identity, and so $\gamma = e$. □

The following is a standard property of the holonomy cover (cf. Connes [19], Haefliger [35]):

LEMMA 5.7. *If α is locally free at x, then the holonomy cover $\widetilde{M}_x$ of the leaf L_x of $\mathcal{F}_\alpha$ through x is the full Galois cover $\widetilde{M} \to M$. If α is essentially free, then all holonomy covers are equal to $\widetilde{M}$.* □

COROLLARY 5.8. *If α is locally free at $x \in X$, then there is a restriction map of exact sequences:*

$$(5.15)\qquad \begin{array}{ccccccccc} 0 & \to & \mathcal{U}_{-1}(V/\mathcal{F}_\alpha) & \to & \mathcal{T}(V/\mathcal{F}_\alpha) & \to & \mathcal{C}(V/\mathcal{F}_\alpha) & \to & 0 \\ & & \downarrow r_x & & \downarrow r_x & & \downarrow r_x & & \\ 0 & \to & \mathcal{U}_{-1} & \to & \mathcal{T} & \to & \mathcal{C}(\widetilde{M}) & \to & 0 \end{array}$$

PROOF: Each operator on the top row restricts to an operator on $\widetilde{\mathcal{H}}_x$ which is canonically identified to $\widetilde{\mathcal{H}} = L^2(\widetilde{M}, \widetilde{E})$. The restrictions clearly lie in the appropriate algebras. □

For x such that $\widetilde{M}_x$ is not equal to $\widetilde{M}$, it is not possible to define (in general) the diagram (5.15). The difficulty is that the uniform operators do not behave well when lifted to a covering, for both their domains can drastically change and the decay condition away from the diagonal can fail.

We can now define the generalized Toeplitz index of special multipliers. We say that $\varphi \in U(N, \mathcal{A})$ is *amenable* if φ is Γ-uniform, and a choice of invariant measure $\mathfrak{m}$ on X_φ is called a *mean* fo φ. We say that φ is *uniquely amenable* if φ is Γ-almost periodic. Our previous discussion has shown:

LEMMA 5.9. *If φ is α-related for $\alpha : \Gamma \to U(N)$, then φ is uniquely amenable.* □

Continuing the discussion of the introduction, we set for a choice of mean $\mathfrak{m}$,

$$(5.16)\qquad A_\varphi = C^*(V/\mathcal{F}_\varphi); \mathrm{tr}_A = \mathrm{tr}_{\mathfrak{m}}.$$

For an amenable multiplier φ, the generalized index of the Toeplitz operator $T_\epsilon(\varphi) \in \mathcal{T}$ on $\widetilde{M}$ is

$$(5.17)\qquad \mathrm{Ind}_{\mathfrak{m}}([\varphi]) = \mathrm{tr}_{\mathfrak{m}} \partial_{\mathcal{F}}[\Phi]$$

where Φ is the continuous extension of φ to V.

In general, there is no closed procedure for describing the index (5.17) directly in terms of the kernel $\partial[\varphi] \in \mathcal{U}_{-1}$. Let us recapitulate its construction: The set of Γ-translates of $\partial[\varphi]$ form a subset of the image

$$r_x : \mathcal{U}_{-1}(V/\mathcal{F}_\varphi) \to \mathcal{U}_{-1}.$$

The next step is to define the mean in terms of these translates. Unfortunately, for a Γ-uniform multiplier φ it is possible that the orbit of $\hat{\varphi} \in X_\varphi$ is not even contained in the support of $\mathfrak{m}$. As X_φ consists of the limit sequences of translates of φ, $\Gamma \cdot \hat{\varphi}$ disjoint from support ($\mathfrak{m}$) is equivalent to $\mathfrak{m}$ being defined only on the ∞-points in Γ, or equivalently is determined by the behavior of Γ "at infinity". As such, the abstract mean $\mathfrak{m}$ is then impossible to calculate from the restrictions of $\partial[\varphi]$ to compact sets in $\mathcal{M}$.

Suppose that Γ is amenable, and φ is uniquely amenable and Γ acts freely on $\hat{\varphi}$, then we can derive an explicit recipe for $\mathrm{Ind}_{\mathfrak{m}}(T_\epsilon(\varphi))$. Moreover, this formula calculates the index for $\epsilon = 0$. By Lemma 5.6, Γ acting on X_φ is essentially free, so that for $x = \hat{\varphi} \in X_\varphi$ we have the restriction

$$\mathrm{tr}_{\mathfrak{m}} \circ r_x : \mathcal{U}_{-\ell}(V/\mathcal{F}_\varphi) \to \mathcal{U}_\ell \to \mathbf{C}. \tag{5.18}$$

Uniqueness of measure $\mathfrak{m}$ implies the uniqueness of the trace on $\mathcal{U}_{-\ell}(V/\mathcal{F}_\varphi)$, so that $\mathrm{tr}_{\mathfrak{m}} \circ r_x$ must agree (up to a scale factor) with the traces on $U_{-\ell}$ constructed in §3, restricted to this subalgebra. Moreover, by uniqueness the lim*-construction reduces to an ordinary limit on these kernels. Thus we have:

THEOREM 5.10. *Let Γ be amenable and φ a uniquely amenable multiplier on which Γ acts freely. Then for any Fölner sequence $\{\Gamma_n\} \subset \Gamma$,*

$$Ind_{\mathfrak{m}}([\varphi]) = vol(M)^{-1} \lim_{n\to\infty} \frac{1}{\#\Gamma_n} \int_{M_n} Tr(\partial[\varphi]) \cdot dy. \tag{5.19}$$

Moreover, there is a topological formula

$$= vol(M)^{-1} \lim_{n\to\infty} \frac{1}{\#\Gamma_n} \int_{M_n} ch^*(\varphi) \cup \psi^{-1} \cdot ch(\sigma_D) \cup Td(\widetilde{M}) \tag{5.20}$$

where $ch^*(\varphi)$ is the pull-back to $\widetilde{M}$ of the universal Chern form on $U(N)$. □

The identification of (5.20) with the right-hand-side of (5.12) follows from the usual description of the transverse invariant measure for $\mathcal{F}_\varphi$ obtained from $\mathfrak{m}$ as an asymptotic cycle [42], which is the domain of integration M_n viewed as a subset of a typical leaf in V.

Theorem 5.10 is the Toeplitz index analogue of Roe's Index Theorem (8.2, [44]) for open manifolds with mean derived from Γ. Deriving this formula from the foliation index theorem has the additional advantage that we can embed the problem into a von Neumann context. As proven in [30], for $\epsilon = 0$ the Toeplitz operator $T_0(\varphi)$ has an index class in $W^*(V/\mathcal{F}_\varphi)$ the von Neumann algebra of $(\mathcal{F}_\varphi, \mathfrak{m})$, and its $\mathfrak{m}$-index equals the expressions (5.19) and (5.20). In the special case that φ is α-related, the extended multiplier $\widetilde{\Phi}$ is related to φ by multiplication by $h \in \mathrm{Hull}(\alpha)$. Thus, the corresponding leafwise Toeplitz operators are similarly related, so their index forms (2.22) are independent of $x \in X_\varphi$. Thus, we conclude the limits $T_\epsilon(\varphi)$ as $\epsilon \to 0$ are leafwise uniform, so that for Γ amenable formula (5.20) remains true for $T_0(\varphi)$.

Let us describe the main application we have for formula (5.12) and Theorem 5.10. Fix a closed subgroup $H \subset U(N)$ and a faithful representation $\alpha : \Gamma \to H$ with dense image. Form the principal bundles

$$\widehat{P}_\alpha = \widetilde{M} \times H/(y,h) \sim (y \cdot \gamma^{-1}, \alpha(\gamma)h)$$
$$P_\alpha = \widetilde{M} \times U(N)/(y,g) \sim (y \cdot \gamma^{-1}, \alpha(\gamma)g).$$

There is an inclusion of bundles

$$\begin{array}{ccc} H & \subseteq & U(N) \\ \downarrow & & \downarrow \\ \widehat{P}_\alpha & \to & P_\alpha \\ \searrow & & \swarrow \\ & M & \end{array}$$

A section $s : M \to P_\alpha$ is equivalent to an α-related multiplier $\varphi : \widetilde{M} \to U(N)$ by Proposition 4.2, and the above construction yields all of the multipliers in $\mathcal{R}_\alpha$ for $H = \text{Hull}(\alpha)$. The symbol φ is Γ-almost periodic, with $\mathfrak{m}$ the left-invariant Haar measure on H. The generalized index, $\text{Ind}_{\mathfrak{m}}([\varphi])$ is defined by (5.11) with Φ given by $\widetilde{\Phi}(y \cdot \gamma^{-1}, \alpha(\gamma) \cdot h) = h \cdot \varphi(y)$ for $(y,h) \in \widetilde{M} \times H$. The topological formula (5.12) simplifies to (5.20) if Γ is amenable.

The remarkable property of the generalized Toeplitz index of $\varphi \in \mathcal{R}_\alpha$ is that the topological formula for $\text{Ind}_{\mathfrak{m}}([\varphi])$ has an altenative description in terms of the Cheeger-Simons class of the flat bundle $P_\alpha \to M$ with section s, [16]. This is an odd dimensional cohomology class which we denote by

$$\text{Tch}(\varphi) \in H^{\text{odd}}(M; \mathbf{R}),$$

as both α and s are determined by φ. The identification $H = \text{Hull}(\alpha) = X_\varphi$ identifies $\widehat{P}_\alpha = V_\varphi$. The fibration map $\pi : \widehat{P}_\alpha \to M$ lifts $\text{Tch}(\varphi)$ to a class $\pi^*(\text{Tch}(\varphi))$ on $\widehat{P}_\alpha$.

THEOREM (THEOREM 4.8, [31]). *Let* $\varphi \in \mathcal{R}_\alpha$. *Then*

(5.21) $$\text{Ind}_{\mathfrak{m}}([\varphi]) = (-1)^m \langle \pi^*(\text{Tch}(\varphi) \cup \text{Td}(M)) \cup \psi^{-1}\text{ch}(\sigma_D), C_{\mathfrak{m}} \rangle \qquad \square$$

For Γ amenable, the content of (5.21) is to equate the leafwise differential forms representing $\pi^*(\text{Tch}(\varphi))$ with the form $\text{ch}^*(\varphi)$ appearing in (5.20).

We conclude this section by deriving the proof of Theorem 1 of the introduction. Introduce the topological space

$$X = \times_{\delta \in \Gamma} \{0,1\}_\delta$$

with typical point $x = \{x_\delta \mid \delta \in \Gamma\}$. The δ-coordinate of x is x_δ. The basic open sets are the cylinders: for $\Delta = \{\delta_1, \dots, \delta_p\} \subset \Gamma$ and $n_1, \dots, n_p \in \{0,1\}$

$$X_\Delta = \{x \in X \mid x_{\delta_i} = n_i\}.$$

Γ acts on X via translation:

$$\gamma \cdot \{x_\delta\} = \{x_{\gamma \cdot \delta}\}$$

which is continuous. Give each factor $\{0,1\}_\delta$ the $(1/2, 1/2)$-measure and X the resulting product measure, $\mathfrak{m}$. Observe, for example, that $\mathfrak{m}(X_\Delta) = 2^{-p}$. The measure $\mathfrak{m}$ is Γ-invariant. It is an easy exercise to show the Γ-action is essentially-free, although it is never free. (The constant sequences provide two fixed points, and if $\Gamma = \mathbf{Z}^m$ then the periodic orbits are dense.)

Form the topological space $V = \Gamma \backslash (\widetilde{M} \times X)$ as above, where all of the holonomy covers $\widetilde{M}_x = \widetilde{M}$ for $\mathcal{F}$. Observe that $\varphi : M \to U(N)$ lifts to $\varphi : V \to U(N)$ which is smooth on leaves. We can thus define the Toeplitz foliation index for $T_\epsilon(\varphi)$, and (5.12) gives a topological formula for $\mathrm{Ind}_{\mathfrak{m}}([\varphi])$. The key point is that both the lifts $D_{\mathcal{F}}$ and φ are independent of x so that φ, $T_\epsilon(\varphi)$ and $\partial[\varphi]$ are independent of x under the identification $\widetilde{M}_x \cong \widetilde{M}$. Consequently, the restriction to the diagonal of the pointwise trace of (2.22) descends to a function on M. The $\mathfrak{m}$-trace used in $\mathrm{tr}_{\mathfrak{m}} \partial[\varphi]$ is thus calculated by restricting $\mathrm{Tr}(\partial[\varphi])$ to a fundamental domain $M_0 \subset \widetilde{M} \cong \widetilde{M}_x$ in a typical leaf as the fiber X has $\mathfrak{m}$-mass 1. Hence

$$\mathrm{Ind}_{\mathfrak{m}}[\varphi] = \mathrm{vol}(M)^{-1} \int_{M_0} \mathrm{Tr}(\partial[\varphi]) dy. \tag{5.22}$$

The right side of (5.22) is precisely the Γ-trace of the difference of projections $\partial[\varphi]$. By (§7, Kasparov [40]), this is equal to the left-hand-side of (1.20).

The topological formula (5.12) for $\mathrm{Ind}_{\mathfrak{m}}[\varphi]$ also reduces to an integral over M as all of the symbol data is lifted from M. The topological index (5.12) thus agrees with the topological index for $T_D(u)$, defined in the introduction. Combining the foliation index theorem with the odd index theorem for compact manifolds yields equality (1.21).

§6. The Relative Eta Invariant for Coverings.

A principal theme of this paper is that the odd index theorem for coverings applies to more than just periodic multipliers on the covering. The α-related multipliers provide the next largest natural class, and for these the analytic index is calculated via a trace on a foliation C^*-algebra associated to the hull closure $\mathrm{Hull}(\alpha)$ of the unitary representation α. It was observed in section 4 that there is a close relationship between α-multipliers and sections of associated flat bundles. The purpose of this section is to prove that correspondingly, the Breuer index $\mathrm{Ind}_{\mathfrak{m}}(\varphi)$ for $\varphi \in \mathcal{R}_\alpha$ is equal to an analytic invariant defined directly on M in terms of D and the flat bundle data (P_α, s) determined by φ. This is the content of Theorem 2 of the Introduction for which we give a direct analytic proof below.

Theorem 2 was first proven in the special case Γ is abelian in [29] and in full generality in [31]. Both papers use the method of renormalization applied in cyclic cohomology, but their essence was to reduce the equality between the two sides of (1.27) to an equality between two integer-valued index problems, and then observe that the symbol data in both Fredholm contexts were homotopic. In contrast, we use here the definition of the eta-invariant in terms of an integral involving the heat kernel to directly relate the relative eta invariant to a spectral flow on $\widetilde{M}$ involving P_ϵ and φ.

More precisely, it is possible to define a "type II eta invariant" of $\widetilde{D}$ directly on $\widetilde{M}$. For the signature operator this was done by Cheeger and Gromov [14], and for more general geometric (Dirac) operators independently by G. Peric [53] and M. Ramachandran [43]. This type II eta invariant is the same for $\widetilde{D}$ and its unitary conjugate $\varphi^* \widetilde{D} \varphi$, as φ extends to a unitary Φ on the full C^*-algebra A_φ associated to φ. Thus, a path between $\widetilde{D}$ on $\varphi^* \widetilde{D} \varphi$ yields a family of eta invariants for which the integral of its distributional derivative is zero. This distributional derivative is the difference of two terms, the first is smooth and locally calculable from $\widetilde{D}$ and $\varphi^* \widetilde{D} \varphi$. Its integral is equal to the relative eta-invariant on M. The second term is non-local, and integrates to the spectral flow between $\widetilde{D}$ and $\varphi^* \widetilde{D} \varphi$. In the type II context, this spectral flow is again equal to the Breuer index of $T_\epsilon(\varphi)$, proving Theorem 2. The rest of this section is devoted to filling in the details of this sketch of proof.

Fix the geometric operator D, a representation $\alpha : \pi_1(M) \to U(N)$ with image isomorphic to a quotient Γ of $\pi_1(M)$, and section s of the associated flat principal $U(N)$-bundle P_α. The vector bundle E_α of the Introduction is associated to P_α via the natural action of $U(N)$ on $\mathbf{C}^N$. This bundle lifts to product $\widetilde{M} \times \mathbf{C}^N$ over the Γ-covering $\widetilde{M} \to M$, and s lifts to an α-related multiplier $\varphi : \widetilde{M} \to U(N)$. Note that the section s induces a trivialization

$$\begin{aligned} M \times U(N) &\cong P_\alpha \\ (x, g) &\to s(x) \cdot g \end{aligned}$$

which in turn induces Θ of the Introduction.

We have two extension of D to $C^\infty(M, E \otimes \mathbf{C}^N)$: the product extension $D_0 = D \otimes \nabla^0$, and that defined via the flat connection $\nabla^{\overline{\alpha}}$ from E_α, $D_1 = D \otimes \nabla^{\overline{\alpha}}$. The operator D_0 lifts to $\widetilde{D}_0 = \widetilde{D} \otimes \nabla^0$ on $C^\infty(\widetilde{M}, \widetilde{E} \otimes \mathbf{C}^N)$.

LEMMA 6.1. *The operator D_1 lifts to $\widetilde{D}_1 = \varphi^* \widetilde{D}_0 \varphi$ on $C^\infty(\widetilde{M}, \widetilde{E} \otimes \mathbf{C}^N)$.*

PROOF: The map induced by Θ on sections is implemented by φ when lifted to $\widetilde{M}$. Given $f \in C^\infty(M, E \otimes \mathbf{C}^N)$, the lift $\tilde{f}$ is Γ-periodic, so the product $\varphi \cdot \tilde{f}$ is α-related so descends to a section $\varphi(f) \in C^\infty(M, E_\alpha)$. Clearly, the inverse is multiplication by the adjoint φ^*. By definition, the extension $D \otimes \nabla^\alpha$ on $C^\infty(M, E_\alpha)$ lifts to the product extension $\widetilde{D} \otimes \nabla^0$ on $\widetilde{M}$, and D_1 is the push-forward of $D \otimes \nabla^\alpha$ via Θ, so that lifting to $\widetilde{M}$ we have

$$\widetilde{D}_1 = \varphi^* (\widetilde{D} \otimes \nabla^0) \varphi = \varphi^* \widetilde{D}_0 \varphi. \tag{6.1}$$

□

For $0 \le t \le 1$ we define

$$\begin{cases} D_t &= (1-t) D_0 + t \cdot D_1 = D_0 + t \cdot G \\ \widetilde{D}_t &= (1-t) \widetilde{D}_0 + t \cdot \widetilde{D}_1 = \widetilde{D}_0 + t \cdot \widetilde{G} \end{cases} \tag{6.2}$$

where $\widetilde{G} = \varphi^*[D, \varphi]$ is a matrix-valued Γ-periodic smooth function on $\widetilde{M}$, which descends to G on M. The first-order elliptic operators D_t and $\widetilde{D}_t$ are of geometric type as D is assumed to be a geometric operator (cf. §2, [44]). In particular, they are essentially self-adjoint. As M is compact, each D_t has isolated pure-point spectrum, while the spectrum of each $\widetilde{D}_t$ can be a mixture of continuous and pure-point. In particular, 0 need not be isolated in $\mathrm{Spec}(\widetilde{D}_t)$. A geometric operator, A, has the Bismut-Freed local cancellation property ([11], cf. also [43]) that the local trace of $Ae^{-\xi A^2}$ has an asymptotic expansion holomorphic in ξ, beginning with $\xi^{1/2}$. Define the eta function

$$\eta(D_t, s) = \frac{1}{2} \cdot \Gamma((s+1)/2)^{-1} \int_0^\infty \xi^{(s-1)/2} \mathrm{tr}_M\{D_t e^{-\xi D_t^2}\} d\xi. \tag{6.3}$$

As 0 is isolated in $\mathrm{Spec}(D_t)$, this defines a holomorphic function for $s > m$. However, the Bismut-Freed cancellation property implies that $\eta(D_t, s)$ is in fact holomorphic for $s > -1/2$, a considerable sharpening of the usual results. In particular, we define

$$\begin{aligned} \eta(D_t) &= \eta(D_t, 0) \\ &= \frac{1}{2\sqrt{\pi}} \int_0^\infty \xi^{-1/2} \mathrm{tr}_M\{D_t e^{-\xi D_t^2}\} d\xi \end{aligned} \tag{6.4}$$

the latter being an improper, conditionally convergent integral at $\xi = 0$. Let us also introduce the truncated integral

$$\eta^a(D_t) = \frac{1}{2\sqrt{\pi}} \int_0^a \xi^{-1/2} \mathrm{tr}_M\{D_t e^{-\xi D_t^2}\} d\xi. \tag{6.5}$$

We would also like to define $\eta_\Gamma(\widetilde{D}_t, s)$ by replacing the trace on M with the Γ-trace on $\widetilde{\mathcal{H}}^N$ in (6.3). However, this requires that 0 be isolated in $\mathrm{Spec}(\widetilde{D}_t)$, a condition that patently fails in spectral flow applications. We adopt an alternative strategy, and introduce for $a > 0$

$$\eta^a_\Gamma(\widetilde{D}_t, s) = \frac{1}{2} \cdot \Gamma((s+1)/2)^{-1} \cdot \int_0^a \xi^{(s-1)/2} \mathrm{tr}_\Gamma\{\widetilde{D}_t e^{-\xi \widetilde{D}_t^2}\} d\xi. \tag{6.6}$$

By the Bismut-Freed local cancellation, this is holomorphic for $s > -1/2$. Moreover, for $s = 0$ and D the signature operator, Cheeger and Gromov [14] proved that the limit exists as $a \to \infty$. Ramachandran extended this to include all geometric operators, using a clever trick based on defining a tempered distribution associated to D and tr_Γ ([43]). So we define for D geometric:

$$\eta_\Gamma(\widetilde{D}_t) = \lim_{a \to \infty} \eta^a_\Gamma(\widetilde{D}_t, 0). \tag{6.7}$$

We next make a simple observation that is the basis for our proof of Theorem 2.

LEMMA 6.2.

$$\eta_\Gamma(\widetilde{D}_0) = \eta_\Gamma(\widetilde{D}_1). \tag{6.8}$$

PROOF: The operators $\widetilde{D}_0$ and $\widetilde{D}_1$ are unitarily conjugate by Lemma 6.1, but the unitary φ is not Γ-periodic so does not a priori satisfy $\mathrm{tr}_\Gamma(\varphi A) = \mathrm{tr}_\Gamma(A\varphi)$ for a Γ-trace-class operator A. However, for the foliated space $(V, \mathcal{F}_\varphi)$ obtained from the hull construction of section 5, the leafwise operators $D_{\mathcal{F},0}$ and $D_{\mathcal{F},1}$ are independent of $x \in X_\varphi$. Thus, the Γ-trace in (6.4) can be replaced with the foliation trace, $\mathrm{tr}_{\mathfrak{m}}$, and $\widetilde{D}_t$ with the corresponding leafwise operator $D_{\mathcal{F},t}$ lifted from D_t. The unitary φ extends to a unitary Φ defined on all of V, and acts as an outer automorphism of the algebra $\mathcal{U}_{-\infty}(\mathcal{F}_\varphi)$, so we have $\mathrm{tr}_{\mathfrak{m}}(\Phi \circ A) = \mathrm{tr}_{\mathfrak{m}}(A \circ \Phi)$ for $A \in \mathcal{U}_{-\infty}(\mathcal{F}_\varphi)$. Taking $A = \Phi^* \circ \widetilde{D}_{\mathcal{F},t} e^{-\xi \widetilde{D}^2_{\mathcal{F},t}}$ and integrating yields (6.8). □

It is standard fact that if 0 is not an eigenvalue of D_t for $t = t_0$, then the function $t \mapsto \eta(D_t)$ is differentiable at t_0, and the derivative $\dot{\eta}(D_t)$ is given by a local formula derived from the total symbol of D_t. In particular, if the spectrum of all D_t is discrete, then the derivative extends to a smooth function for all $0 \le t \le 1$ and we can set

$$\eta(D, \alpha, \Theta) = \int_0^1 \dot{\eta}(D_t)dt. \tag{6.9}$$

The local formula for $\dot{\eta}(D_t)$ can be derived directly from the definition (6.3) as in Gilkey (page 83, [34]).

Our task is to make sense of $\frac{d}{dt}\{\eta_\Gamma(\widetilde{D}_t)\}$, then use (6.8) and the fundamental theorem of calculus to derive the identity which implies (1.27). The difficulty is that zero need not be isolated in $\mathrm{Spec}(D_t)$ for a continuum of values of t, so the "derivative" will be distribution-valued in an essential way. The key observation which overcomes the obstacle is that $\eta^a_\Gamma(\widetilde{D}_t, s)$ is holomorphic for $s > -1/2$ and all $a > 0$. We prove that it has a derivative, given by a local term coinciding with $\dot{\eta}(D_t)$, and a second boundary-value term that will be equated to the spectral flow interpretation of $\mathrm{Ind}_{\mathfrak{m}}(\varphi)$. Moreover,

$$0 = \lim_{a\to\infty} \int_0^1 \frac{d}{dt}\{\eta^a_\Gamma(\widetilde{D}_t, 0)\}dt \tag{6.10}$$

by (6.8) so that we obtain (1.27). It remains to justify these claims.

PROPOSITION 6.3. *For all $a > 0$, $\eta^a_\Gamma(\widetilde{D}_t)$ is differentiable as a function of t, and we have*

$$\int_0^1 \frac{d}{dt}\{\eta^a_\Gamma(\widetilde{D}_t)\}dt = \eta(D, \alpha, \Theta) - SF_\Gamma(\widetilde{D}, \varphi, a) \tag{6.11}$$

$$SF_\Gamma(\widetilde{D}, \varphi, a) = \sqrt{\frac{a}{\pi}} \cdot \int_0^1 \mathrm{tr}_\Gamma(\widetilde{G} e^{-a\widetilde{D}^2_t})dt. \tag{6.12}$$

PROOF: For $s > m+2$, the integrand in (6.6) is uniformly differentiable in t for $0 < \xi \le a$, so that the derivative of $\eta^a_\Gamma(\widetilde{D}_t, s)$ exists and is equal to the integral of the derivative inside. Then compute as in (page 83, [34]) to obtain

$$\frac{d}{dt}\eta^a_\Gamma(\widetilde{D}_t, s) =$$

$$\frac{1}{2}\cdot\Gamma((s+1)/2)^{-1}\int_0^a \xi^{(s-1)/2}\left(1+2\xi\frac{d}{d\xi}\right)\mathrm{tr}_\Gamma(\widetilde{G}e^{-\xi\widetilde{D}_t^2})d\xi. \tag{6.13}$$

Integrate by parts to obtain

$$\left\{\begin{array}{l} -\frac{s}{2}\cdot\Gamma((s+1/2))^{-1}\int_0^a \xi^{(s+1)/2}\mathrm{tr}_\Gamma(\widetilde{G}e^{-\xi\widetilde{D}_t^2})d\xi \\ +\Gamma((s+1)/2)^{-1}\cdot\xi^{(s+1)/2}\mathrm{tr}_\Gamma(\widetilde{G}e^{-\xi\widetilde{D}_t^2})|_{\xi=0}^a \end{array}\right. \tag{6.14}$$

The heat equation asymptotics for the compact case, M, carry over to $\widetilde{M}$ as well (cf. Chapter 13, [47]), so that the first term in (6.14) is a meromorphic function of s with no pole at $s=0$. Its value at $s=0$ is determined by a local expression derived from the complete symbol of $\widetilde{G}e^{-\xi\widetilde{D}_t^2}$, which is equal to the local expression for $Ge^{-\xi D_t^2}$ on M. Thus, the Γ-trace of it reduces to an integral over M which equals $\dot{\eta}(D_t)$.

The second term in (6.14) for $s > m-1$ reduces to

$$\Gamma((s+1)/2)^{-1}\cdot a^{(s+1)/2}\cdot\mathrm{tr}_\Gamma(\widetilde{G}e^{-a\widetilde{D}_t^2}) \tag{6.15}$$

as the limit in the evaluation for $\xi \to 0$ vanishes. The term (6.15) is holomorphic for $s > -1/2$, so its value at $s=0$ is defined. Integrate (6.15) for $s=0$ to obtain (6.12), proving (6.11). □

The next step is to identify (6.12) with the Breuer index $\mathrm{Ind}_{\mathfrak{m}}(\varphi)$. Consider the function

$$\sigma^a(\lambda) = \frac{1}{\sqrt{\pi}}\int_0^a \xi^{-1/2}\lambda e^{-\xi\lambda^2}d\xi \tag{6.16}$$

which satisfies

$$\lim_{\lambda\to\pm\infty}\sigma^a(\lambda) = \pm 1 \tag{6.17}$$

$$\frac{d}{d\lambda}\sigma^a(\lambda) = 2\cdot\sqrt{\frac{a}{\pi}}\cdot e^{-a\lambda^2} > 0 \tag{6.18}$$

and hence

$$\left\{\begin{array}{ll} |\sigma^a(\lambda)-1| < c(\epsilon)e^{-(a\lambda^2)/2} & \text{for } \lambda > \epsilon \\ |\sigma^a(\lambda)+1| < c(\epsilon)e^{-(a\lambda^2)/2} & \text{for } \lambda < -\epsilon \end{array}\right. \tag{6.19}$$

For each $E > 0$ and $0 \le t \le 1$ introduce the spectral projection on $\widetilde{\mathcal{H}}^N$,

$$\chi(E,t) = \chi_{[-E,E]}(\widetilde{D}_t) \tag{6.20}$$

associated to $\widetilde{D}_t$ and the interval $[-E, E]$.

LEMMA 6.4.

$$SF_\Gamma(\widetilde{D}, \varphi, a) = \lim_{E\to\infty} -\frac{1}{2}\int_0^1 \mathrm{tr}_\Gamma\Big\{\Big(\frac{d}{dt}\sigma^a(\widetilde{D}_t)\Big)\circ \chi(E,t)\Big\}dt. \tag{6.21}$$

PROOF: $\chi(E,t)$ is Γ-trace class for all E and commutes with D_t, so that we can repeat the integration-by-parts method of Proposition 6.3 to reduce the right-hand-side of (6.21) to the expression:

$$\lim_{E\to\infty} \sqrt{\frac{a}{\pi}}\cdot \mathrm{tr}_\Gamma\{\widetilde{G}\cdot e^{-a\widetilde{D}_t^2}\circ \chi(E,t)\}. \tag{6.22}$$

The proof is completed by noting that $\widetilde{G}$ is a bounded operator, and so for some constant

$$\mathrm{tr}_\Gamma\{\widetilde{G}\circ e^{-a\widetilde{D}_t^2}\circ (I-\chi(E,t))\} \le c\cdot ||G||\cdot e^{-aE^2}. \tag{6.23}$$ □

The final step uses standard spectral properties of geometric (first-order) operators:

PROPOSITION 6.5.

$$\begin{aligned}&\mathrm{Ind}_{\mathfrak{m}}(\varphi) = \\ &\lim_{a,E\to\infty} -\frac{1}{2}\int_0^1 \mathrm{tr}_\Gamma\Big\{\Big(\frac{d}{dt}\sigma^a(\widetilde{D}_t)\Big)\circ \chi(E,t)\Big\}dt.\end{aligned} \tag{6.24}$$

PROOF: For an α-related multiplier φ, we noted that the $\mathrm{Hull}(\alpha)\cong \overline{\alpha(\Gamma)}\subset U(N)$, and the restrictions of Φ to leaves is given by the translates $g\cdot\varphi$ of φ, for $g\in U(N)$. Thus, the leafwise operators appearing in (2.22) are leaf-independent after taking the pointwise trace. We can therefore let $\epsilon\to 0$ in the index construction of section 2, use normality of $\mathrm{Tr}_{\mathfrak{m}}$ and leaf independence to identify

$$\mathrm{Ind}_{\mathfrak{m}}(\varphi) = \mathrm{tr}_\Gamma\{\Pi(\varphi)\} - \mathrm{tr}_\Gamma\{\Pi(\varphi^*)\} \tag{6.25}$$

where $\Pi(\varphi)$ is projection onto the kernel of $P_0^+\circ\varphi\circ P_0^+ - P_0^-$, and $\Pi(\varphi^*)$ the corresponding projection for the adjoint.

The geometric operator D satisfies an elementary commutation rule with functions (viewed as diagonal multipliers) $[D,f]=\nabla f$, where ∇ is a first-order differential operator constructed from the $\mathrm{Spin}^{\mathbf{C}}$-structure and covariant differentiation on $E\to M$. In particular, for φ unitary, this implied that $\widetilde{G}=\varphi^*[D,\varphi]$ is a uniformly bounded operator on $\widetilde{\mathcal{H}}^N$. Let $c_2 = ||\widetilde{G}||$. This gives the standard estimates.

LEMMA 6.6. *There are operator inequalities*

$$\widetilde{D} - c_2 \leq \varphi^* \widetilde{D} \varphi \leq \widetilde{D} - c_2 \tag{6.26}$$

and hence for $E > c_2$,

$$\chi_{[-E+c_2,E-c_2]}(\widetilde{D}) \leq \chi_{[-E,E]}(\varphi^* \widetilde{D} \varphi) \leq \chi_{[-E-c_2,E+c_2]}(\widetilde{D}). \tag{6.27}$$

□

As the range of $\Pi(\varphi)$ is the closure of the subspace range $(P_0^+) \cap$ range $(\varphi^* P_0^-)$, and $\Pi(\varphi^*)$ the closure of range $(P_0^+) \cap$ range (φP_0^-), we deduce

COROLLARY 6.7. *For* $E > c_2$,

$$\begin{cases} \Pi(\varphi) \circ \chi(E,0) &= \Pi(\varphi) \\ \Pi(\varphi^*) \circ \chi(E,0) &= \Pi(\varphi^*). \end{cases} \tag{6.28}$$

□

Combine (6.28) with the equality (6.25) and $P_0^+ + P_0^- = \mathrm{Id}$ to obtain

COROLLARY 6.8. *For* $E > c_2$

$$\begin{aligned} \mathrm{Ind}_m(\varphi) = -\frac{1}{2}[&\mathrm{tr}_\Gamma\{(\varphi^* P_0^+ \varphi - P_0^+) \circ \chi(E,0)\} \\ &- \mathrm{tr}_\Gamma\{(\varphi^* P_0^- \varphi - P_0^-) \circ \chi(E,0\}]. \end{aligned} \tag{6.29}$$

□

To conclude the proof of Proposition 6.5, we must show that the right-hand-sides of (6.24) and (6.29) are equal. First note that by the estimate (6.23) and operation inequalties (6.27), we can replace $\chi(E,t)$ in (6.24) by $\chi(E,0)$. Then integrate the right-side of (6.24) to obtain

$$\lim_{a,E\to\infty} -\frac{1}{2}\mathrm{tr}_\Gamma\{(\sigma^a(D_1) - \sigma^a(D_0)) \circ \chi(E,0)\}. \tag{6.30}$$

Then combine the exponential estimate (6.19) with (6.26) and the Sobolev estimate for an open manifold (cf. Chapter 5, [47]) to deduce that (6.30) converges uniformly to

$$-\frac{1}{2} \cdot \lim_{E\to\infty} \mathrm{tr}_\Gamma\left\{\left(\varphi^* \lim_{a\to 0} \sigma^a(\widetilde{D})\varphi - \lim_{a\to 0} \sigma^a(\widetilde{D})\right) \circ \chi(E,0)\right\} \tag{6.31}$$

which is equal to the right-hand-side of (6.29). □

REMARK 6.9. The above analytic proof of Theorem 2 also applies for Γ finite, even trivial. The multiplier φ is then a unitary over a finite cover of M, and we identify the spectral flow for a family of eta-invariants on M with a Toeplitz index. This equality is asserted in a preliminary form in (§7, [3]), and Booss-Wojciechowski gave a topological proof in [12]. As is evident from the results of this section, the analytic proof extends to the Γ-almost periodic case. □

Let us conclude this paper with the remark that our study of Toeplitz indices for *normal* multipliers φ on $\widetilde{M}$ could equally well be carried out for non-self-adjoint D coupled to *normal vector bundles* over $\widetilde{M}$. In this case, the approach of this paper yields index problems similar to those which arise for $C_r^*(\Gamma)$-bundles over M in Kasparov-Mischenko approach to the Novikov Conjecture (cf. [32]). On the other hand, the even case generalizes the almost periodic index theory on $\mathbf{R}^m$, [51], so deserves further investigation.

References

1. M. F. Atiyah. Elliptic operators, discrete groups and von Neumann algebras. *Astérisque,* vol. 32/33 (1976), 43-72.

2. M. F. Atiyah, V. Patodi & I. M. Singer. Spectral asymmetry and Riemannian Geometry, I. *Math. Proc. Camb. Phil. Soc.,* **77**(1975), 43-69.

3. M. F. Atiyah, V. Patodi & I. M. Singer. Spectral asymmetry and Riemannian Geometry, III. *Math. Proc. Camb. Phil. Soc.,* **79**(1975), 71-99.

4. M. F. Atiyah & W. Schmid. A geometric construction of the discrete series for semi-simple Lie groups. *Invent. Math.,* **42**(1977), 1-62.

5. M. F. Atiyah & I. M. Singer. The index of elliptic operators, I. *Annals of Math.,* vol. 87 (1968), 484-530.

6. M. F. Atiyah & I. M. Singer. Index theory for skew-adjoint Fredholm operators. *Publ. Inst. Hautes Etudes Sci.* no. 37 (1969), 305-326.

7. P. Baum & A. Connes. Geometric K-theory for Lie groups and foliations. *Preprint Inst. Hautes Etudes Sci.* (1982).

8. P. Baum & R. G. Douglas. K-homology and index theory. In "*Operator Algebras and Applications*", Proc. Symp. Pure Math., vol. 38, part I. Amer. Math. Soc., Providence (1982), 117-173.

9. P. Baum & R. G. Douglas. Toeplitz operators and Poincaré duality. in "*Proc. Toeplitz Memorial Conf., Tel Aviv 1981*", Birkhauser, Basel (1982), 137-166.

10. J. Bellisard. K-theory of C^*-algebras in solid state physics. *Lecture Notes in Physics* vol. 257, Springer-Verlag, Berlin (1987), 99-156.

11. J.-M. Bismut & D. Freed. The analysis of elliptic families: II. Dirac operators, eta invariants and the holonomy theorem. *Commun. Math. Phys.* **107**(1986), 103-163.

12. B. Booss & K. Wojciechowski. Desuspensions of splitting elliptic symbols, I. *Ann. Global Analysis and Geometry* **3**(1985), 330-363.

13. J. Cheeger & M. Gromov. On the characteristic numbers of complete manifolds of bounded curvature and finite volume. In "*Differential geometry and Complex Analysis. Volume dedicated to H. Rauch,*" Ed. by I. Chavel & H. Farkas, Springer-Verlag, Berlin (1985).

14. J. Cheeger & M. Gromov. Bounds on the von Neumann dimension of L^2-cohomology and the Gauss-Bonnet theorem for open manifolds. *Jour. Diff. Geom.* **21**(1985), 1-34.

15. J. Cheeger, M. Gromov & M. Taylor. Finite propagation speed, kernel estimates for functions of the Laplace operator, and the geometry of complete Riemannian manifolds. *Jour. Diff. Geom.* **17**(1982), 15-54.

16. J. Cheeger & J. Simons. Differential characters and geometric invariants, *Preprint* (1972). Appeared in "*Geometry and Topology. Proceedings, Univ. of Maryland 1983-84*", Lecture Notes in Math., vol. 1167, Springer-Verlag, Berlin (1986).

17. A. Connes. The von Neumann algebra of a foliation. *Lecture Notes in Physics* vol. 80. Springer-Verlag, Berlin (1978), 145-151.

18. A. Connes. Sur la théorie non-commutative de l'integration. *Lecture Notes in Math.* vol. 725. Springer-Verlag, Berlin (1979), 19-143.

19. A. Connes. A survey of foliations and operator algebras. In "*Operator Algebras and Applications,*" Proc. Symp. Pure Math., vol. 38, part I. Amer. Math. Soc., Providence (1982), 521-628.

20. A. Connes. Non-commutative differential geometry. I. The Chern character in K-homology. *Publ. Inst. Hautes Etudes Sci.* no. 62 (1986), 41-93.

21. A. Connes. Non-commutative differential geometry. II. deRham homology and non-commutative algebra. *Publ. Inst. Hautes Etudes Sci.* no. 62 (1986), 91-144.

22. A. Connes. Colloquium Lecture. Univ. California at Berkeley. (May, 1985).

23. A. Connes & M. Karoubi. Caractère multiplicatif d'un module de Fredholm. *C. R. Acad. Sci. Paris.* t. 299 (1984), 963-968.

24. A. Connes & H. Moscovici. The L^2-index theorem for homogeneous spaces. *Bulletin Amer. Math. Soc.* vol. 1 (1987), 688-690.

25. A. Connes & H. Moscovici. Conjecture de Novikov et groupes hyperboliques. *C. R. Acad. Sci. Paris.* t. 307 (1988), 475-480.

26. A. Connes & H. Moscovivi. Cyclic cohomology, the Novikov Conjecture and hyperbolic groups. *Preprint Inst. Hautes Etudes Sci.* (Oct. 1988), IHES/M/88/51.

27. R. Curto, P. Muhly & J. Xia. Toeplitz operators on flows. *University of Iowa preprint.* (1987).

28. R. G. Douglas, S. Hurder & J. Kaminker. Eta invariants and von Neumann algebras. *Bulletin Amer. Math. Soc.*, vol. 21 (1988), 83-87.

29. R. G. Douglas, S. Hurder & J. Kaminker. Toeplitz operators and the eta invariant: the case of S^1. In *"Index Theory of Elliptic Operators, Foliations, and Operator Algebras,"* Contemp. Math. vol. 70, Amer. Math. Soc., Providence (1988), 11-41.

30. R. G. Douglas, S. Hurder & J. Kaminker. The longitudinal cocycle and the index of Toeplitz operators. *I.U.-P.U.@ Indianapolis, Preprint* (1988).

31. R. G. Douglas, S. Hurder & J. Kaminker. Cyclic cocycles, renormalizaiton and eta-invariants. *University of Illinois at Chicago, Preprint* (1989).

32. T. Fack. Sur la conjecture de Novikov. in *"Index Theory of Elliptic Operators, Foliations, and Operator Algebras."* Contemp. Math. vol. 70, Amer. Math. Soc., Providence (1988), 43-102.

33. J. Feldman. *Lectures on Orbit Equivalence for Amenable Group Actions.* Birkhauser, to appear.

34. P. B. Gilkey. *Invariant Theory, the Heat Equation and the Atiyah-Singer Index Theorem.* Publish or Perish, vol. 11, Wilmington (1984).

35. A. Haefliger. Groupoides d'holonomie et classifiants. In *"Structures Transverse des Feuilletages,"* Asterisque, vol. 116 (1984), 70-97.

36. N. Higson. A primer on Kasparov's KK-theory. *University of Pennsylvania, Preprint.* (1988).

37. S. Hurder. *Analysis and Geometry of Foliations.* Research Monograph based on Ulam Lectures at University of Colorado, 1988-89.

38. R. Ji & J. Kaminker. The K-theory of Toeplitz extensions, *Jour. Operator Theory* **19**(1988), 347-354.

39. R. Ji & J. Xia. On the classification of commutator ideals. *Jour. Functional Anal.* **78**(1988), 208-232.

40. G. G. Kasparov. K-functor and extension of C^*-algebras. *Izv. Akad. Nayk SSSR,* Ser. Mat. 44 (1980), 571-636.

41. C. C. Moore & C. Schochet. *Analysis on Foliated Spaces.* Math. Sciences Research Inst. Publ. No. 9, Springer-Verlag, Berlin (1988).

42. J. Plante. Foliations with measure-preserving holonomy. *Annals of Math.* vol. 102 (1975), 327-361.

43. M. Ramachandran. Cheeger-Gromov inequality for type-II eta invariants. *Univ. of Colorado, Preprint* (May 1989).

44. J. Roe. An index theorem on open manifolds. I. *Jour. Diff. Geom.* **27**(1988), 87-113.

45. J. Roe. An index theorem on open manifolds. II. *Jour. Diff. Geom.* **27**(1988), 115-136.

46. J. Roe. Finite propagation speed and Connes foliation algebra. *Proc. Camb. Phil. Soc.* **102**(1987), 459-466.

47. J. Roe. *Elliptic Operators, Topology and Asymptotic Methods.* Pitnam Research Notes in Math. vol. 179, Longman Scientific. New York (1988).

48. D. Ruelle & D. Sullivan. Currents, flows and diffeomorphisms. *Topology* **14**(1975), 319-327.

49. I. M. Singer. Recent applications of index theory for elliptic operators. *Proc. Symp. Pure Math.* vol. 23, Amer. Math. Soc., Providence (1971), 11-31.

50. I. M. Singer. Some remarks on operator theory and index theory. In *"K-theory and Operator Algebras. Proceedings, Univ. of Georgia 1975,"* Lecture Notes in Math. vol. 575, Springer-Verlag, Berlin (1977), 128-138.

51. M. A. Subin. The spectral theory and the index of elliptic operators with almost periodic coefficients. *Uspehi Mat. Nauk* **34**(1979), 95-135. English transl.: *Russian Math. Surveys* 34:2 (1979), 109-157.

52. K. Wojciechowski. A note on the space of pseudo-differential projections with the same principal symbol. *Jour. Operator Theory* **15**(1986), 207-216.

53. G. Peric. *The Eta Invariant of Foliated Spaces*. Thesis, Ohio State University (1989).

Department of Mathematics, Statistics, and Computer Science
University of Illinois at Chicago, Chicago, Illinois 60680

Contemporary Mathematics
Volume **105**, 1990

The Lefschetz Fixed Point Theorem for Foliated Manifolds

CONNOR LAZAROV[1] AND JAMES HEITSCH[2]

1. Introduction. In this paper we outline our results on the Lefschetz fixed point theorem for foliations of a compact manifold and some applications. The details will appear in [HL1] and [HL2].

We briefly recall the classical Lefschetz fixed point theorem [AB, I & II]. Assume that we are given an elliptic complex $\{E_i, d_i\}$ on a compact C^∞ manifold and a geometric endomorphism $T = \{T_i\}$ of this complex which covers a smooth map f of M with isolated non-degenerate fixed points. The trace of T_i on $H^i(E, d)$ is defined and

$$\sum(-1)^i TR(T_i) = \sum_p \sigma(p)$$

where $\sum_p$ is a sum over the fixed points of f and $\sigma(p)$ is a number which depends only on infinitesimal data of $\{E_i, d_i\}$, f, and T at p.

In our theorem we assume that we are given a foliation F of a compact manifold with an invariant transverse measure ν, a leaf preserving diffeomorhism f with nondegenerate transversal fixed point set, a leafwise elliptic complex, and a geometric enodmorphism T. We then define the global object $TR_\nu(T_i)$ coming from the action of T_i on the reduced L^2 cohomology of the complex restricted to the leaves. Our main theorem describes $\Sigma(-1)^i TR_\nu(T_i)$ in terms of the integral, with respect to $d\nu$, of a family of leafwise measures on the fixed point set which, again, depend only upon infinitesimal data on the fixed point set.

In the cases of the classical leafwise elliptic complexes, the leafwise measures are given by familiar expressions. In particular if the fixed point set on each leaf is a set of isolated points we recover the measures of [AB, II]. At the opposite extreme, when T is the identity, we get an index theorem for foliations and more generally, a version of the G index theorem for foliations.

For foliations by orientable surfaces we get, as an application, a generalization of the classical finiteness theorem of the automorphism group of a compact Riemann surface with genus greater than one. And for foliations by spin manifolds we get a generalization of the rigidity theorem of [AH] for compact spin manifolds with non-vanishing $\hat{A}$ genus.

1980 *Mathematics Subject Classification* (1985 *Revision*). Primary 57R30, 58C30.

[1] Partially supported by PSC-CUNY Grant.

[2] Partially supported by NSF Grant DMS 8503029.

2. Dirac Operators and Geometric Endomorphisms. Let M be a compact oriented Riemannian manifold of dimension m and F a codimension q oriented foliation of M. A *Dirac complex* (E, d) *along* F consists of

1. A family $E = (E_0, \dots, E_k)$ of Hermitian vector bundles on M.
2. A family $d = (d_0, \dots, d_{k-1})$ of differential operators where $d_i : C^\infty(E_i) \to C^\infty(E_{i+1})$. The operators d_i differentiate only in leaf directions and the complex (E, d) is a generalized Dirac complex [GL]. This means that $E = \bigoplus_i E_i$ is a module over the bundle of Clifford algebras $C(T^*F)$ and E has a connection ∇ which is compatible with Clifford multiplication and the Riemannian connection on T^*F and that the operator $D = \oplus(d_i + d_{i-1}^*) : C^\infty(E) \to C^\infty(E)$ is given by

$$C^\infty(E) \xrightarrow{\nabla} C^\infty(T^*M \otimes E) \xrightarrow{r} C^\infty(T^*F \otimes E) \xrightarrow{m} C^\infty(E)$$

where m is Clifford multiplication and r is induced by the restriction map.

Let f be a diffeomorphism of M which takes each leaf of F to itself. A *geometric endormorphism* of (E, d) over f is a family $T = (T_0, \dots, T_k)$ of complex linear maps

$$T_i : C^\infty(E_i) \to C^\infty(E_i)$$

such that $d_i T_i = T_{i+1} d_i$ and for each i there is a smooth bundle map $A_i : f^*E_i \to E_i$ so that T_i is given by

$$T_i(s)(x) = A_{i,x}(s(f(x))).$$

Let $E_i^L = E_i$ restricted to the leaf L. T_i induces a map T_i^L on $L^2(E_i^L)$ and a map

$$T_i^{L,\#} : H_L^i(E, d) \to H_L^i(E, d)$$

where $H_L^i(E, d)$ is reduced L^2 cohomology $\ker(d_i^L)/\overline{\operatorname{im}(d_{i-1}^L)}$ (see [HL1] Proposition (2.2.1)). Here d_i^L is the restriction of d_i to sections of E_i^L. Let $d_i^{L,*}$ be the adjoint of d_i^L, and

$$\Delta_i^L = d_{i-1}^L d_{i-1}^{L,*} + d_i^{L,*} d_i^L.$$

Using Hodge theory we can identify $H_L^i(E, d)$ with $\ker(\Delta_i^L)$ and $T_i^{L,\#}$ passes to a map $T_i^{L,*}$ on $\ker(\Delta_i^L)$ given by compression of T_i^L to this kernel. (See [HL1] Proposition (2.2.2)). We denote the family $\{T_i^{L,*}\}$ by T_i^*.

3. The Global $TR_\nu(T_i^*)$. We now assume that we are given a holonomy invariant Radon transverse measure ν. Let $\lambda = \{\lambda^L\}$ be a family of measures on the leaves $\{L\}$ of F which vary measureably in the direction transverse to the leaves. Using ν we can integrate λ to arrive at a measure $\int \lambda \, d\nu$ on M. (See [MS IV]). We describe this construction briefly.

Let R be a complete Borel transversal to F. That is, a Borel subset of M which intersects each leaf in a non empty countable set. ν is then a measure on R. Let $\pi : M \to R$ be a measurable map with the property that for each x, $\pi(x)$ lies on the leaf through x. Let A be a Borel subset of M with characteristic function χ_A.

Definition. $\int_A \lambda d\nu = \int_R(\int_{\pi^{-1}(t)} \chi_A d\lambda^t) d\nu(t)$. Here t ranges over R and λ^t is the measure λ^L where L is the leaf containing t. It is a consequence of the invariance of ν that this is independent of R and π. It is easy to construct such π and R. (See [HL1] (2.3)).

Let $P_i^L : L^2(E_i^L) \to \ker(\Delta_i^L)$ be the projection. Then $T_i^{L,*} = P_i^L T_i^L P_i^L$. $T_i^{L,*}$ is a smoothing operator. Let $k_L^{i,T}(x,y)$ be its Schwartz kernel. $k_L^{i,T}(x,x) \in C^\infty(E_i^L \otimes (E_i^L)^*)$ so the pointwise trace tr $k_L^{i,T}(x,x)$ makes sense. Let λ^L be the volume form on L coming from M. Then tr $k^{i,T}(x,x)\lambda = \{\text{tr } k_L^{i,T}(x,x)\lambda^L\}$ is a family of transversely measurable leafwise measures. (See [HL1] (2.3.27)).

Definition. $TR_\nu(T_i^*) = \int_M (tr\ k^{i,T}(x,x)\lambda) d\nu$.

Definition. $L_\nu(T) = \sum_{i=0}^{k} (-1)^i TR_\nu(T_i^*)$.

4. Fixed Point Indices and the Lefschetz Theorem. In order to state our Lefschetz theorem we must make a restriction on the fixed point set and on the diffeomorphism f. Let N be the fixed point set. We assume that N is a closed submanifold which is transverse to F. Let $N^L = N \cap L$. We assume that f is non-degenerate on each N^L. This means the following. Let η^L denote the bundle $(TL \mid N^L)/TN^L$ on N^L. $df \mid L$ induces a map $df_\eta : \eta^L \to \eta^L$ covering the identity map on N^L. To say that f is non degenerate means that for all x in N, $\mathrm{Det}(I_x - df_{\eta,x}) \neq 0$. Note that the identify map of M satisfies this condition.

Lefschetz Fixed Point Theorem for Foliations. ([HL1] (3.1.1)). Let M, F, (E,d), ν, f, A, T and N be as above. To each N^L we can assign a smooth measure a^L which depends only on f, A, the symbols of Δ_i, the metrics, and the derivatives of these, to a finite order on N^L. The family $a = \{a^L\}$ varies smoothly transversely, and

$$L_\nu(T) = \int_N a\ d\nu.$$

Suppose (E,d) is one of the classical complexes (DeRham, Dolbeault, Signature, Spin) and $T = f^*$ acting on forms. Then a^L is the usual integrand of the Atiyah-Singer G index theorem for the relevant complex. If f is the identity, a^L is then the integrand of the Connes index theorem (see [ML]) so $L_\nu(I)$ is equal to the Connes index.

If each component of N^L is an isolated point x, then for an arbitrary Dirac complex we can identify a^L as in [AB, II] to be

$$a^L(x) = \sum_{j=0}^{k} (-1)^j \frac{tr\ A_{j,x}}{|det(I_x - df_{L,x})|}.$$

For the classical complexes with $T = f^*$, the $a^L(x)$ can be further explicated. E.g. for the DeRham complex,

$$a^L(x) = \text{ sign } (det(I_x - df_{L,x})).$$

Expressions for $a^L(x)$ when N^L is a union of isolated fixed points in the case of the signature, spin, and Dolbeault complexes can be found in [AB, II] and [HL1].

5. Applications.

Analogue of the Atiyah L^2 Covering Theorem. If we take f equal to the identity, $L_\nu(I)$ can be interpreted as an index of the leafwise elliptic complex (E, d). The Connes index $I^C(E, d)$ of this complex is defined using the lift of (E, d) to the holonomy groupoid of the foliation, whereas $L_\nu(I)$ is defined on M. The holonomy groupoid can be thought of as the union of the holonomy covering spaces of the leaves. Thus the equality of $L_\nu(I)$ and $I^C(E, d)$ can be thought of as an analogue of the covering theorem of [A] which relates the index of an elliptic operator on a compact manifold to the von Neumann index of the operator lifted to a Galois covering space.

Rigidity Theorems for Foliations by Surfaces and Spin Manifolds. Let M be a compact Riemannian manifold and F and ν as above. We begin by defining leafwise characteristic numbers using ν.

Assume that F has leaves of dimension $2k$. Let K^L be the curvature form of the Riemannian connection on the leaf L coming from the metric induced on L from M. Let Pf be the Pfaffian polynomial function on matrices. $\{Pf(K^L)\}$ is then a smooth family of leafwise measures.

Definition. $\chi_\nu(F) = \int_M Pf(K^L)d\nu$.

Let $\hat{A}_k(p_1, \ldots, p_k)$ be the $\hat{A}$ polynomial of [AS, III] p. 570, and $\hat{A}_k(K^L)$ the $2k$ form we get by replacing $p_1, \ldots, p_k$ by the Pontrjagin polynomial functions ([KN, II], p. 302) in the curvature K^L.

Definition. $\hat{A}_\nu(F) = \int_M \hat{A}_k(K^L)d\nu$.

Our first rigidity theorem is a generalization of the classical theorem of the finiteness of the automorphism group of a compact Riemann surface of genus greater than one.

THEOREM A. *Let M be a compact orientable Riemannian manifold and F an orientable foliation by surfaces. Let ν be a non-negative invariant transverse measure. If $\chi_\nu(F) < 0$ then no compact connected Lie group can act nontrivially on M as a group of isometries which take each leaf of F to itself.*

We note that if M has dimension three, we can state a slightly stronger theorem. Let $O(M; F)$ be the group of isometries of M which take each leaf of F to itself. If $\chi_\nu(F) < 0$ then $O(M; F)$ is totally pathwise disconnected.

Our next rigidity theorem is a generalization of the main theorem of [AH]. Assume the leaves of F have dimension $2k$ and that the principal $SO(2k)$ bundle Q of the cotangent bundle along the leaves of F has a reduction to a Spin $(2k)$ bundle P. We say that P is a spin structure on F. Let G act by isometries of M taking each leaf of F to itself. We will say that G preserves the spin structure on F if the action of G on Q lifts to an action on P compatible with the projection on Spin $(2k)$ to $SO(2k)$.

THEOREM B. *Let M be a compact orientable Riemannian manifold and F a foliation with a spin structure. Let ν be an invariant transverse measure for F. If $\hat{A}_\nu(F) \neq 0$ then no compact connected Lie group can act non-trivially as a group of isometries which take each leaf to itself and which preserves the spin structure on F.*

6. An example. We present an example of a foliated manifold and a non-trivial geometric endomorphism T for which $L_\nu(T) \neq L_\nu(I)$ for the leafwise DeRham complex. We consider a surface Σ of genus 3. We take the complex structure on Σ given by a symmetric polygonal fundamental domain D centered at 0 in the Poincaré disc H. Let $\{a, a'\}$, $\{b, b'\}$, $\{c, c'\}$, $\{d, d'\}$, $\{e, e'\}$, $\{g, g'\}$ be opposite sides of D. Let A, B, C, D, E, G be the oriented isometries of H with no fixed points, which take, respectively, a to a', b to b', c to c', etc. Γ, the fundamental group of Σ, is the group generated by these isometries and their inverses subject to $GE^{-1}DC^{-1}BA^{-1}G^{-1}ED^{-1}CB^{-1}A = I$. Let v_1 and v_2 be independent irrational numbers. We define a homomorphism $h : \Gamma \to$ translations $(\mathbf{R})$ by sending

$$\begin{array}{rcl} A, E, C^{-1} & \text{to} & x \to x + v_1 \\ B, G, D^{-1} & \text{to} & x \to x + v_2 \\ A^{-1}, E^{-1}, C & \text{to} & x \to x - v_1 \\ B^{-1}, G^{-1}, D & \text{to} & x \to x - v_2. \end{array}$$

The relation is preserved by h. We let $M = H \times_\Gamma S^1$ where the identifications are given by $(u, y) \sim (\gamma u, h(\gamma)y)$, for $\gamma \in \Gamma$. M is a flat S^1 bundle over Σ with a foliation F given by the images of $H \times \{\text{point}\}$ in M. The action of Γ on S^1 preserves the standard measure on S^1 and so gives rise to an invariant measure ν for F. It is easy to see that $\chi_\nu(F) < 0$. Rotation of H by $-2\pi/3$ about the center preserves the identifications which define M and induces a diffeomorphism f of M that takes each leaf to itself. One can show that the fixed points of f are isolated and non-degenerate on each leaf, the fixed point set is a closed submanifold transverse to F, at each fixed point the fixed point index sign $det(I - df) = +1$, and finally for the leafwise DeRham complex $L_\nu(f) = 2\nu(S^1) > 0$. Since $L_\nu(I) = \chi_\nu(F)$ this shows $L_\nu(f) \neq L_\nu(I)$. (We are using $L_\nu(f)$ for $L_\nu(T)$ where T is the endomorphism of the leafwise DeRham complex induced by df). See [HL1] for examples of non-trivial $L_\nu(T)$ for the other classical complexes.

We should note that for this example $O(M; F)$ is infinite. We can construct an isometric leaf preserving action of $Z \times Z$ on M by

$$(m, n)(u, y) = (u, y + mv_1 + nv_2).$$

REFERENCES

[A] ATIYAH, M.F., *Elliptic Operators, Discrete Groups and Von Neumann Algebras*, Soc. Mathematique de France Asterisque, **32-33** (1976), pp. 43-72.

[AB] ATIYAH, M.F. AND BOTT, R., *A Lefschetz Fixed Point Formula for Elliptic Differential Operators I and II*, Ann. Math., **86** (1967), pp. 374-407, Ann. Math., **88** (1968), pp. 451-491.

[ABP] ATIYAH, M.F., BOTT, R., AND PATODI, V.K., *On the Heat Equation and the Index Theorem*, Invent. Math., **19** (1973), pp. 279-330; Errata ibid **28** (1975), pp. 277-280.

[AH] ATIYAH, M.F. AND HIRZEBRUCH, F., *Spin Manifolds and Group Actions*, Essays on Topology and Related Topics, Memoirs dedies a' George DeRham, Springer Verlag, N.Y., 1970, pp. 18-27.

[AS] ATIYAH, M.F. AND SINGER, I.M., *The Index of Elliptic Operators: II, III*, Ann. Math., **87** (1968), pp. 484-530, pp. 546-604.

[ASe] ATIYAH, M.F. AND SEGAL, G., *The Index of Elliptic Operators: II*, Ann. Math., **87** (1968), pp. 531-545.

[C] CHERNOFF, P., *Essential Self Adjointness of Powers of Generators of Hyperbolic Equations*, J. Funct. Anal., **12** (1973), pp. 401-404.

[ChGT] CHEEGER, J., GROMOV, M. AND TAYLOR, M., *Finite Propogation Speed, Kernal Estimates for Functions of the Laplace Operator and the Geometry of Complete Riemannian Manifolds*, J. Diff. Geo., **17** (1982), 15-53.

[Co1] CONNES, A., *Sur La Theorie Non Commuative de L'Integration*, Lecture Notes in Math., vol. 755, Springer-Verlag, Spring 1979.

[Co2] CONNES, A., *A Survey of Foliations and Operator Algebras*, Proc. Symp. Pure Math., vol. 38 part 1, AMS, Providence, R.I., 1982, pp. 521-628.

[DS] DUNFORD, N. AND SCHWARTZ, J., *Linear Operators*, part 2, Spectral Theory, Interscience (John Wiley), 1963.

[G1] GILKEY, P., *Leftschetz Fixed Point Theorem and the Heat Equation*, Partial Diff. Equations and Geometry - Proc. of the Park City Conference, Ed. Christopher Byrnes, Marcel Dekker, (1979), pp. 91-94.

[G2] GILKEY, P., *Invariance Theory, The Heat Equation, and the Atiyah-Singer Theorem*, Math. Lec. Series, **11**, Publish or Perish Press, Wilmington, Del., 1984.

[GL] GROMOV, M. AND LAWSON, B., *Positive Scalar Curvature and the Dirac Operator on Complete Riemannian Manifolds*, Publications IHES, **56** (1983), pp. 83-196.

[HL1] HEITSCH, J. AND LAZAROV, C., *A Lefschetz Fixed Point Theorem for Foliated Manifolds*, (to appear, Topology).

[HL2] HEITSCH, J. AND LAZAROV, C., *Rigidity Theorems for Foliations by Surfaces and Spin Manifolds*, (to appear).

[KN] KOBAYASHI, S. AND NOMIZU, K., *Foundations of Differential Geometry*, vol. II, Interscience Press, 1969.

[MS] MOORE, C. AND SCHOCHET, C., *Global Analysis of Foliated Spaces*, MSRI Publications, 1988.

[Q] QUILLEN, D., *Superconnections and the Chern Character*, Topology, **24** (1985), pp. 89-95.

[RS] REED, M. AND SIMON, B., *Methods of Mathematical Physics*, vol. 1, Academic Press (1972).

[Ro1] ROE, J., *An Index Theorem on Open Manifolds*, J. Diff. Geo., **27** (1988), pp. 87-113.

[Ro2] ROE, J., *Finite Propogation Speed and Connes' Foliation Algebra*, Math. Proc. Camb. Phil. Soc., **102** (1987), pp. 459-466.

[T] TAYLOR, M., *Pseudo-differential Operators*, Princeton University Press, 1982.

Herbert Lehman College - CUNY - Bronx, New York 10468
University of Illinois at Chicago - Chicago, Illinois 60680

Contemporary Mathematics
Volume **105**, 1990

L^2-ACYCLICITY AND L^2-TORSION INVARIANTS

Varghese Mathai and **Alan L. Carey**

Abstract:

This paper discusses a generalisation of Reidemeister-Franz torsion to smooth closed oriented manifolds which are L^2-acyclic. The definition exploits the theory of finite von Neumann algebras in an essential way.

Introduction

We study a class of smooth closed oriented manifolds, called L^2-acyclic manifolds, which have the property that the complex of L^2-differential forms on the universal cover has trivial L^2-cohomology. By results of Dodzuik [9], this notion depends only on the oriented homotopy type of the smooth, closed manifold.

We invent a new differential invariant of an L^2-acyclic manifold which we call L^2-RF torsion. The theory of von Neumann algebras with a faithful finite trace (i.e. finite von Neumann algebras) is used in an essential way in our work.

In section 1, we introduce several new concepts and we develop the algebraic theory of L^2-RF torsion. We have tried to highlight the main differences between L^2-RF torsion and ordinary RF torsion.

In section 2, we study L^2-acyclic representations and it is observed that the class of L^2-acyclic manifolds contains all odd dimensional closed oriented hyperbolic manifolds and also all closed, connected oriented flat manifolds. We also show how to construct new L^2-acyclic manifolds from known L^2- acyclic manifolds.

In section 3, we define the L^2-RF torsion of an L^2-acyclic representation. We prove that it is a differential invariant and we also establish some of its other properties. The technical details of the invariance are new though some results depend on earlier work on ordinary RF torsion.

This work is still in an early stage, much remains to be done in the way of calculations, examples and so on. We present some simple computations in section 4 (for the circle and torus).

Our exposition of L^2-RF torsion is similar in spirit to the expositions of ordinary RF torsion by J. Milnor [12], M. Cohen [7], D. Ray and I. Singer [14].

There arose a number of very interesting questions, conjectures and open problems as a result of our investigations:

1980 *Mathematics Subject Classification* (1985 *Revision*). Primary 47, 53, 57.

(1) **Problem**: Compute the L^2-RF torsion for closed oriented hyperbolic manifolds (i.e. manifolds of constant negative sectional curvature).

By analogy with de Rham's classification theorem of spherical manifolds [12] a natural question is to ask if a closed oriented hyperbolic manifold of odd dimension is characterised up to its length spectrum by its L^2-RF torsion i.e. if M and N are closed oriented hyperbolic manifolds of dimension $2k-1$, and if M and N have the same L^2-RF torsion, then is M isospectral to N?

(2) It is natural to look for an L^2-analytic torsion which should equal our L^2-RF torsion. The second author has made some progress on this and will report on it in the near future.

(3) **Question**: Let M be a smooth, closed, oriented, connected three dimensional manifold such that $\pi_1(M)$ is infinite. When is M an L^2-acyclic manifold? We give partial results in section 2.

Many of the results of this paper extend to non-compact Cheeger-Gromov manifolds and will be discussed in a future paper. The ideas discussed here require mathematical tools from a variety of areas and so, in the interests of accessability, we have included some expository material.

Acknowledgement: The second named author would like to thank the organisers of the 1988 AMS summer conference at Brunswick, Maine for an invitation to participate.

1. L^2-RF Torsion

In this section we study Hilbert modules and define a L^2-Reidemeister Franz torsion for a class of complexes over a finite von Neumann algebra $\mathcal{U}$.

1.1: Hilbert $\mathcal{U}$ modules

Definition 1.1: Let $\mathcal{U}$ be a finite von Neumann algebra. This means $\mathcal{U}$ has a finite faithful normal trace τ. Then $\mathcal{U}$ is naturally a pre-Hilbert space with inner product given by

$$(a,b) \equiv \tau(a^*b) \qquad \text{for all } a,b \in \mathcal{U}.$$

We shall denote the L^2-completion of $\mathcal{U}$ by $\ell^2(\mathcal{U})$. There is a natural representation of $\mathcal{U}$ on $\ell^2(\mathcal{U})$ given by left multiplication, which is faithful and normal. This will be called the left regular representation.

Definition 1.2: $\mathcal{M}$ is said to be a Hilbert $\mathcal{U}$ module if $\mathcal{M}$ is a closed $\mathcal{U}$ invariant subspace of $\oplus_{x\in X}\ell^2(\mathcal{U})$ where X is a finite set.

Definition 1.3: The direct sum of Hilbert $\mathcal{U}$ modules $\mathcal{M}$ and $\mathcal{N}$ is the direct sum $\mathcal{M}\oplus\mathcal{N}$ of $\mathcal{U}$ modules equipped with the direct sum inner product. $\mathcal{M}\oplus\mathcal{N}$ is easily seen to be a Hilbert $\mathcal{U}$ module.

Definition 1.4: A free Hilbert $\mathcal{U}$ module $\mathcal{F}$ is a module of the form $\oplus_{x\in X}\ell^2(\mathcal{U})$. A Hilbert $\mathcal{U}$ module $\mathcal{M}$ is said to be stably free if the direct sum of $\mathcal{M}$ with some free Hilbert $\mathcal{U}$ module is a free Hilbert $\mathcal{U}$ module.

Remark 1.5: Observe that a Hilbert $\mathcal{U}$ module $\mathcal{M}$ is actually finitely generated and projective, since it is easily seen that $\mathcal{M}^\perp$ is also a Hilbert $\mathcal{U}$ module and $\mathcal{M} \oplus \mathcal{M}^\perp = \oplus_{x \in X} \ell^2(\mathcal{U})$.

Definition 1.6: Let $\mathcal{M}$ and $\mathcal{N}$ be Hilbert $\mathcal{U}$ modules. Let $L(\mathcal{M}, \mathcal{N})$ denote the set of all continuous, linear $\mathcal{U}$-module maps.

Note that $L(\ell^2(\mathcal{U}), \ell^2(\mathcal{U}))$ is naturally a finite von Neumann algebra. This follows from the fact that it is isomorphic to the commutant $\mathcal{U}'$ of the left regular representation which in turn is anti-isomorphic to $\mathcal{U}$ (Dixmier [8,pp80,99]). Furthermore the commutant of the $\mathcal{U}$ action on a free Hilbert $\mathcal{U}$ module is isomorphic to $\mathcal{U}' \otimes M_n$ where M_n denotes the $n \times n$ matrices and n is the cardinality of the index set of the free module. (This is a consequence of [8, pp116]). This algebra is also a left Hilbert $\mathcal{U}$ module, the Hilbert space structure coming from the trace $\tau \otimes tr_n$ where tr_n is the usual matrix trace and τ denotes the trace on $\mathcal{U}'$ as well (note that the commutant is also a finite von Neumann algebra as well). Henceforth we normalise τ by $\tau(Id) = 1$ where Id is the identity operator on $\ell^2(\mathcal{U})$. Now τ is defined for subspaces $\mathcal{L}$ of $\ell^2(\mathcal{U})$ such that the orthogonal projection, $P_{\mathcal{L}} : \ell^2(\mathcal{U}) \to \hat{\mathcal{L}}$ ($\hat{\mathcal{L}}$ is the closure of $\mathcal{L}$ in $\ell^2(\mathcal{U})$), belongs to $\mathcal{U}'$ for we can define $\tau(\mathcal{L}) \equiv \tau(P_{\mathcal{L}})$.

We extend the definition of τ to subspaces $\mathcal{L}$ of $\oplus_{x \in X} \ell^2(\mathcal{U}))$ such that the orthogonal projection, $P_{\mathcal{L}} : \oplus_{x \in X} \ell^2(\mathcal{U}) \to \hat{\mathcal{L}}$ belongs to $\mathcal{U}' \otimes \operatorname{End}(\mathbb{C}^n)$ using the trace introduced above on the latter. To lighten the notation we will simply denote by τ the trace $\tau \otimes tr_n$ for any n although the normalisation will be such as to give the value n on the identity element. Note that $\tau(\mathcal{L}) < \infty$ as $\mathcal{L}$ is finitely generated over $\mathcal{U}$. This means in particular that the dimension of any Hilbert $\mathcal{U}$ module $\mathcal{M}$ is well defined and we write it as $\tau(\mathcal{M}) = \tau(P_{\mathcal{M}})$.

Definition 1.7: $f : \mathcal{M} \to \mathcal{N}$ is a homomorphism of Hilbert $\mathcal{U}$ modules if f is a $\mathcal{U}$ module map which is a bounded linear operator. Note that $\ker f$ and $\overline{\text{range } f}$ are also Hilbert $\mathcal{U}$ modules. $f : \mathcal{M} \to \mathcal{N}$ is said to be a weak isomorphism if $\ker f = 0$ and $\overline{\text{range } f} = \mathcal{N}$.

Proposition 1.8 : If $f : \mathcal{M} \to \mathcal{N}$ is a weak isomorphism of Hilbert $\mathcal{U}$ modules then $f = g \circ h$ where $h : \mathcal{M} \to \mathcal{M}$ is a self adjoint weak automorphism, of Hilbert $\mathcal{U}$ modules, and $g : \mathcal{M} \to \mathcal{N}$ is an isomorphsim of Hilbert $\mathcal{U}$ modules.

This last result is just the polar decomposition restated for module maps. Note in particular that the isomorphism of the proposition is thus an isometry. One application is the following result.

Proposition 1.9 (J. Cohen [6]): Every Hilbert $\mathcal{U}$ module $\mathcal{M}$ is isometric to $\oplus_{k=1}^n I_k$ where I_k is a Hilbert $\mathcal{U}$ submodule of $\ell^2(\mathcal{U})$.

Definition 1.10: Let $\mathcal{M}$ be a Hilbert $\mathcal{U}$ submodule of $\oplus_{x \in X} \ell^2(\mathcal{U})$, and let p be the orthogonal projection onto $\mathcal{M}$. Then $p \in \mathcal{U}' \otimes \operatorname{End}(\mathbb{C}^n)$ where n is the cardinality of X. We say

$$0 \to \mathcal{L} \to \mathcal{M} \xrightarrow{f} \mathcal{N} \to O$$

is a weakly short exact sequence of Hilbert $\mathcal{U}$ modules if $\ker f = \mathcal{L}$ and $\overline{\text{Image } f} = \mathcal{N}$.

We can exploit the previous results to show:

Lemma 1.11: If $0 \to \mathcal{L} \to \mathcal{M} \xrightarrow{p} \mathcal{N} \to 0$ is a weakly short exact sequence of Hilbert

$\mathcal{U}$-modules then there is a short exact sequence of Hilbert $\mathcal{U}$-modules

$$\begin{array}{ccccccccc} 0 & \rightarrow & \mathcal{L} & \rightarrow & \mathcal{M} & \xrightarrow{p} & \mathcal{N} & \rightarrow & 0 \\ & & \updownarrow 1 & & \downarrow h & & & & \\ 0 & \rightarrow & \mathcal{L} & \rightarrow & \mathcal{M} & \xrightarrow{q} & \mathcal{N} & \rightarrow & 0 \end{array}$$

where h is a self adjoint weak automorphism such that $h|_{\mathcal{L}}$ is the identity.

Proof: Firstly $\mathcal{M} = \ker p \oplus (\ker p)^{\perp}$. Let $p' \equiv p|_{(\ker p)^{\perp}}$, $p' = (\ker p)^{\perp} \rightarrow \mathcal{N}$ be a weak isomorphism. By proposition 1.9, there is a self adjoint weak automorphism h' of $(\ker p)^{\perp}$ and an isomorphism $q' : (\ker p)^{\perp} \rightarrow \mathcal{N}$ such that $p' = q' \circ h'$. Define $h \equiv 1 \oplus h'$ and $q \equiv 0 \oplus q'$.

Lemma 1.12: Let $O \rightarrow \mathcal{L} \rightarrow \mathcal{M} \xrightarrow{p} \mathcal{N} \rightarrow O$ be a weakly short exact sequence of Hilbert $\mathcal{U}$ modules. If $\mathcal{M}, \mathcal{N}$ are stably free Hilbert $\mathcal{U}$ modules, then so is $\mathcal{L}$.

Proof: By the above lemma there is a short exact sequence

$$O \rightarrow \mathcal{L} \rightarrow \mathcal{M} \xrightarrow{q} \mathcal{N} \rightarrow O$$

of Hilbert $\mathcal{U}$ modules where q is surjective. Assume that $\mathcal{N} \oplus \mathcal{F}$ is a free Hilbert $\mathcal{U}$ module, where $\mathcal{F}$ is a free Hilbert $\mathcal{U}$ module. Then $q \oplus 1 : \mathcal{M} \oplus \mathcal{F} \rightarrow \mathcal{N} \oplus \mathcal{F}$ has a section $s : \mathcal{N} \oplus \mathcal{F} \rightarrow \mathcal{M} \oplus \mathcal{F}$ (obtained by mapping each basis element to an arbitrary element in its inverse image). Then $s : \mathcal{N} \rightarrow \mathcal{M}$ defined by $s \equiv pr_1 \circ S \circ i_1$ where $i_1 : \mathcal{N} \rightarrow \mathcal{N} \oplus \mathcal{F}$ is the inclusion map and $pr_1 : \mathcal{M} \oplus \mathcal{F} \rightarrow \mathcal{M}$ is the projection map, is a section of the map q i.e.$q \circ s = 1$. Hence $\mathcal{M} = s(\mathcal{N}) \oplus \mathcal{L}$. If $\mathcal{M} \oplus \mathcal{F}_1$ is a free Hilbert $\mathcal{U}$ module, where $\mathcal{F}_1$ is a free Hilbert $\mathcal{U}$ module, it follows that $s(\mathcal{N}) \oplus \mathcal{F} \oplus \mathcal{F}_1 \oplus \mathcal{L}$ is a free Hilbert $\mathcal{U}$ module i.e. $\mathcal{L}$ is a stably free Hilbert $\mathcal{U}$ module.

Proposition 1.13 (See Cheeger and Gromov [5, section 1]): (1) Let $\mathcal{M}$ be a Hilbert $\mathcal{U}$ module. Then $\tau(\mathcal{M}) = 0$ if and only if $\mathcal{M} = 0$.

(2) $\tau(\ell^2(\mathcal{U}) \otimes \mathbf{C}^k) = k$.

(3) Let $f : \mathcal{M} \rightarrow \mathcal{N}$ be a Hilbert $\mathcal{U}$ module homomorphism. Then $\tau(\mathcal{M}) = \tau(\ker f) + \tau(\overline{\text{range } f})$.

(4) (Continuity) Let $\mathcal{M}_1 \supseteq \mathcal{M}_2 \ldots \supseteq \mathcal{M}_j \supseteq \ldots$ be Hilbert $\mathcal{U}$ submodules. Then $\tau(\cap_{j=1}^{\infty} \mathcal{M}_j) = \lim_{j \rightarrow \infty} \tau(\mathcal{M}_j)$.

(5) If $O \rightarrow \mathcal{L} \rightarrow \mathcal{M} \rightarrow \mathcal{N} \rightarrow O$ is a weakly short exact sequence of Hilbert $\mathcal{U}$ modules, then $\tau(\mathcal{M}) = \tau(\mathcal{N}) + \tau(\mathcal{L})$.

The special case of the above which is of greatest interest for us is where π is a discrete group and $\mathcal{U}(\pi)$ is the von Neumann algebra generated by the left regular representation of π. Then $\mathcal{U}(\pi)$ is a finite von Neumann algebra [8, pp319]. We let τ be the faithful normal trace on $\mathcal{U}(\pi)$. The left regular representation acting on $\ell^2(\pi)$ is a homomorphism $\rho : \pi \rightarrow \mathcal{U}(\pi)$. More generally we will refer to any homomorphism of π into the unitary group of a finite von Neumann algebra as a unitary representation.

1.2: $\mathbf{L^2}$-homological algebra

Let $C : 0 \to C^0 \xrightarrow{d} C^1 \xrightarrow{d} \dots \xrightarrow{d} C^n \to 0$ be a complex over $\mathcal{U}$ where each C^j is a Hilbert $\mathcal{U}$ module, and $d \in L(C^j, C^{j+1})$. We shall call (C, d) a Hilbert $\mathcal{U}$ complex. Let δ denote the Hilbert adjoint of d and $\Delta_j = d\delta + \delta d$ acting on C^j. Let $\mathcal{H}^j_{(2)}(C)$ denote the kernel of Δ_j.

We will prove that the Hodge theorem holds i.e.

Proposition 1.14 (Hodge theorem): Let (C, d) be a Hilbert $\mathcal{U}$ complex. Then

$$C^j = \mathcal{H}^j_{(2)}(C) \oplus \overline{dC^{j-1}} \oplus \overline{\delta C^{j+1}} \text{ for all } j \geq 0.$$

Proof: Since $\Delta_j : C^j \to C^j$ is a Hilbert $\mathcal{U}$-module homomorphism, it follows from proposition 1.14(3) that

$$C^j = \mathcal{H}^j_{(2)}(C) \oplus \overline{\text{range } \Delta_j}.$$

Let $f = \Delta_j\, r = d(\delta r) + \delta(dr)$ i.e. range $\Delta_j \subseteq dC^{j-1} \oplus \delta C^{j+1}$ which implies that $\overline{\text{range } \Delta_j} \subseteq \overline{dC^{j-1}} \oplus \overline{\delta C^{j+1}}$.

Let $f \in \mathcal{H}^j_{(2)}(C) \cap \overline{dC^{j-1}} \oplus \overline{\delta C^{j+1}}$, then $f \in \mathcal{H}^j_{(2)}(C)$ implies that $df = 0 = \delta f$, while $f \in \overline{dC^{j-1}} \oplus \overline{\delta C^{j+1}}$ implies that there are sequences u_j and w_k such that $f = \lim_j du_j + \lim_k \delta w_k$. Now $df = 0$ implies that $\lim_k d\delta w_k = 0$, since $d \in L(C^j, C^{j+1})$. Thus by Cauchy-Schwartz $0 = (\lim_k d\delta w_k, w_k) = \lim_k ||\delta w_k||^2$, hence $\lim_k \delta w_k = 0$. Similarly one can show that $\delta f = 0$ implies that $\lim_j du_j = 0$. Hence we see that $f = 0$. In particular

$$\overline{dC^{j-1}} \oplus \overline{\delta C^{j+1}} \subseteq \overline{\text{range } \Delta_j}.$$

We can define the L^2-cohomology groups of the Hodge complex (C, d) as follows

$$H^j_{(2)}(C) \equiv \ker d_j / \overline{\text{image } d_{j-1}}.$$

The L^2-Betti numbers are then defined to be

$$b_j^{(2)}(C) \equiv \tau(H^j_{(2)}(C)).$$

For a Hilbert $\mathcal{U}$ complex (C, d) it is clear that we have a natural isomorphism (called the Hodge isomorphism)

$$H^j_{(2)}(C) \cong \mathcal{H}^j_{(2)}(C).$$

An isometry F between Hilbert $\mathcal{U}$ complexes (C, d) and (C_1, d^1) is defined to be a sequence of isometries $F_j : C^j \to C_1^j$ which are $\mathcal{U}$ module maps and commute with differentials i.e.

$$F_{j+1} d_j = d_j^1 F_j.$$

For such an isometry between Hilbert $\mathcal{U}$ complexes there is an induced isometry between the L^2-cohomology groups of the complexes i.e.

$$F_{j*} : H^j_{(2)}(C) \to H^j_{(2)}(C_1) \text{ for all } j \geq 0$$

In particular the L^2-Betti numbers of the complexes are equal i.e.

$$b_j^{(2)}(C) = b_j^{(2)}(C_1) \text{ for all } j \geq 0$$

Proposition 1.15 (Cheeger and Gromov [4, theorem 2.1]): Let

$$0 \to (C', d') \xrightarrow{\alpha} (C, d) \xrightarrow{\beta} (C'', d'') \to 0$$

be a short exact sequence of Hilbert $\mathcal{U}$ complexes. Then there is a weakly long exact sequence in L^2-cohomology

$$\ldots \to H^j_{(2)}(C') \xrightarrow{\alpha} H^j_{(2)}(C) \xrightarrow{\beta} H^j_{(2)}(C'') \to H^{j+1}_{(2)}(C') \to \ldots$$

Proof: If (C, d) is a Hilbert $\mathcal{U}$ complex, then the spectral projections $E(\lambda)$ of the self adjoint operators $d\delta$ and δd satisfy $\tau(E(\lambda)) < \alpha$, as the range of $E(\lambda)$ is a subspace of $\oplus_j C^j$. The proposition then follows from theorem 2.1 in [4].

1.3: $\mathcal{U}$-RF complexes:

Definition 1.16: The rank (over $\mathcal{U}$) of a free Hilbert $\mathcal{U}$ module $\mathcal{F}$ is defined to be $\tau(\mathcal{F})$. Note that the rank of $\mathcal{F}$ is always a non-negative integer. The rank of a stably free Hilbert $\mathcal{U}$ module $\mathcal{M}$ is also defined to be $\tau(\mathcal{M})$ which is also a non-negative integer. This can be seen as follows: if $\mathcal{M} \oplus \mathcal{F}$ is isometric to $\mathcal{F}_1$, where $\mathcal{F}$ and $\mathcal{F}_1$ are free Hilbert $\mathcal{U}$ modules, then the rank of $\mathcal{M}$ is also equal to the difference rank $\mathcal{F}_1$ - rank $\mathcal{F}$, which is a non-negative integer.

Definition 1.17: Let $\mathcal{F}$ be a free Hilbert $\mathcal{U}$ module of rank k. We say $e = (e_1, e_2, \ldots, e_k), e_j \in \mathcal{F}$ pairwise orthogonal is a basis for $\mathcal{F}$ over $\mathcal{U}$ if $\oplus_{j=1}^{r} \mathcal{U} e_j$ is dense in $\mathcal{F}$. Two bases e and f of $\mathcal{F}$ are said to be $\mathcal{U}$ related if there is an invertible element S in the commutant of the $\mathcal{U}$ action which maps e to f. It is clear that $\mathcal{U}$ related is an equivalence relation. By the remarks after definition 1.7 we see that such an operator can be thought of as an invertible element of $\mathcal{U} \otimes M_k$, that is as a $Gl(k, \mathcal{U})$ matrix. We denote this matrix of S by e/f.

For example let $\mathcal{F}_k$ denote the standard free Hilbert $\mathcal{U}$ module $\ell^2(\mathcal{U})^k \cong \ell^2(\mathcal{U}) \otimes \mathbf{C}^k$. Then $\mathcal{F}_k$ has the canonical basis $f_1 = (1, 0, \ldots, 0)$, $f_2 = (0, 1, \ldots, 0) \ldots$, $f_k = (0, 0, \ldots, 1)$.

Lemma 1.18: Let

$$0 \to (\mathcal{M}, e) \xrightarrow{i} (\mathcal{N}, f) \xrightarrow{p} (\mathcal{L}, g) \to 0$$

be a short exact sequence of (free) Hilbert $\mathcal{U}$ modules where e, f and g are bases of the respective modules with $i(e)$ $\mathcal{U}$ related to f. Then there is a basis eg of $\mathcal{N}$ which is $\mathcal{U}$ related to f.

Proof: Lift each element g_j of the basis g of $\mathcal{L}$ to an element g'_j in $\oplus_j \mathcal{U} f_j$. Then $i(e) \oplus g'$ is $\mathcal{U}$ related to f (as bases of $\mathcal{N}$). It is easy to see that any other similar lift g'' is $\mathcal{U}$ related to g', completing the argument.

Definition 1.19: Let (C, d) be a Hilbert $\mathcal{U}$ complex of free Hilbert $\mathcal{U}$ modules C^j. Choose a basis e_j of each C^j. Then $d : C^j \to C^{j+1}$ is represented by a matrix $m(d, e)$ with entries from $\mathcal{U}'$ (cf remarks after 1.7) and the triple (C, d, e) is called a $\mathcal{U}$-RF complex.

Remark 1.20: These ideas also work for complexes of stably free Hilbert $\mathcal{U}$ modules however we will not have need of this extension here.

1.4. The Fuglede-Kadison determinant:

We will now discuss the Fuglede-Kadison [11, 8] determinant of a finite Von Neumann algebra $\mathcal{U}$.

We shall stick to our earlier convention and denote the induced trace on $M_n(\mathcal{U})$ by τ. If $A \in Gl(n, \mathcal{U})$, then $A^*A \in Gl(n, \mathcal{U})$ is positive definite and $\log(A^*A) \in Gl(n, \mathcal{U})$ is self-adjoint. The Fuglede-Kadison determinant

$$|\mathrm{Det}_\tau| : Gl(n, \mathcal{U}) \to \mathbb{R}_+^*$$

is defined by

$$|\mathrm{Det}_\tau|(A) \equiv \exp[1/2\tau(\log(A^*A))]$$

and it has the following properties:

(1) $|\mathrm{Det}_\tau|(H) = \exp(\tau(\log(H)))$ if $H \in Gl(n, \mathcal{U})$, $H = H^*$ and $H \geq 0$.

(2) $|\mathrm{Det}_\tau|(\lambda I) = |\lambda|^n, \quad \lambda \neq 0$.

(3) $|\mathrm{Det}_\tau|(A) = |\mathrm{Det}_\tau|(A^*) = |\mathrm{Det}_\tau|((A^*A)^{1/2})$ if $A \in Gl(n, \mathcal{U})$.

(4) $|\mathrm{Det}_\tau|(AB) = |\mathrm{Det}_\tau|(A)|\mathrm{Det}_\tau|(B)$ if $A, B \in Gl(n, \mathcal{U})$.

(5) $|\mathrm{Det}_\tau|(A) = |\mathrm{Det}_\tau|(UAU^*)$ if $A, U \in Gl(n, \mathcal{U})$ and $\mathcal{U}$ is unitary.

(6) $|\mathrm{Det}_\tau|(A) \leq |A|$ if $A \in Gl(n, \mathcal{U})$.

For us the most important fact about this determinant is that it has a non-trivial extension (i.e. not the algebraic one) to certain singular operators, namely those which are injective but whose range is not closed. If it were not for this the ensuing discussion would have little content. The point is that using the spectral representation $A^*A = \int \lambda d\mu(E_\lambda)$ the preceding definition still applies with the understanding that $|\mathrm{Det}_\tau|(A) = 0$ when $\int \ln\lambda d\mu(E_\lambda)$ is divergent. With obvious modifications all of the preceding properties continue to hold together with the following computationally useful result: for $B \in M_n(\mathcal{U})$, $B \geq 0$,

(7) $|\mathrm{Det}_\tau|(B) = \lim_{\epsilon \to 0} |\mathrm{Det}_\tau|(B + \epsilon)$

Note that [11] discusses the preceding results only for II_1 factors. The generalisation to the finite case is given in [8] except for the discussion of singular operators. However it is not hard to see that the argument in [11] for this extension holds also for finite von Neumann algebras.

We conclude this subsection with a final definition.

Definition 1.21: We introduce an equivalence relation $\sim$ on $\mathcal{U}$ related bases by defining $e \sim f$ if $|\mathrm{Det}_\tau|(e/f) = 1$.

Remark 1.22: In lemma 1.19 we note that the $\sim$ class of the basis eg is well defined.

1.5: Definition of L^2-RF torsion

In this section we give our definition of L^2-RF torsion.

To each $\mathcal{U}$-RF complex (C, d, e) we can associate the triple (E, D, e) given by

$$E^0 \equiv C^{od} \equiv \oplus_{j \geq 0} C^{2j+1}$$

$$E^1 \equiv C^{ev} \equiv \oplus_{j\geq 0} C^{2j}$$

and

$$D : E^0 \to E^1, \quad D \equiv d + \delta.$$

Clearly E^j are free Hilbert $\mathcal{U}$ modules with bases $c_0 \equiv \cup_{j\geq 0} e_{2j+1}$ for E^0 and $c_1 \equiv \cup_{j\geq 0} e_{2j}$ for E^1.

Then $D : E^0 \to E^1$ and recalling the remarks after definition 1.7 and definition 1.20 we may regard D as acting via a matrix with entries from $\mathcal{U}'$.

Definition 1.23:If the L^2-Betti numbers of the complex (C, d, e) vanish we say it is L^2-acyclic. In this case D is a weak isomorphism so E^0 and E^1 are isomorphic free $\mathcal{U}$ modules. The choice of bases c_0 and c_1 makes this identification explicit so that D is represented by a matrix $m(D, e)$ as an operator on say E^0 and we may define the L^2-RF torsion to be

$$T_{(2)}(C, d, e) \equiv |\mathrm{Det}_\tau|(m(D, e)).$$

Note that while D is injective it may not have closed range thus a complex may be L^2-acyclic without being acyclic. This is one important difference between L^2-RF torsion and ordinary torsion. A second stems from the fact that the Fuglede-Kadison determinant is non-trivial on a wider class of operators than just the invertibles. This is what makes this L^2-RF torsion interesting for a wider class of manifolds.

If c_j' is an alternate set of basis elements for E^j induced from an alternate $\mathcal{U}$ related basis a_g for C^g, then we have

Lemma 1.24: $T'_{(2)}(C, d, e) = T'_{(2)}(C, d, a) \prod_{i=0}^{n} |\mathrm{Det}_\tau|(a_i/e_i)^{(-1)^i}$.

Proof: $m(D, e) = c_0/c_0' m(D, a) c_1'/c_1$ i.e.

$$\begin{pmatrix} e_1/a_1 & & 0 \\ & e_3/a_3 & \\ 0 & & \ddots \end{pmatrix} \quad m(D, a) \quad \begin{pmatrix} a_0/e_0 & & 0 \\ & a_2/e_2 & \\ 0 & & \ddots \end{pmatrix}$$

Hence we have

$$T'_{(2)}(C, d, e) = \prod_{j\geq 0} |\mathrm{Det}_\tau|(e_{2j+1}/a_{2j+1}) T'_{(2)}(C, d, a) \prod_{j\geq 0} |\mathrm{Det}_\tau|(a_{2j}/e_{2j})$$

using the fact that e/f is the inverse of f/e we obtain the result.

1.6: L²-RF torsion via chain complexes.

Instead of studying Hilbert $\mathcal{U}$ (cochain) complexes, we could just as well study Hilbert $\mathcal{U}$ chain complexes (C, ∂) where

$$0 \to C_n \xrightarrow{\partial_n} C_{n-1} \xrightarrow{\partial_{n-1}} \ldots \xrightarrow{\partial_1} C_0 \to 0$$

is a chain complex over $\mathcal{U}$ and each C_j is a Hilbert $\mathcal{U}$ module with $\partial_j \in L(C_j, C_{j-1})$. The L^2-homology groups of this chain complex (C, ∂) are defined to be

$$H_j^{(2)}(C) = \ker \partial_j / (\overline{\mathrm{range}\ \partial_{j+1}}).$$

Definition 1.25: Let (C, ∂) be a Hilbert $\mathcal{U}$ chain complex of free Hilbert $\mathcal{U}$-modules C_j, and $f_j = (f_j^1, \ldots, f_j^k)$ be a basis for C_j over $\mathcal{U}$. If ∂ in the basis f_j is represented by a matrix with coefficients in $\mathcal{U}'$ then (C, ∂, f) is called a $\mathcal{U}$-RF chain complex.

Now let (C, ∂, f) be a L^2-acyclic $\mathcal{U}$-RF chain complex (i.e. $H_j^2(C) = 0$ for all $j \geq 0$). Let $\rho : \pi \to \mathcal{U}$ be a unitary representation of π. We can define the L^2-RF torsion of (C, ∂, f) as follows:

Definition 1.26: Let D' denote $\partial + \partial^*$ restricted to the odd part of the chain complex (C_{od}). Then as in definition 1.24 we let

$$T_{(2)}(C, \partial, f) \equiv |\mathrm{Det}_\tau|(m(D', f)).$$

Let (C, d, e) be a L^2-acyclic $\mathcal{U}$-RF (cochain) complex. Its dual chain complex

$$0 \to C_n \xrightarrow{\partial_n} C_{n-1} \xrightarrow{\partial_{n-1}} \ldots \xrightarrow{\partial_1} C_0 \to 0$$

is defined by: C_j is the topological dual space to C^j and ∂_j is the adjoint map to d_{j-1} and f_j is the dual basis to e_j. Since C^j is a Hilbert space, we can identity C_j with C^j. Under the identification, ∂_j becomes δ_j and f_j becomes e_j. Then $T_{(2)}(C, \partial, f)$ can be seen to be equal to $T_{(2)}(C, \delta, e)$.

Proposition 1.27: $T_{(2)}(C, d, e) = T_{(2)}(C, \delta, e) = T_{(2)}(C, \partial, f)$.

Proof. If D denotes $d + \delta$ restricted to C^{od} then by the remarks above there is a natural isomorphism taking bases to bases:

$$\begin{array}{ccc} C_{od} & \xrightarrow{D'} & C_{ev} \\ \downarrow \wr & & \wr \downarrow \\ C^{od} & \xrightarrow{D} & C^{ev} \end{array}$$

Hence $m(D', f)^t = m(D, e)$ where t denotes the transpose relative to the preceding isomorphism. But then

$$T_{(2)}(C, \partial, f) = |\mathrm{Det}_\tau|(m(D', f)) = |\mathrm{Det}_\tau|(m(D, e)^t) = T_{(2)}(C, d, e)$$

as required.

1.7: L^2-RF torsion for cochain homotopy equivalences

Definitions 1.28: Let $f : (C, d, e) \to (C', d', e')$ and $g : (C, d, e) \to (C', d', e')$ be maps of $\mathcal{U}$-RF complexes (by this we mean cochain maps which are also bounded $\mathcal{U}$ module maps). Then f is said to be L^2 homotopic to g if there is a sequence of maps $D^j \in L(C^j, C'^{j-1})$ such that $d'_{j-1} D^j + D^{j+1} d_j = f_j - g_j$ for all j. Also a map of $\mathcal{U}$-RF complexes $f : (C, d, e) \to (C', d', e')$ is said to be a (cochain) L^2 homotopy equivalence if there is a map of $\mathcal{U}$-RF complexes $g : (C', d', e') \to (C, d, e)$ such that both $f \circ g$ and $g \circ f$ are L^2 homotopic to the identity map. Another way of saying this is the following: if Hf, Hg are induced maps in L^2 cohomology, then $Hf \circ Hg = 1$ and $Hg \circ Hf = 1$.

Let $f : (C, d, e) \to (C', d', e')$ be a map of $\mathcal{U}$-RF complexes. Then we can define the mapping cone (C_f, d_f, e_f) as follows

$$C_f^j \equiv C^j \oplus C'^{j-1},$$

$$(d_f)_j \equiv (d_j + (f_j - d'_{j-1})),$$
$$(e_f)_j \equiv e_j \cup e'_{j-1}.$$

Clearly (C_f, d_f, e_f) is a $\mathcal{U}$-RF complex.

Lemma 1.29: If $f : (C, d, e) \to (C', d', e')$ is a L^2 homotopy equivalence of $\mathcal{U}$-RF complexes, then (C_f, d_f, e_f) is an L^2-acyclic $\mathcal{U}$-RF complex.

Proof: Define $(\hat{C}, \hat{d}, \hat{e})$ by

$$\hat{C}^j \equiv C'^{j-1}, \quad \hat{d}_j \equiv -d'_{j-1} \text{ and } \hat{e}_j \equiv e_{j-1}.$$

Then $(\hat{C}, \hat{d}, \hat{e})$ is a $\mathcal{U}$-Rf complex and we have the weakly short exact sequence of $\mathcal{U}$-RF complexes

$$0 \to (\hat{C}, \hat{d}, \hat{e}) \xrightarrow{\alpha} (C_f, d_f, e_f) \xrightarrow{\beta} (C, d, e) \to 0$$

which gives rise to a weakly long exact sequence in cohomology by proposition 1.16

$$\ldots \to H^{j-1}_{(2)}(C') \xrightarrow{H\alpha} H^j_{(2)}(C_f) \xrightarrow{H\beta} H^j_{(2)}(C) \xrightarrow{Hf} H^j_{(2)}(C') \to \ldots$$

where we have used the fact that $H^j_{(2)}(\hat{C}) = H^{j-1}_{(2)}(C')$ and $H\alpha, H\beta, Hf$ are the induced maps in L^2 cohomology. Hf is an isomorphism since f is a L^2 homotopy equivalence, and so from the long exact sequence we see that $H^j_{(2)}(C_f) = 0$ for all $j \geq 0$.

Definition 1.30: The L^2-RF torsion of a L^2 homotopy equivalence of $\mathcal{U}(\pi)$-RF complexes $f : (C, d, e) \to (C', d', e')$ is defined to be

$$T_{(2)}(f, e, e') \equiv T_{(2)}(C_f, d_f, e_f).$$

We have two comments on this definition.

Remark 1.31: Let $f : (C, d, e) \to (C', d', e')$ be a map of L^2-acyclic $\mathcal{U}$-RF complexes. Then f is trivially a L^2 homotopy equivalence.

Definition 1.32: A map $f : (C, d, e) \to (C', d', e')$ is said to be a L^2 simple isomorphism of $\mathcal{U}$-RF complexes if $T_{(2)}(f, e, e') = 1$.

Remark 1.33: Let $(\mathcal{M}, e)$ and $(\mathcal{N}, g)$ be free Hilbert $\mathcal{U}$-modules with bases e of $\mathcal{M}$ and g of $\mathcal{N}$. Let $F : \mathcal{M} \to \mathcal{N}$ be an isomorphism of Hilbert $\mathcal{U}$-modules. Then we can form the mapping cone complex (C_F, d_F, e_F) where

$$C^0_F = \mathcal{M}, \quad C^1_F = \mathcal{N}, \quad C^j_F = 0 \text{ for } j \neq 0, 1$$
$$(d_F)_0 = F, \quad (d_F)_j = 0 \text{ for } j \neq 0$$
$$(e_F)_0 = e, \quad (e_F)_1 = g$$

and the L^2-RF torsion of F, $T_{(2)}(F, e, g)$ is then equal to $T_{(2)}(C_F, d_F, e_F)$ Note that if the matrix of F in the above bases $m(F, e, g)$ is the identity matrix, then F is a simple L^2 isomorphism.

1.8: The Euler Characteristic:

Let $C : 0 \to C^0 \xrightarrow{d} C^1 \xrightarrow{d} \dots \xrightarrow{d} C^n \to 0$ be a Hilbert $\mathcal{U}$ complex. Then the Euler characteristic of C, $\chi(C)$, is defined by $\chi(C) = \sum_{j=0}^{n}(-1)^j\tau(C^j)$. Observe that

$$\tau(C^j) = \tau(\overline{dC^{j-1}}) + \tau(\overline{\delta C^{j+1}}) + \tau(\mathcal{H}^j_{(2)}(C)).$$

Also $d : \overline{\delta C^{j+1}} \to \overline{dC^j}$ is a weak isomorphism. By proposition 1.9 $\overline{\delta C^{j+1}}$ is isometric to $\overline{dC^j}$, so $\tau(\overline{\delta C^{j+1}}) = \tau(\overline{dC^j})$ and we have

$$\chi(C) = \sum_{j=0}^{n}(-1)^j\tau(C^j) = \sum_{j=0}^{n}(-1)^j[\tau(\overline{dC^{j-1}}) + \tau(\overline{dC^j}) + \tau(\mathcal{H}^j_{(2)}(C))]$$

where $C^{-1} = C^{n+1} = 0$, that is $\chi(C) = \sum_{j=0}^{n}(-1)^j\tau(\mathcal{H}^j_{(2)}(C))$.

If (C, d, e) is a $\mathcal{U}$-RF complex then $\chi(C)$ is an integer since $\tau(C^j)$ is free for all $j \geq 0$.

2. L^2-acyclicity

Let M be a closed manifold of dimension n, and $\rho : \pi_1(M) \to \mathcal{U}$ a unitary representation of the fundamental group of M. Let $E_\rho \equiv \widetilde{M} \times_\rho \ell^2(\mathcal{U}) \to M$ be the associated $\mathcal{U}$ finite Hilbert bundle over M and $H^*_{(2)}(M, E_\rho))$ denote the L^2-cohomology groups of M with values in E_ρ, given a Riemannian metric on M.

Definition 2.1: A unitary representation $\rho : \pi_1(M) \to \mathcal{U}$ is said to be L^2-acyclic if $H^*_{(2)}(M, E_\rho) = 0$ for all $*$.

Remark 2.2: Since the condition $H^*_{(2)}(M, E_\rho) = 0$ depends only on the quasi-isometry class of the Riemannian metric on M, and since on a compact manifold all Riemannian metrics are quasi-isometric to each other, it follows that the condition $H^*_{(2)}(M, E_\rho) = 0$ depends only on the differential structure on M and on the unitary representation ρ.

We consider the special case of the above when $E_\rho \equiv \widetilde{M} \times_\rho \ell^2(\pi) \to M$ is the associated Hilbert $\mathcal{U}(\pi)$ bundle over M, where $\widetilde{M} \xrightarrow{pr} M$ denotes the universal cover of M. For the rest of the paper we assume that the Riemannian metric on $\widetilde{M}$ is induced from a Riemannian metric on M.

Consider the following complexes:
(i) $(\Omega^*_{(2)}(\widetilde{M}), d)$, which denotes the complex of L^2 differential forms on $\widetilde{M}$ and
(ii) $(\Omega^*_{(2)}(M, E_\rho), d)$, which denotes the complex of L^2 differential forms on M with values in E_ρ.

There is a canonical isometry of complexes given by

$$\Omega^*_{(2)}(\widetilde{M}) \to \Omega^*_{(2)}(M, E_\rho),$$

$$\omega_x \to \sum_{\gamma \equiv \pi} \omega_{\gamma x}.\gamma, \qquad x \in \widetilde{M} \text{ and } \omega \in \Omega^*_{(2)}(\widetilde{M})$$

where we canonically identify the spaces $T^*_{\gamma x}\ \widetilde{M}$ with $T^*_{pr(x)}M$. This induces an isometry on the L^2 cohomology groups

$$H^*_{(2)}(\widetilde{M}) \to H^*_{(2)}(M, E_\rho) \text{ for all } * .$$

The L^2-Betti numbers of M, $b_k^{(2)}(M)$ are defined to be

$$b_k^{(2)}(M) \equiv \tau(H_k^{(2)}(\widetilde{M})) = \tau(H_{(2)}^k(M, E_\rho)).$$

Definition 2.3: A closed, oriented connected manifold M is said to be a L^2-acyclic manifold if the left regular representation of the fundamental group of M is a L^2-acyclic representation i.e. M is a L^2-acyclic manifold if all the L^2-Betti numbers of $\widetilde{M}$ vanish.

Remark 2.4: Dodziuk [9] has proved that the L^2-Betti numbers of the universal cover of a closed oriented manifold M are homotopy invariants of M. This implies that any closed oriented manifold which is homotopy equivalent to a L^2-acyclic manifold is also a L^2-acyclic manifold.

Although the condition of L^2-acyclicity is a homotopy invariant of closed manifolds, the following provides obstructions in terms of classical homotopy invariants.

Proposition 2.5: Let M be a L^2-acyclic manifold. Then:
(1) The fundamental group $\pi_1(M)$ is infinite.
(2) The Euler characteristic $\chi(M)$ is zero.
(3) The signature $\sigma(M)$ is zero.

Proof: (1) Since M is connected, a function f on $\widetilde{M}$ satisfying $df = 0$ is a constant and hence $b_{(2)}^0(\widetilde{M}) = 0$ if and only if $\pi_1(M)$ is infinite.

We establish (2) and (3) by the Atiyah and Singer L^2-index theorem for coverings [1], [16], which says that the Euler characteristic of M is equal to the von Neumann dimension of $\sum_j (-1)^j H_{(2)}^j(\widetilde{M})$ which is zero since M is L^2-acyclic, and the signature of M is equal to the von Neumann dimension of $H_{(2)}^+(\widetilde{M})$ − the dimension of $H_{(2)}^-(\widetilde{M})$ (where $+, -$ is the decomposition of $H_{(2)}^*(\widetilde{M})$ given by the signature involution) which is zero since M is L^2-acyclic, completing the proof.

We recall the definition of a rotationally symmetric (RS) Riemannian manifold (alias a "model" cf [10]). X is said to be an RS Riemannian manifold if
(1) there is a point $p \in X$ such that the exponential map is a diffeomorphism of T_pX onto X,
(2) every linear isometry of T_pX can be realised as the differential of an isometry of X.

Examples 2.6: Both complete, simply connected, hyperbolic manifolds (i.e. manifolds of constant negative sectional curvature) and complete, simply connected flat manifolds (i.e. manifolds of constant zero sectional curvature) are rotationally symmetric.

Proposition 2.7 (Dodzuik [10]): (a) Let M be a closed, oriented, connected manifold of odd dimension. Assume that there is a Riemannian metric on M which induces a RS Riemannian metric of infinite volume on the universal cover of M. Then M is a L^2-acyclic manifold.

(b) If M is a closed, connected, oriented, flat manifold then M is L^2-acyclic.

Proof: If M satisfies the hypotheses of the theorem, then by the main theorem in Dodzuik [10], $H_{(2)}^*(\widetilde{M}) = 0$ for all $*$. i.e. M is a L^2-acyclic manifold.

Corollary 2.8: Let G be a connected non-compact semisimple Lie group and let π be a torsion free, cocompact, discrete subgroup of G. Let K be a maximal compact subgroup of G and let G/K have odd dimension. Then if $\pi \backslash G/K$ is orientable it is L^2-acyclic.

We now recall the definition of an acyclically foliated (A-foliated) manifold. A closed, connected manifold X is said to be an A-foliated manifold if there is a nowhere zero, closed 1-form on X. On such a manifold the second author has proved that there are acyclic representations.

We record for the reader's convenience some well known concepts. Let $E\pi \to B\pi$ be a principal bundle with structure group π and contractible total space $E\pi$ over the paracompact space $B\pi$. Then $B\pi$ is called the classifying space of the group π. A countably generated discrete group π is said to be amenable if there is a finitely additive, left invariant measure on π.

Proposition 2.9: Let M be a smooth three dimensional manifold which is closed, connected and oriented. Assume that $\pi_1(M)$ is an infinite, amenable group. Then M is a L^2-acyclic manifold.

Proof: As $\pi_1(M)$ is infinite it follows that $b^0_{(2)}(\widetilde{M}) = 0$. Also Cheeger and Gromov [5] prove that the natural map

$$\rho : H^j_{(2)}(\widetilde{M}) \to H^j_{deRham}(\widetilde{M})$$

is injective for all $j \geq 0$ whenever $\pi_1(M)$ is infinite and amenable. Hence $b^1_{(2)}(\widetilde{M}) = 0$. By the L^2-Poincare duality (see Atiyah [1]) we have $b^2_{(2)}(\widetilde{M}) = b^1_{(2)}(\widetilde{M}) = 0$ and similarly $b^3_{(2)}(\widetilde{M}) = 0$ completing the proof.

A complete Riemannian metric g on M is said to satisfy $geo(M, g) \leq 1$ if the injectivity radius of g is greater than or equal to 1 and the sectional curvatures of g have absolute value less than or equal to 1.

The following provides more examples of L^2-acyclic manifolds.

Proposition 2.10 (Cheeger and Gromov [3], [4], [5]):
(1) Let π be a discrete group which contains an infinite and amenable normal subgroup. Then $B\pi$ is an L^2 acyclic manifold, if $B\pi$ is closed.
(2) Let M be a closed, oriented manifold which admits a sequence of Riemannian metrics g_j such that $geo(\widetilde{M}, \widetilde{g}_j) \leq 1$ and Volume $(M, g_j) \to 0$ as $j \to \infty$, then M is an L^2 acyclic manifold.
(3) Let M be a closed, oriented manifold. If $\mathbb{R}_+ \times M$ admits a complete Riemannian metric g such that Volume $(\mathbb{R}_+ \times M, g) < \infty$ and $geo(\mathbb{R}_+ \times \widetilde{M}, \widetilde{g}) \leq 1$. Then M is an L^2-acyclic manifold.

Lemma 2.11: If M is an L^2-acyclic manifold and N is a closed, oriented manifold, then $M \times N$ is an L^2-acyclic manifold.

Proof: Since the L^2 cohomology of the universal cover of $M \times N$ is independent of the choice of Riemannian metric on $M \times N$, we can choose the product metric on $M \times N$, which induces the product metric on $\widetilde{M} \times \widetilde{N}$, to calculate the L^2 cohomology of $\widetilde{M} \times \widetilde{N}$.

According to Zucker [20], we have a Kunneth theorem in this situation i.e.

$$H^k_{(2)}(\widetilde{M} \times \widetilde{N}) \cong \oplus_{i+j=k} H^j_{(2)}(\widetilde{M}) \otimes H^i_{(2)}(\widetilde{N}).$$

Question 2.12: Let $F \to M \to B$ be a fibre bundle such that M and B are closed, oriented manifolds and F is an L^2-acyclic manifold. Then under what additional conditions is M an L^2-acyclic manifold? One way to study this is to study the L^2 version of the Leray theorem for spectral sequences proved by Cheeger and Gromov [4, section 2]. However we choose to study this question via more elementary methods. First we need to make some definitions.

Definition 2.13: A fibre bundle $F \to M \xrightarrow{p} B$ is said to have quasi-totally geodesic fibres if there are Riemannian metrics on F, M, B and a locally finite open covering $\{U_\alpha\}$ of B such that $p^{-1}(U_\alpha)$ is quasi-isometric to $U_\alpha \times F$ where the Riemannian metrics on $p^{-1}(U_\alpha)$ and U_α (for all α) are the restrictions of the Riemannian metrics on M and B respectively.

Example 2.14: Any fibre bundle which has totally geodesic fibres.

Let (M, g_M) and (N, g_N) be Riemannian manifolds (g_M and g_N denote the corresponding metrics) and let $f : M \to N$ be a diffeomorphism. Then f is said to be a quasi-isometry if $f^* g_N$ is quasi-isometric to g_M, that is there is a constant K such that

$$g_M/K \leq f^* g_N \leq K g_M.$$

We note that if M is compact (with or without boundary), then any diffeomorphism is a quasi-isometry.

Let $f : M \to N$ be a quasi-isometry. Then clearly $\widetilde{f} : \widetilde{M} \to \widetilde{N}$ is a quasi-isometry if the metrics on $\widetilde{M}$ and $\widetilde{N}$ are induced from M and N respectively (where as usual the tilde denotes the universal cover).

Lemma 2.15: Let $F \to M \xrightarrow{p} B$ be a fibre bundle of closed oriented manifolds. Choose Riemannian metrics on F and B and a p-compatible metric on M. Then the fibre bundle has quasi-totally geodesic fibres.

Proof : Assume that the minimum of the injectivity radius and the convexity radius is $\geq \epsilon$. Choose a finite cover $(B_{\epsilon/2}(x_\alpha))_\alpha$ of B consisting of geodesically convex normal balls of radius $\epsilon/2$ and centre $x_\alpha \in B$. Then $p^{-1}(B_\epsilon(x_\alpha))$ is a tubular neighbourhood of $p^{-1}(x_\alpha)$ and by local triviality of the fibre bundle there is a diffeomorphism

$$\phi_\alpha : p^{-1}(B_\epsilon(x_\alpha)) \xrightarrow{\sim} B_\epsilon(x_\alpha) \times F.$$

Clearly ϕ_α restricts to a map

$$\phi_\alpha : p^{-1}(\overline{B_{3\epsilon/4}(x_\alpha)}) \xrightarrow{\sim} \overline{B_{3\epsilon/4}(x_\alpha)} \times F$$

of smooth compact manifolds with boundary. By the remarks before the lemma ϕ_α is a quasi- isometry and $U_\alpha \equiv B_{\epsilon/2}(x_\alpha)$ is the desired open cover of B with the property that $(p^{-1}(U_\alpha, \phi_\alpha)_\alpha$ trivialises the fibre bundle and

$$\phi_\alpha : p^{-1}(U_\alpha) \xrightarrow{\sim} U_\alpha \times F$$

is a quasi-isometry.

Definition 2.16: A fibre bundle of connected manifolds $F \xrightarrow{i} M \xrightarrow{p} B$ is said to be special if the following holds. Let

$$\ldots \to \pi_2(B) \xrightarrow{\partial} \pi_1(F) \xrightarrow{i^*} \pi_1(M) \xrightarrow{p^*} \pi_1(B) \to 0$$

be part of the long exact sequence of homotopy groups. Then the fibre bundle above is said to be special if ker i^* is trivial, or equivalently, if image ∂ is trivial.

Example 2.17. Any fibre bundle $F \to M \to B$ such that $\pi_2(B)$ is trivial is special.

Lemma 2.18: Let $F \to M \xrightarrow{p} B$ be a special fibre bundle of closed connected oriented Riemannian manifolds. Then the fibre bundle $\widetilde{F} \to \widetilde{M} \xrightarrow{\widetilde{p}} \widetilde{B}$ has quasi-totally geodesic fibres (where the covering spaces all have the induced metrics).

Proof: We choose the finite open cover (U_α) of B as in the proof of lemma 2.15 such that $(p^{-1}(U_\alpha), \phi_\alpha)_\alpha$ is a local trivialisation of the fibre bundle M and the ϕ_α are quasi-isometries. Let $U'_\alpha \subset \widetilde{B}$ be a lift of U_α. Clearly $\gamma.U'_\alpha$ is isometric to U_α for all $\gamma \in \pi_1(B)$. Hence

$$(*) \qquad \widetilde{p}^{-1}(\gamma U'_\alpha) \text{ is isometric to } \widetilde{p^{-1}(U_\alpha)}$$

for all $\gamma \in \pi_1(B)$ and for all α. Now ϕ_α is a quasi-isometry and so $\widetilde{\phi}_\alpha$: $\widetilde{p^{-1}(U_\alpha)} \xrightarrow{\sim} U_\alpha \times \widetilde{F}$ is a quasi-isometry since the metrics on $\widetilde{p^{-1}(U_\alpha)}$ and $U_\alpha \times \widetilde{F}$ are induced from those in $p^{-1}(U_\alpha)$ and $U_\alpha \times F$ respectively. Now using $(*)$ we have quasi-isometries

$$\phi^\gamma_\alpha : \widetilde{p}^{-1}(\gamma U'_\alpha) \xrightarrow{\sim} U_\alpha \times \widetilde{F}$$

for all $\gamma \in \pi_1(B)$ and for all α. Hence $(\widetilde{p}^{-1}(\gamma U'_\alpha), \phi^\gamma_\alpha)_{\alpha,\gamma}$ is a local trivialisation of $\widetilde{F} \to \widetilde{M} \xrightarrow{\widetilde{p}} \widetilde{B}$ such that ϕ^γ_α is a quasi-isometry for all $\gamma \in \pi_1(B)$ and for all α.

Proposition 2.19. If F is an L^2-acyclic manifold and M and B are closed, connected, oriented manifolds and $F \xrightarrow{i} M \xrightarrow{p} B$ is a special fibre bundle then M is an L^2-acyclic manifold.

Proof. Since the fibre bundle is special, there is a short exact sequence of homotopy groups

$$0 \to \pi_1(F) \xrightarrow{i^*} \pi_1(M) \xrightarrow{p^*} \pi_1(B) \to 0.$$

Hence there is a fibre bundle

$$\widetilde{F} \to \widetilde{M} \xrightarrow{\widetilde{p}} \widetilde{B},$$

where $\widetilde{F} \to F$, $\widetilde{M} \to M$ and $\widetilde{B} \to B$ are the universal coverings. By lemma 2.15 the fibre bundle $F \xrightarrow{i} M \xrightarrow{p} B$ has quasi-totally geodesic fibres. Endowing $\widetilde{F}, \widetilde{M}, \widetilde{B}$ with the induced Riemannian metrics, one can see by lemma 2.18 that $\widetilde{F} \to \widetilde{M} \xrightarrow{\widetilde{p}} \widetilde{B}$ also has quasi-totally geodesic fibres i.e. there is a locally finite open covering $\{U_\alpha\}_\alpha$ of $\widetilde{B}$ such that $\widetilde{p}^{-1}(U_\alpha)$ is quasi-isometric to $U_\alpha \times \widetilde{F}$. In particular

$$H^*_{(2)}(\widetilde{p}^{-1}(U_\alpha)) \cong H^*_{(2)}(U_\alpha \times \widetilde{F}) \cong \oplus_{i+j=*} H^i_{(2)}(U_\alpha) \otimes H^j_{(2)}(\widetilde{F})$$

by Zucker's theorem [19]. Since F is L^2-acyclic, it follows that $H^*_{(2)}(\widetilde{p}^{-1}(U_\alpha)) = 0$ for all $*$. In a similar way, one can prove that $\widetilde{p}^{-1}(\cap U_\alpha) = \cap\, \widetilde{p}^{-1}(U_\alpha)$ for finite intersections, has trivial L^2-cohomology. By Cheeger and Gromov [4, theorem 2.1], there is Mayer-Vietoris long exact sequence in L^2-cohomology for open coverings (by scaling the Riemannian metrics on the closed manifolds F, M, B we can assume that conditions $geo(F), geo(M), geo(B) \leq 1$ are satisfied). Applying this sequence to the covering $\widetilde{p}^{-1}(U_\alpha)$ of $\widetilde{M}$, we see that $\widetilde{M}$ has trivial L^2- cohomology i.e. M is an L^2-acyclic manifold.

Corollary 2.20: Let G be any of the following Lie groups $Gl(n,\mathbb{C})$, $U(p,q)$, $PSl(2,\mathbb{R})$, $Gl_0(2,\mathbb{R}), SO_0(2,n)$ and let K be a maximal compact subgroup of G. Let π be a closed, torsion-free, discrete, cocompact subgroup of G such that

$$(*) \qquad 0 \to K \to \pi\backslash G \to \pi\backslash G/K \to 0$$

is a fibre bundle of closed oriented manifolds. Then $\pi\backslash G$ is an L^2-acyclic manifold.

Proof: When G is any one of the Lie groups in the list above, we see that K is an L^2 acyclic manifold. Since $\pi\backslash G/K$ is the Eilenberg-Maclane space $K(\pi, 1)$, then $\pi_2(\pi\backslash G/K)$ is trivial, i.e. the fibre bundle (*) is special. Hence by proposition 2.19 $\pi\backslash G$ is an L^2-acyclic manifold. (Note that $U(n)$ is L^2-acyclic because the fibration $U(n) \to U(n+1) \to S^{2n+1}$ may be shown to be special using the fact that $\pi_2(S^{2n+1}) = 0$ for all $n \geq 1$, proposition 2.19 and induction on n starting with the fact that $U(1)$ is L^2-acyclic. Finally note that by a theorem of Borel [2], examples of subgroups $\pi \subset G$ considered in the corollary always exist.

Proposition 2.21: Let $M_1 \to M$ be a finite normal cover, where M is an oriented manifold. Then M is L^2-acyclic if and only if M_1 is L^2-acyclic.

Proof: Clearly M_1 is a closed oriented manifold such that $\pi_1(M_1)$ is a normal subgroup of $\pi_1(M)$ of finite index r. Also note that $\widetilde{M}_1 = \widetilde{M}$. Hence by [6, p133] we see that

$$b^j_{(2)}(\widetilde{M}_1) = r b^j_{(2)}(\widetilde{M}).$$

Remark 2.22: Let $t : |K| \to M$ be a C^1 triangulation of the manifold M and if $\rho : \pi_1 M \to \mathcal{U}$ is a unitary representation, let $\rho_t \equiv (t_*^{-1})\rho$ where t_* is the induced isomorphism on fundamental groups. Then it follows from theorem of Dodzuik [9] that

$$H^j_{(2)}(K, \rho_t) \cong \mathcal{H}^j_{(2)}(M, \rho) \cong H^j_{(2)}(M, \rho)$$

canonically. (Dodzuik proves this in the case when ρ is the left regular representation. However, his proof is valid in general). It follows that ρ is L^2-acyclic if and only if ρ_t is L^2-acyclic.

Definition 2.23: A fibre bundle of closed oriented manifolds $F \xrightarrow{i} M \xrightarrow{p} B$ is said to be flat if there is a representation

$$\rho : \pi_1(B) \to \text{Diff}(F)$$

where Diff(F) is the group of all diffeomorphisms of F and $M = \widetilde{B} \times_\rho F$.

Since

$$\widetilde{M}=\widetilde{B}\times\widetilde{F}$$

we see that

$$\pi_j(\widetilde{M})=\pi_j(\widetilde{B}\times\widetilde{F}),\ \ j\geq 2$$

$$=\pi_j(\widetilde{B})\times\pi_j(\widetilde{F}),\ \ j\geq 2$$

i.e.

$$\pi_j(M)=\pi_j(B)\times\pi_j(F),\ \ j\geq 2.$$

So from the long exact sequence in homotopy for the fibration $F\to M\to B$, we get a short exact sequence of fundamental groups

$$0\to\pi_1(F)\to\pi_1(M)\to\pi_1(B)\to 0.$$

In particular, a flat fibre bundle is a special fibre bundle.

Corollary 2.24. If $F\to M\to B$ is a flat fibre bundle of closed, oriented manifolds and if F is a L^2-acyclic manifold, then M is a L^2-acyclic manifold.

Proof. The fibre bundle is special by the remarks above and so the hypotheses of proposition 2.19 hold.

We digress at this point to discuss geometric structures on a manifold. A Riemannian metric on a connected closed manifold M is said to be locally homogeneous if given any two points p and q in M there are open neighbourhoods of U and V of p and q respectively and an isometry $(U,p)\to(V,q)$. A closed manifold M admits a geometric structure if M admits a locally homogeneous Riemannian metric. Singer has shown that [17] the induced locally homogeneous Riemannian metric on $X\equiv\widetilde{M}$ has a group of isometries G which acts transitively on X with a compact isotropy group. We then say (following Thurston) that M admits a geometric structure modelled on (X,G). Thurston [18] has classified all the three dimensional geometries; there are eight of them:

1. E^3 2. H^3 3. $S^2\times E^1$ 4. $H^2\times E^1$ 5. $\widetilde{SL(2,\mathbb{R})}$ 6. Heis (the three dimensional Heisenberg group) 7. Sol (the three dimensional solvable Lie group) 8. S^3

In each case G is the group of isometries of X.

Proposition 2.25: Let M be a closed, connected, oriented, three dimensional manifold which admits a geometric structure and suppose $\pi_1(M)$ is infinite. Then M is an L^2-acyclic manifold.

Proof: Since $\pi_1(M)$ is infinite X cannot be S^3 but all other possibilities could occur. Now a case by case study gives

1. $E^3=\widetilde{T}^3$ as Riemannian manifolds and so L^2-acyclicity follows by proposition 2.7.
2. Dodzuik's theorem (proposition 2.7) gives L^2 acyclicity of H^3.
3. From Zucker's theorem [20] $H^j_{(2)}(S^2\times E^1)=\oplus_{i+k=j}H^i_{(2)}(S^2)\otimes H^k_{(2)}(E^1)$ and as $E^1\cong\widetilde{S}^1$ we have by proposition 2.7 $H^k_{(2)}(E^1)=0$ giving L^2-acyclicity.

4. L^2 acyclicity for $H^2 \times E^1$ follows by an argument similar to 3.

5. Using $PSL(2,\mathbb{R}) = SL(2,\mathbb{R})/\{-1,1\}$ we obtain $\widetilde{PSL(2,\mathbb{R})} = \widetilde{SL(2,\mathbb{R})}$ as Riemannian manifolds. If π is the fundamental group of a closed Riemann surface then $\pi\backslash\widetilde{PSL}(2,\mathbb{R}) = \widetilde{PSL(2,\mathbb{R})}$. By proposition 2.19, $\pi\backslash PSL(2,\mathbb{R})$ is L^2 acyclic. Hence $\widetilde{SL(2,\mathbb{R})}$ is L^2 acyclic.

6. $\underline{\text{Heis}} = \left\{ \begin{pmatrix} 1 & x & z \\ & 1 & y \\ & & 1 \end{pmatrix} : x, y, z \in \mathbb{R} \right\}$. Let $\Gamma \subset \underline{\text{Heis}}$ be the subgroup consisting of matrices with integer entries. Then $M \equiv \Gamma\backslash\underline{\text{Heis}}$ is a closed, oriented connected manifold and we have a fibration $S^1 \to M \to \mathbb{T}^2$. By proposition 2.19, M is an L^2-acyclic manifold.

7. Let $\alpha : \mathbb{R} \to Aut(\mathbb{R}^2)$ be given by $\alpha_t(x,y) = (e^t x, e^{-t} y)$ where $(x,y) \in \mathbb{R}^2$ and $t \in \mathbb{R}$. Then $Sol = \mathbb{R}^2 \times_\alpha \mathbb{R}$. Let $\pi \subset Sol$ be the subgroup $\mathbb{Z}^2 \times_\alpha \mathbb{Z}$ obtained by restricting the elements of $\mathbb{R}^2$ to the integers and taking $\mathbb{Z}$ to be generated by a fixed non-zero element t_0. Then $M \equiv \pi\backslash Sol$ is a closed, oriented connected manifold and we have the flat torus bundle: $\mathbb{T}^2 \to M \to S^1$ with monodromy given by α (restricted to $\mathbb{Z}$). By proposition 2.19 M is L^2-acyclic.

3. L^2-torsion for L^2-acyclic manifolds

Let M be a smooth, closed, oriented Riemannian manifold and $\widetilde{M}$ the universal cover of M with the induced Riemannian metric. Then π acts by orientation preserving isometries on $\widetilde{M}$. Let K denote the simplicial complex of a smooth triangulation $t : |K| \to M$ and $\widetilde{K}$ the simplicial complex of the induced triangulation of $\widetilde{M}$. Let $C^j(\widetilde{K})$ denote the space of oriented cochains of $\widetilde{K}$ of degree j. If we identify a j-simplex σ of $\widetilde{K}$ with the corresponding cochain defined by it, then an element $f \in C^j(\widetilde{K})$ can be written as a formal linear combination $\Sigma f_\sigma \sigma$ where the sum is taken over all j-simplices σ of $\widetilde{K}$. Define the space of L^2 cochains of degree j to be

$$C^j_{(2)}(\widetilde{K}) \equiv \{f \in C^j(\widetilde{K}) : \Sigma |f_\sigma|^2 < \infty\}$$

It can be quite easily seen that $C^j_{(2)}(\widetilde{K}) \cong C^j_{(2)}(K) \otimes \ell^2(\pi)$ i.e. the space of L^2 cochains of K is a free, finitely generated Hilbert $\mathcal{U}(\pi)$ module.

Let $\mathcal{U}$ be a finite von Neumann algebra and τ be the faithful normal trace on $\mathcal{U}$ normalised by $\tau(Id) = 1$ where Id is the identity operator on $\ell^2(\mathcal{U})$. Let $\pi = \pi_1(M)$ and $\rho : \pi \to \mathcal{U}$ a unitary representation of the fundamental group of M. Assume that ρ is an L^2-acyclic representation. Then denote by ρ_t the representation $(t_*)^*\rho : \pi' = \pi_1|K| \to \mathcal{U}$, where t_* is the isomorphism induced by t on fundamental groups.

Let $E_{\rho_t} = \widetilde{K} \times_{\rho_t} \ell^2(\mathcal{U}) \to K$ be the associated $\mathcal{U}$ finite Hilbert bundle over K, and L_{ρ_t} the corresponding local coefficient system consisting of the flat sections of E_{ρ_t}. We define the cochain complex $C^j_{(2)}(K, L_{\rho_t})$ as follows:

The group π' acts on $C^j_{(2)}(\widetilde{K})$ unitarily on the right, and on $\ell^2(\mathcal{U})$ unitarily on the left via the representation ρ_t. Define

$$C^j_{(2)}(K, L_{\rho_t}) \equiv C^j_{(2)}((\widetilde{K}), \ell^2(\mathcal{U}))^{\pi'} \cong C^j_{(2)}(K) \otimes \ell^2(\mathcal{U}).$$

Hence the L^2-cochains of K with local coefficients in L_{ρ_t} form a free, finitely generated Hilbert $\mathcal{U}$ module. Each $f \in C^j_{(2)}(K, L_{\rho_t})$ can be written as a sum $\Sigma f_\sigma \sigma$ where σ runs over all the j-simplices of K, and f_σ is a flat section of E_{ρ_t} over the set Star(σ) and $\Sigma |f_\sigma|^2 < \infty$. The coboundary operator

$$d^c : C^j_{(2)}(K, L_{\rho_t}) \to C^{j+1}_{(2)}(K, L_{\rho_t})$$

defined by

$$d^c(\Sigma f_\sigma \sigma) \equiv \Sigma f_\sigma d^c \sigma$$

where $d^c\sigma = \Sigma_{\lambda > \sigma} \lambda$ is the usual coboundary of σ. Observe that d^c is a bounded operator. An extension of a result of Dodzuik [9] shows that $C_{(2)}(K, L_{\rho_t}), d^c, e)$ is L^2-acyclic and so we can define the L^2-RF torsion with respect to K as

$$T_{(2)}(K, \rho_t) \equiv T_{(2)}(C_{(2)}(K, L_{\rho_t}), d^c, e).$$

The notation indicates that the choice of ordering for the basis of the complex is not important. This is the case because another choice of basis changes the matrix of $d^c + \delta^c$ by pre- and post-multiplication by matrices with integral entries and determinant one. We can also define the L^2-RF torsion

$$T_{(2)}(M, \rho) \equiv T_{(2)}(K, \rho_t)$$

and will show below that this is indeed independent of the triangulation. In particular, if we take $\rho : \pi \to \mathcal{U}(\pi)$ to be the left regular representation, we obtain a canonical isometry

$$C^j_{(2)}(\widetilde{K}) \to C^j_{(2)}(K, L_{\rho_t})$$

$$f_\sigma \to \sum_\gamma f_{\gamma\sigma}\gamma$$

of cochain complexes.

Assume that M is an L^2-acyclic manifold and that $t : |K| \to M$ is a C^1-triangulation of M. Then by Dodzuik's theorem [9] we see that $(C_{(2)}(\widetilde{K}), d^c, e)$ is a L^2-acyclic $\mathcal{U}(\pi')$-RF complex, where e_j is the natural basis of C^j which are the j-simplices of the triangulation. Hence it has a L^2-RF torsion $T_{(2)}(C_{(2)}(\widetilde{K}), d^c, e)$. We define the L^2-RF torsion of K to be

$$T_{(2)}(K) \equiv T_{(2)}(C_{(2)}(\widetilde{K}), d^c, e)$$

and define the L^2-RF torsion of M to be

$$T_{(2)}(M) \equiv T_{(2)}(C_{(2)}(\widetilde{K}), d^c, e).$$

The rest of the section is devoted to proving that $T_{(2)}(M)$ is well defined i.e. it is independent of the choice of triangulation $t : |K| \to M$ of M.

Henceforth K and L will denote simplicial complexes which are finite triangulations of a closed oriented manifold.

Let K be a simplicial complex and $L \subset K$ a subcomplex such that $i : L \to K$ is a homotopy equivalence.

Then we can form

$$C^j_{(2)}(\widetilde{K},\widetilde{L}) \cong C^j_{(2)}(K,L)\otimes \ell^2(\pi')$$

which is a free Hilbert $\mathcal{U}(\pi')$ module. Here $\pi' \equiv \pi_1|K|$, and standard arguments show that $(C^j_{(2)}(\widetilde{K},\widetilde{L}), d^c, e)$ is a L^2-acyclic $\mathcal{U}(\pi')$-RF complex where e is the canonical basis defined by the pair $(\widetilde{K},\widetilde{L})$. We can define the torsion

$$T_{(2)}(K,L) \equiv T_{(2)}(C_{(2)}(\widetilde{K},\widetilde{L}), d^c, e).$$

Let $S : C^*(L) \to C^*(K)$ be a continuous map. If $K = \lambda_0 \cup \lambda_1 \cup \ldots \cup \lambda_n$ and $L = \sigma_0 \cup \sigma_1 \cup \ldots \cup \sigma_\ell$ and the 'dual' simplices defined by

$$\sigma^*_j(\sigma_s) = \delta_{js} \text{ and } \lambda^*_j(\lambda_s) = \delta_{js}$$

then the map $\widetilde{S}$: $C^*(\widetilde{L}) \to C^*(\widetilde{K})$ is defined by

$$\widetilde{S}\left(\sum_{\gamma,j} a_{\gamma j}\gamma\sigma^*_j\right) = \sum_{\gamma,r}\left(\sum_j a_{\gamma j}q_{jr}\right)\gamma\lambda^*_r$$

where

$$\widetilde{S}\,(\sigma^*_j) = S(\sigma^*_j) = \sum_r q_{jr}\lambda^*_r.$$

It follows that S induces a bounded linear map $S_{(2)}$ (the restriction of $\widetilde{S}$ to L^2 cochains)

$$S_{(2)} : C^*_{(2)}(\widetilde{L}) \to C^*_{(2)}(\widetilde{K})$$

such that

$$||S_{(2)}||_{(2)} \le \max_{j,r} |q_{jr}|.$$

Remark 3.1: If f is a simplicial map from K to L then we simplify the notation for the induced map on L^2-cohomology $f^*_{(2)}$ to $f_{(2)}$.

Lemma 3.2: If $f, g : K \to L$ are simplicial maps which are homotopic such that f and g induce isomorphisms on fundamental groups. Then $f_{(2)}$ is chain homotopic to $g_{(2)}$.

Proof: We know that if f is homotopic to g, then there is a cochain homotopy D : $C^*(L) \to C^*(K)$ such that

$$d_K D + D d_L = f^* - g^*$$

We also have that

$$\widetilde{d}_K\widetilde{D} + \widetilde{D}\widetilde{d}_L = \widetilde{f}^* - \widetilde{g}^* \qquad (*)$$

where $\widetilde{D}, \widetilde{f}, \widetilde{g}, \widetilde{d}$ are the lifts of D, f, g, d respectively to the universal cover. By the above discussion, (*) becomes

$$d_{K(2)}D_{(2)} + D_{(2)}d_{L(2)} = f_{(2)} - g_{(2)}.$$

Lemma 3.3: If $\phi : K \to L$ is an isomorphism of simplicial complexes and if $(C_{(2)}(\widetilde{K}), d^c, e)$ and $(C_{(2)}(\widetilde{L}), d^c, e')$ are L^2-acyclic $\mathcal{U}$-RF complexes then we have $T_{(2)}(K) = T_{(2)}(L)$.

Proof: It is enough to show that ϕ induces a simple isometry of complexes. Note that if e_j is a canonical basis for $C_{(2)}(\widetilde{K})$ over $\mathcal{U}(\pi')$ where $\pi' \equiv \pi_1|K|$, then $\phi(e_j)$ is the canonical basis for $C_{(2)}(\widetilde{L})$ over $\mathcal{U}(\pi)$, where $\pi \equiv \pi_1|L|$. Hence the matrix of ϕ with respect to the preferred bases is the identity and it is an isomorphism of complexes so the result follows.

Lemma 3.4: Let $L \subset K$ be a subcomplex of a finite simplex K such that the inclusion map $i : L \to K$ is a homotopy equivalence. Then the complex $(C_{(2)}(\widetilde{K}, \widetilde{L}), d_{(2)})$ is acyclic.

Proof: We know that the complex $(C(K, L), d)$ is acyclic. So there is a contraction operator T (i.e. $dT + Td = 1$) on $C(K, L)$. Hence T defines a bounded operator on the relative L^2 cochains $T_{(2)}$ as in the discussion preceding remark 3.1, such that $d_{(2)}T_{(2)} + T_{(2)}d_{(2)} = 1$. It follows from this that we have an acyclic complex and in particular the range of $d^j_{(2)}$ is closed. So

$$H^j_{(2)}(C_{(2)}(\widetilde{K}, \widetilde{L}), d_{(2)}) = H^j(C_{(2)}(\widetilde{K}, \widetilde{L}), d_{(2)}) = 0$$

for all $j \geq 0$.

Lemma 3.5: The operator $d_{(2)} + \delta_{(2)}$ has bounded inverse as an operator from $C^{od}_{(2)}$ to $C^{ev}_{(2)}$.

Proof: By the preceding lemma we see that the range of $d^j_{(2)}$ is closed and the same argument applied to the adjoint gives the range of $\delta^j_{(2)}$ closed for all $j \geq 0$. So the Hodge theorem in this situation is

$$C^j_{(2)}(\widetilde{K}, \widetilde{L}) = d_{(2)}(C^{j-1}_{(2)}(\widetilde{K}, \widetilde{L}) \oplus \delta_{(2)}(C^{j+1}_{(2)}(\widetilde{K}, \widetilde{L}).$$

That is $d_{(2)} + \delta_{(2)}$ restricted to $C^{od}_{(2)}$ is an isomorphism. Hence $m(d_{(2)} + \delta_{(2)}, e)$ which is in $M_n \otimes \mathcal{U}$ is invertible in this algebra.

Lemma 3.6: Let G be an element of $\mathcal{U}' \otimes M_{r+s}$ which has the form (with respect to some basis) $\begin{pmatrix} A & 0 \\ B & C \end{pmatrix}$ with $A \in \mathcal{U}' \otimes M_r$, $C \in \mathcal{U}' \otimes M_s$ an invertible element and $B \in \mathcal{U}' \otimes M_{r \times s}$. Then G is the product $\begin{pmatrix} A & 0 \\ 0 & C \end{pmatrix} \begin{pmatrix} 1 & 0 \\ C^{-1}B & 1 \end{pmatrix}$. The latter operator is a commutator.

Proof: Only the last statement does not follow by multiplying out the product. It follows however from the observation:

$$\begin{pmatrix} 1 & 0 \\ X & 1 \end{pmatrix} = \begin{pmatrix} 1 & 0 \\ 2X & 1 \end{pmatrix} \begin{pmatrix} 1 & 0 \\ 0 & \frac{1}{2} \end{pmatrix} \begin{pmatrix} 1 & 0 \\ -2X & 1 \end{pmatrix} \begin{pmatrix} 1 & 0 \\ 0 & 2 \end{pmatrix}.$$

Proposition 3.7: Let K and L be triangulations of an L^2-acyclic manifold. Let L be a subcomplex of K such that the inclusion map is a homotopy equivalence. Then

$$T_{(2)}(K) = T_{(2)}(K, L) T_{(2)}(L)$$

Proof: By Dodzuik[9] K and L are L^2-acyclic. We have

$$C_{(2)}(\widetilde{K}) = C_{(2)}(\widetilde{L}) \oplus H$$

where $H \cong C_{(2)}(\widetilde{L})^{\perp}$ is a graded Hilbert space which has basis e_H chosen to satisfy the conditions: (i) that it extend a basis of $C_{(2)}(\widetilde{L})$ to a basis of $C_{(2)}(\widetilde{K})$ $\mathcal{U}(\pi)$-related to the canonical basis, and hence (ii) that e_H projects onto a basis of the relative complex $C_{(2)}(\widetilde{K},\widetilde{L})$ Now the differential

$$d^K_{(2)} : C^{od}_{(2)}(\widetilde{K}) \to C^{ev}_{(2)}(\widetilde{K})$$

has matrix relative to this basis of the form

$$d^K_{(2)} = \begin{pmatrix} d^L_{(2)} & 0 \\ pd^K_{(2)} & d^H \end{pmatrix}.$$

Here $p : C^{ev}_{(2)}(\widetilde{K}) \to C^{ev}_{(2)}(\widetilde{L})$ is the orthogonal projection. As $(d^K_{(2)})^2 = 0$ we see that $(d^H)^2 = 0$ and hence (H, d^H, e_H) is a Hilbert $\mathcal{U}(\pi)$-RF complex. The following diagram of $\mathcal{U}(\pi)$-RF complexes is commutative:

$$\begin{array}{ccccccccc} 0 & \to & C_{(2)}(\widetilde{L}) & \to & C_{(2)}(\widetilde{K}) & \to & C_{(2)}(\widetilde{K},\widetilde{L}) & \to & 0 \\ & & \| & & \| & & \downarrow & & \\ 0 & \to & C_{(2)}(\widetilde{L}) & \to & C_{(2)}(\widetilde{K}) & \to & H & \to & 0 \end{array}$$

It follows that there is a natural isomorphism, taking bases to bases, of (H, d^H) and $(C_{(2)}(\widetilde{K},\widetilde{L}), d_{(2)})$. By lemma 3.4 $(C_{(2)}(\widetilde{K},\widetilde{L}), d_{(2)})$ is acyclic and lemma 3.5 implies that $d^H + \delta^H$ is in $GL(r, \mathcal{U}(\pi)')$. Also, $T_{(2)}(K, L) = T_{(2)}(H)$ by the remarks above. Similarly

$$\delta^K_{(2)} : C^{od}_{(2)}(\widetilde{K}) \to C^{ev}_{(2)}(\widetilde{K})$$

has a decomposition

$$\delta^K_{(2)} = \begin{pmatrix} \delta^L_{(2)} & 0 \\ p\delta^K_{(2)} & \delta^H \end{pmatrix}$$

as L is a subcomplex of K. It follows that

$$m\left(d^K_{(2)} + \delta^K_{(2)}, e_K\right) = \begin{pmatrix} m\left(d^L_{(2)} + \delta^L_{(2)}, e_L\right) & 0 \\ m(p\left(d^K_{(2)} + \delta^K_{(2)}\right), e_L \cup e_H) & m(d^H + \delta^H, e_H) \end{pmatrix}$$

is a matrix with coefficients in $\mathcal{U}' \otimes M_{r+s}$ (where we have chosen r, s to match with our notation in lemma 3.6) as all maps are $\mathcal{U}$ module maps. Now we can apply lemma 3.6 so that

$$m(d^K_{(2)} + \delta^K_{(2)}, e_K) = m(d^L_{(2)} + \delta^L_{(2)}, e_L) \oplus m(d^H + \delta^H, e_H)$$

modulo commutators. Hence

$$|\mathrm{Det}_\tau|(m(d^K_{(2)} + \delta^K_{(2)}, e_K)) = |\mathrm{Det}_\tau|(m(d^L_{(2)} + \delta^L_{(2)}, e_L))|\mathrm{Det}_\tau|(m(d^H + \delta^H, e_H))$$

which suffices to prove the result.

Proposition 3.8: Retain the hypotheses of proposition 3.7. If $K - L$ is simply connected then $T_{(2)}(K, L) = 1$.

Proof: Assume that $K - L$ has a single component Γ. Choose a representative component $\widetilde{\Gamma}$ of $\widetilde{K} - \widetilde{L}$. Then $\widetilde{\Gamma}$ projects homeomorphically onto Γ. Now for each j- simplex σ in

$K - L$ there is a unique simplex $\tilde{\sigma}$ in $\tilde{\Gamma}$ which lies over $\tilde{\sigma}$. Observe that no representative simplex $\tilde{\sigma}$ of $\tilde{K} - \tilde{L}$ can intersect a proper translate $\gamma\,\tilde{\sigma}$ for any γ differing from the identity. This means that $d_{(2)}\sigma$ can be expressed as a linear combination of representative $j+1$ simplices with integer coefficients.

Now $(C(\tilde{K}, \tilde{L})\ \tilde{d})$ is an acyclic complex with coefficients over $\mathbb{C}$ [12]. By lemma 3.4 $(C_{(2)}(\tilde{K}, \tilde{L}), d_{(2)})$ is acyclic. Also these complexes may be chosen to have the same bases say $e = (e_j)$ (but over different algebras). We first note that the matrix of $\tilde{d}$ which equals that of $d_{(2)}$ with this basis choice, has all its entries in $\mathbb{Z}$. Let C'^j be the free $\mathbb{Z}$ module defined by the basis e. Then since $m(\tilde{d}, e) = m(d_{(2)}, e)$ are matrices over $\mathbb{Z}$ both $\tilde{d}$ and $d_{(2)}$ restrict to a coboundary operator $d' : C'^j \to C'^{j+1}$ for all $j \geq 0$. From [7, theorem 18.2] we know that the complex (C', d', e') is also acyclic. Introduce $\delta' : C'^{od} \to C'^{ev}$ the transpose of d' with respect to an identification of C'^j and its dual given by choosing bases appropriately. We claim that D' (which is $d' + \delta'$ restricted to C'^{od}) is an isomorphism of $\mathbb{Z}$ modules. (This assertion needs some comment. Acyclicity of (C', d', e') gives exact sequences $0 \to \ker d'^{i+1} \to C'^i \to \ker d'^{i+2} \to 0$ for each i of free $\mathbb{Z}$ modules, as submodules of free $\mathbb{Z}$ modules are free. Hence C'^i splits into a direct sum of $\ker d'^{i+1}$ and a complementary module isomorphic to $\ker d'^{i+2}$. Now the transpose $\delta'^{i+1} : \ker d'^{i+2} \to C'^i/\ker d'^{i+1}$ is an isomorphism which combined with the previous remark implies that $\delta'^{i+1} : C'^{i+1} \to C'^i/\ker d'^{i+1}$ is surjective. Surjectivity of D' now follows.) In particular both $m(D', e)$ and $m(D', e)^{-1}$ are matrices over $\mathbb{Z}$. This forces $|\det|(m(D', e)) = 1$. Finally notice that $m(D_{(2)}, e) = m(D', e)$ (here $D_{(2)}$ is the restriction of $d_{(2)} + \delta_{(2)}$ to $C^{od}_{(2)}$) and the Fuglede-Kadison determinant of a matrix with entries in $\mathbb{Z}$ is the absolute value of the ordinary determinant and hence by the above is one.

If K has several components, then we go through the above argument for each component and prove the result.

We recall some familiar concepts. Let K and L be simplicial complexes such that L is a subcomplex of K and $K = L \cup v\sigma$ where v is a vertex of K and σ is a simplex of K. We then say that K is an elementary expansion of L or equivalently, L is an elementary collapse of K. Since $K - L = v\sigma$ is simply connected, we see that $T_{(2)}(K, L) = 1$.

Next let L be a subcomplex of K. A simplicial map $\phi : K \to L$ is said to be simple homotopy equivalence if there is a finite sequence

$$K = K_0 \to K_1 \to \ldots \to K_n = L$$

where each arrow represents either an elementary expansion or an elementary collapse.

Lemma 3.9: Let K and L be triangulations of an L^2-acyclic manifold. Let L be a subcomplex of K such that the inclusion map is a simple homotopy equivalence, then $T_{(2)}(K) = T_{(2)}(L)$.

Proof: By our assumption, there is a finite sequence

$$K = K_0 \to K_1 \to \ldots \to K_n = L$$

where each arrow represents either an elementary expansion or an elementary collapse. Using proposition 3.7 and induction we see that $T_{(2)}(K) = \{$ products of terms of the type $T_{(2)}(K_j, K_{j+1}) = 1$ if $K_j \to K_{j+1}$ is an elementary collapse and $T_{(2)}(K_{j+1}, K_j) = 1$ if $K_j \to K_{j+1}$ is an elementary expansion $\} \times T_{(2)}(L)$. By the discussion preceding the lemma $T_{(2)}(K) = T_{(2)}(L)$.

Proposition 3.10 (Invariance under subdivision): Let K and K' be finite triangulations of an L^2-acyclic manifold such that K' is a subdivision of K. Then $T_{(2)}(K') = T_{(2)}(K)$.

Proof: It has been shown by M. Cohen [7,theorem 25.1] that the inclusion map of K in K' is a simple homotopy equivalence. By lemma 3.9, $T_{(2)}(K) = T_{(2)}(K')$.

Proposition 3.11 (Combinatorial Invariance): The torsion of an L^2 acyclic manifold M, $T_{(2)}(M)$, does not depend on the choice of C^1 triangulation $t : |K| \to M$.

Proof: Let $u : |L| \to M$ be another C^1 triangulation. By a fundamental theorem of Whitehead and Munkres there exist C^1 subdivisions $h : |K''| \to |K|$ and $\ell : |L''| \to |L|$ which are isomorphic via the isomorphism

$$\phi : K'' \to L''$$

of simplicial complexes. So we have

$$\begin{aligned} T_{(2)}(K) &= T_{(2)}(K'') \text{ by invariance under subdivision} \\ &= T_{(2)}(L'') \text{ by lemma 3.3} \\ &= T_{(2)}(L) \text{ by invariance under subdivision,} \qquad \text{q.e.d.} \end{aligned}$$

There is an analogous theorem in the induced case:

Proposition 3.12 (Combinatorial invariance of $\mathbf{L^2}$-RF torsion): Let $\rho : \pi \to \mathcal{U}$ be an L^2-acyclic representation. Then the L^2-RF torsion $T(M, \rho)$ does not depend on the choice of C^1 triangulation $t : |K| \to M$.

Proof: Exactly as before.

3.1: $\mathbf{L^2}$-RF torsion via Homology theory

We give an alternate but equivalent definition of the L^2-RF torsion of an L^2-acyclic representation.

Let (t, K^*) be the cell complex formed from a cell decomposition of the closed smooth manifold M. If ρ is an L^2-acyclic representation, we can form the chain complex $(C_j^{(2)}(K^*, \rho_t), \partial, \varepsilon)$ where ∂ is the boundary operator and ε_j is the canonical basis for $C_j^{(2)}$. Then we can define the L^2 torsion via Homology theory as follows:

$$T_{(2)}(M, \rho) = T_{(2)}(C_j^{(2)}(K^*, \rho_t), \partial, \varepsilon) = |\mathrm{Det}_\tau| m(D', \varepsilon)$$

where $D' = \partial + \partial^*$ restricted to C_{od} (see section 1.6).

3.2: Properties of L^2-RF torsion

We will prove some of the properties of L^2-RF torsion in this section.

Consider (M_j, ρ_j) where M_j are closed Riemannian manifolds with the same universal cover $\widetilde{M}$ such that the Riemannian metrics on M_1 and M_2 lift to the same metric on $\widetilde{M}$. Let $G_j \equiv \pi_1(M_j)$. Suppose that G_1 is a subgroup of G_2 of index r. If $\rho : G_1 \to \mathcal{U}$ is a unitary representation, then we construct the induced representation $U(\rho) : G_2 \to Gl(r, \mathcal{U})$ as follows. Let $\{a_k\}_{k=1}^r$ be a set of representatives in G_2 for the elements of G_2/G_1 i.e. $G_2 = \cup_{k=1}^r a_k G_1$. Let $\mathcal{H}$ be the vector space generated by all maps $\phi : G_2 \to \mathcal{U}$ satisfying

$$\phi(g_1 g) = \rho(g_1)\phi(g) \quad g_1 \in G_1.$$

The induced representation $U(\rho) : G_2 \to Gl(\mathcal{H})$ is defined by

$$U(\rho)(g_2)\phi(g) \equiv \phi(gg_2) \quad g_2 \in G_2.$$

Since a map ϕ is determined by its values on the a_k, then the map $T\phi \equiv \oplus_{k=1}^r \phi(a_k)$ defines an isomorphism between $\mathcal{H}$ and $\oplus_{k=1}^r \mathcal{U}$. Define $\hat{\rho}\,(g) \equiv \rho(g)$ if $g \in G_1$ and zero otherwise, and define the representation $R : G_2 \to Gl(r, \mathcal{U})$ by

$$R(g)(\oplus_{k=1}^r y_k) \equiv \oplus_{k=1}^r (\sum_{j=1}^r \hat{\rho}\,(a_k g a_j^{-1}) y_j).$$

Then a computation shows that $TU(\rho)(g)\phi = R(g)T\phi$.

Proposition 3.13: $T_{(2)}(M_1, \rho) = T_{(2)}(M_2, U(\rho))^r$

Proof: Let $t_j : |K_j| \to M_j$ be C^1 triangulation such that the universal covers $\widetilde{K}_j$ are isomorphic as simplicial complexes to the universal cover $\widetilde{K}$ of a fixed triangulation $t : |K| \to M$. We define a simple isometry of $\mathcal{U}$-modules

$$S : (C_{(2)}(K_1, L_\rho), d^c, e) \to (C_{(2)}(K_2, L_{U(\rho)}), d'^c, e').$$

as follows (note that the complex on the right is initially to be regarded as a $\mathcal{U} \otimes M_r$ module). Every cochain $c \in C^j_{(2)}(K_1, L_\rho)$ can be considered as an equivariant map c with values in $\ell^2(\mathcal{U})$ i.e. $c(g\sigma) = \rho(g)c(\sigma)$ where $g \in G_1$, σ is a j-simplex in $\widetilde{K}$. Also, every cochain $\hat{c} \in C^j_{(2)}(K_2, L_{U(\rho)})$ can be considered to be an equivariant map c into $\ell^2(\mathcal{U})^r$ i.e. $\hat{c}\,(g\sigma) = R(g)\,\hat{c}\,(\sigma)$ where $g \in G_2$ and σ is a j-simplex in K. Using these identifications define

$$Sc(\sigma) \equiv \oplus_{k=1}^r c(a_k\sigma) \text{ where } \sigma \text{ is a } j\text{-simplex in } K.$$

Since c satisfies $c(g_1\sigma) = \rho(g_1)c(\sigma)$ for $g_1 \in G_1$, it follows that

$$Sc(g\sigma) = R(g)Sc(\sigma) \text{ where } g \in G_2 \text{ and } \sigma \text{ is a } j\text{-simplex in } \widetilde{K}.$$

Obviously S defines a $\mathcal{U}$ module homomorphism between $C^j_{(2)}(K_1, L_\rho)$ and $C^j_{(2)}(K_2, L_{U(\rho)})$. Now for $\tilde{c} = \oplus c_j$ in $C^j_{(2)}(K_2, L_{U(\rho)})$ it follows that $c_1 \in C^j_{(2)}(K_1, L_\rho)$ and that $Sc_1 = \hat{c}$. Injectivity is clear so S is an isometry. Also S commutes with the coboundary operators, since the coboundary operators act componentwise. Also the above proof shows that if e_j is the canonical basis for $C^j_{(2)}(K_1, L_\rho)$, then Se_j is the canonical basis for $C^j_{(2)}(K_2, L_{U(\rho)})$ i.e. the matrix of S in the preferred bases is the identity. So S is a L^2 simple isomorphism of $\mathcal{U}$ modules which we now use to identify these modules. Let τ denote the trace on $\mathcal{U}$ acting on $C^j_{(2)}(K_1, L_\rho)$ constructed as in section 1. Using S we see that the normalised trace on $\mathcal{U} \otimes M_r$ acting on $C^j_{(2)}(K_2, L_{U(\rho)})$ is

$$\tau_2 = \frac{1}{r}\tau.$$

The normalisation of the traces above determines that on the commutants and so we may compare the torsion defined using τ_2 with the torsion calculated using τ to deduce the result.

4. Computation of L²-RF torsion

We compute the L^2-RF torsion of the circle S^1. S^1 has one cell e_0 in dimension zero, and one cell e_1 in dimension one. If we take S^1 to be the quotient $\mathbb{R}/\mathbb{Z}$, then $\{x \in \mathbb{R} : 0 \leq x < 1\}$ is a fundamental domain for S^1. Let γ be the generator of $\pi_1(S^1)$ given by translation by $+1$, then we can define the cells for S^1 as follows:

$$e_1 = \{x \in \mathbb{R} : 0 \leq x \leq 1\}$$

$$e_0 = \{x \in \mathbb{R} : x = 0\}$$

Then the cells of $\tilde{S}^1$ are given by $\gamma^k e_j$ where $k \in \mathbb{Z}$ and $j = 0, 1$ and the boundary operator ∂ acting on the cells of $\tilde{S}^1$ is given by

$$\partial e_1 = \gamma e_0 - e_0 = (\gamma - 1)e_0$$

$$\partial e_0 = 0.$$

We use the definition of L^2-RF torsion via Homology groups.

So

$$T_{(2)}(S^1) = |\mathrm{Det}_\tau|(\partial) = |\mathrm{Det}_\tau|(1 - \gamma)$$

To calculate this we can use the explicit form of the trace for the regular representation of $\mathbb{Z}$. This gives

$$\tau(\ln(1-\gamma)^*(1-\gamma)) = \int_{-\pi}^{\pi} \ln((1 - e^{-i\theta})(1 - e^{i\theta}))d\theta/2\pi = 0$$

Hence we find that the torsion in this case is one. While this is not particularly exciting it is worth commenting that the operator $(1-\gamma)$ is singular so that a naive interpretation of our definition would suggest that the torsion should be zero. A similar calculation (involving a more complicated integral) yields the same result for the 2-torus namely the L^2-RF torsion is one. We have not yet computed more complicated examples.

Proposition 4.1: If $M_1 \to M$ is an r-sheeted normal covering, where M_1, M are closed, connected, oriented manifolds. If M_1 or M is L^2-acyclic, then

$$T_{(2)}(M_1) = T_{(2)}(M)^r$$

Proof: By proposition 2.21 M_1 is L^2-acyclic iff M is L^2 acyclic. The proposition is a corollary of proposition 3.13.

Remark 4.2: Prof M. A. Shubin has kindly sent us a reprint of his article with S. P. Novikov entitled 'Morse inequalities and von Neumann II$_1$ factors' (Doklady (Amer. Math. Soc. translation) 34 (1987) 79-82) in which it is remarked that the methods of that paper may be used to define a II$_1$ factor torsion.

References

1. Atiyah, M., *Elliptic operators, discrete groups and Von Neumann algebras*, Asterisque no. 32-33 (1976) 43-72.

2. Borel, A., *Compact Clifford-Klein forms of locally symmetric spaces*, Topology **2** (1963) 111-122.

3. Cheeger, J. and Gromov, M., *On the characteristic numbers of complete manifolds of bounded curvature and finite volume*, Rauch Mem. vol., Chavel, I. and Farkas H. editors, Springer Verlag (1985) 115-154.

4. Cheeger, J. and Gromov, M., *Bounds on the Von Neumann dimension of L^2-cohomology and the Gauss Bonnett theorem for open manifolds*, Jour. Diff. Geom. **21** (1985) 1-34.

5. Cheeger, J. and Gromov, M., *L^2 cohomology and group cohomology*, Topology **25** (1986) 189-215.

6. Cohen, J., *Von Neumann dimension and the homology of covering spaces*, Quart. Jour. Math. Oxford **30** (1979) 133-142.

7. Cohen, M., *A course in simple homotopy theory*, Springer Verlag GTM **10** (1972).

8. Dixmier, J., *Von Neumann algebras*, North Holland Amsterdam **27** (1981).

9. Dodzuik, J., *DeRham-Hodge theory for L^2-cohomology of infinite coverings*, Topology **16** (1977) 157-165.

10. Dodzuik, J., *L^2-harmonic forms on rotationally symmetric Riemmanian manifolds*, Proc. A.M.S. **77** (Dec 1979) 395-400.

11. Fuglede, B. and Kadison, R.V., *Determinant theory in finite factors*, Annals of Math. **55** (1952) 520-530.

12. Milnor, J., *Whitehead torsion*, Bull. A.M.S. **72** (1966) 358-426.

13. Pazhitnov, A., *An analytic proof of the real part of Novikov's inequalities*, Soviet Math. Dokl. **35** (1987) 456-457.

14. Ray, D. and Singer, I.M. *R-torsion and the Laplacian on Riemannian manifolds*, Advances in Math. **7** (1971) 145-210.

15. Ray, D. and Singer, I.M., *Analytic Torsion*, Proc. Symp. Pure Math. **23** (1972)167-181.

16. Singer, I.M., *Some remarks on operator theory and index theory*, Springer Lecture Notes in Math **575** (1977) 128-137.

17. Singer, I.M., *Infinitesimally homogeneous spaces*, Comm. Pure App. Math. **13** (1960) 685-697.

18. Thurston, W., *Three dimensional manifolds, Kleinian groups and hyperbolic geometry*, Bull. A.M.S. **6** (May 1982) 357-381.

19. Witten, E., *Supersymmetry and Morse theory*, J. Differential Geometry **17** (1982) 661-692.

20. Zucker, S., *L^2 cohomology of warped products and arithmetic groups*, Inv. Math. **70** (1982) 169-218.

Contemporary Mathematics
Volume 105, 1990

SECONDARY CHARACTERISTIC NUMBERS AND LOCALLY FREE S^1-ACTIONS

H. Moriyoshi

0. INTRODUCTION. Quite recently A. Connes has proved remarkable results in [8] which suggest that there might exist a close relationship between secondary characteristic classes and cyclic cohomology. Actually one of his results in [8] is to give well defined additive maps from K-groups of certain C^*-algebras into $\mathbb{C}$ exploiting his idea in [7] to make a pairing between K-theory and cyclic cohomology, which was done in the following geometric situation.

Let V be an oriented smooth manifold and Γ a discrete group acting on V by orientation preserving diffeomorphisms. Let $C_0(V) \rtimes \Gamma$ be the reduced crossed product with this action. We have a natural homomorphism $\mu\colon K^*(V,\Gamma) \longrightarrow K_*[C_0(V) \rtimes \Gamma]$ where $K^*(V,\Gamma)$ denotes the geometric K-group for V. The relevant property is that $\mu(x)$ is described by the longitudinal index of a certain Dirac operator for $x \in K^*(V,\Gamma)$. On the other hand we have the equivariant Chern character $ch_\Gamma\colon K^*(V,\Gamma) \longrightarrow H^*(V,\Gamma)$ which is well understood in a geometric way (it is a $\mathbb{C}$-isomorphism). Suppose that Γ is torsion-free. In this case the equivariant cohomology group $H^*(V,\Gamma)$ is isomorphic to $H_*(V \times_\Gamma E\Gamma;\mathbb{C})$. Then one of the results in [8] is stated as follows: for a given universal secondary characteristic class $\omega \in H^*(V \times_\Gamma E\Gamma;\mathbb{R})$ there exists an additive map $\phi_\omega\colon K_*[C_0(V) \rtimes \Gamma] \longrightarrow \mathbb{C}$ such that

$$\phi_\omega \mu(x) = \langle \omega, ch_\Gamma(x) \rangle$$

for any $x \in K^*(V,\Gamma)$.

1980 Mathematics Subject Classification (1985 Revision). 57R20, 57R30, 58G12.

This result gives us a useful tool to study the behavior of secondary characteristic classes of foliated manifolds since there is a possibility that we can obtain additional information for the secondary characteristic number $\langle \omega, ch_\Gamma(x)\rangle$ by investigating $\mu(x)$. For instance, as a corollary of the above result Connes has proved in [8] the vanishing of certain secondary characteristic numbers for a foliated manifold (W,F) with an integrable spin vector bundle $F \subset TW$ assuming that W admits a Riemannian metric along the leaves with leafwise positive scalar curvature. In this case the longitudinal index of the Dirac operator along the leaves vanishes by a well known result of Lichnerowicz (see [10]) and hence it yields the vanishing of certain secondary characteristic numbers by the above result.

In this article we shall prove a similar vanishing theorem of some secondary characteristic numbers exploiting Connes' result with the idea of perturbing the Dirac operator by a Killing vector field, which was pointed out in [13]. The situation is as follows.

Let V be a closed manifold and $\Gamma \subset \mathrm{Diff}^+(V)$ a discrete subgroup consisting of orientation preserving diffeomorphisms of V. Let M be a closed spin^c manifold and $\tilde{M}$ a Γ-principal bundle over M. Put $N = V \times_\Gamma \tilde{M}$. We note that N is a foliated bundle over M with the projection map $\pi: N \longrightarrow M$. Then one of our main results is stated as follows.

THEOREM A. *Suppose that there exists a locally free* S^1*-action on* M *preserving the* spin^c *structure and that the holonomy of* N *along an* S^1*-orbit in* M *is of finite order. Then for any secondary characteristic class* ω *of the foliated bundle* N *it follows*

$$\langle \omega\pi^*\hat{\mathfrak{A}}(M), |N|\rangle = 0$$

where $\hat{\mathfrak{A}}(M)$ *is the total* $\hat{A}$*-class of* M.

In particular, if we choose a trivial group as Γ and an arbitrary product of Pontrjagin classes of M as ω, then the above theorem shows the vanishing of all Pontrjagin numbers of M with a locally free S^1-action (see also Theorem 4.8). This overlaps one of the results in Bott [4], which is obtained as a corollary of his fixed point formula for Killing vector fields.

THEOREM A is obtained from the following theorem, which corresponds to the result of Lichnerowicz above.

THEOREM B. *Let* W *be an even-dimensional manifold with a proper* Γ*-action. Suppose that* W *admits a* Γ*-invariant* $spin^c$ *structure and that* W/Γ *is compact. Let* $\mathrm{ind}(D)$ *denote the index of the Dirac operator on* W, *which is an element of* $K_0(C^*\Gamma)$. *If there is a locally free* S^1*-action on* W *which preserves the* $spin^c$ *structure and commutes with the* Γ*-action, then we obtain* $\mathrm{ind}(D) = 0$.

This article is organized as follows. Section 1 is devoted to preliminary things. The definition of $C_0(V) \rtimes \Gamma$ and Kasparov modules are stated here. In section 2 we review relevant properties with $K^*(V,\Gamma)$, which reflect the geometric aspects on $K_*[C_0(V) \rtimes \Gamma]$. Then we state precisely one of results in [8] which is needed in our case. In section 3 we give a description of a Kasparov module which represents the index of the Dirac operator mentioned above. Based on these arguments, Theorems A and B are proved in section 4 with some other results.

The author is very grateful to Professor Akio Hattori for his help and encouragement. He also thanks M. Furuta and K. Ono for helpful conversations, and especially M. Hilsum for pointing out an error in the first version of this article.

1. REDUCED CROSSED PRODUCTS AND KASPAROV MODULES.

1.1. Let X be a topological space and Γ an arbitrary discrete group. Let $C_c(X)$ denote the space of all complex-valued continuous functions on X with compact support. Suppose given a right action of Γ on X by homeomorphisms. It induces a left Γ-action on $C_c(X)$ by

$$[g(a)](x) = a(xg) \qquad (x \in X)$$

for $a \in C_c(X)$ and $g \in \Gamma$. Then we define an algebra $C_c(X,\Gamma)$ with an involution $*$ as follows:

i) as a $\mathbb{C}$ vector space

$$C_c(X,\Gamma) = C_c(X\times\Gamma)$$

and an element $a \in C_c(X,\Gamma)$ is viewed as a finite formal sum

$$a = \sum_{g\in\Gamma} a_g U_g$$

where $a_g \in C_c(X)$ is:

$$a_g(x) = a(x,g) \qquad x \in X;$$

ii) the addition and the multiplication are given by

$$a + b = \sum_{g\in\Gamma} (a_g+b_g)U_g$$

$$ab = \sum_{g,h\in\Gamma} a_g g(b_h)U_{gh}$$

for $a = \sum_{g\in\Gamma} a_g U_g$ and $b = \sum_{h\in\Gamma} b_h U_h$;

iii) an involution $*$ is defined by

$$(\sum_{g\in\Gamma} a_g U_g)^* = \sum_{g\in\Gamma} g(\overline{a}_{g^{-1}})U_g.$$

For $x \in X$, put $G_x = \{(y,h) \in X \times \Gamma : yh = x\}$. Let $\ell^2(G_x)$ be the Hilbert spce of all ℓ^2-sequences on G_x. We define a representation π_x of $C_c(X,\Gamma)$ on $\ell^2(G_x)$ to be

$$[\pi_x(a)f](y,h) = \sum_{g\in\Gamma} a_g(y)f(yg,g^{-1}h) \qquad ((y,h) \in G_x)$$

for $a = \sum_{g\in\Gamma} a_g U_g \in C_c(X,\Gamma)$ and $f \in \ell^2(G_x)$. Then we define a C*-norm on $C_c(X,\Gamma)$ by

$$\|a\| = \sup_{x\in V} \|\pi_x(a)\|$$

where in the right hand side $\|\ \|$ denotes the operator norm.

1.2. DEFINITION. *The reduced crossed product* $C_0(X) \rtimes \Gamma$ *is defined to be the completion of* $C_c(X,\Gamma)$ *with respect to the above* C*-*norm. When* X *is* *a point, we call it the reduced group* C^**-algebra for* Γ, *which is denoted by* $C^*\Gamma$.

The rest of this section is devoted to the definition of a Kasparov module, which is needed to describe the K-theory of C^*-algebras. In the following we refer the reader to [5] for the detail.

1.3. Let A be an algebra with an involution $*$ and $\mathcal{E}$ a right A-module. Then $\mathcal{E}$ is called a pre-Hilbert A-module if it is equipped with a map $\langle\ ,\ \rangle: \mathcal{E}\times\mathcal{E}\longrightarrow A$ such that:

i) $\langle\ ,\ \rangle$ is conjugate linear in the first variable and linear in the second;

ii) $\langle\xi,\eta a\rangle = \langle\xi,\eta\rangle a$ for $\xi,\eta\in\mathcal{E}$ and $a\in A$;

iii) $\langle\xi,\eta\rangle^* = \langle\eta,\xi\rangle$ for $\xi,\eta\in\mathcal{E}$;

iv) $\langle\xi,\xi\rangle\geq 0$ for $\xi\in\mathcal{E}$, and $\xi=0$ if and only if $\langle\xi,\xi\rangle=0$.

We call $\langle\ ,\ \rangle$ an A-valued inner product on $\mathcal{E}$.

We further assume that A is a C^*-algebra. Put $\|\xi\| = \|\langle\xi,\xi\rangle\|^{1/2}$ for $\xi\in\mathcal{E}$. Then $\mathcal{E}$ is called a Hilbert A-module if $\mathcal{E}$ satisfies the following additional condition:

v) $\mathcal{E}$ is complete with respect to the above norm $\|\ \|$.

Let A be a C^*-algebra and $\mathcal{E}$ a Hilbert A-module with an inner product $\langle\ ,\ \rangle$. We define $\mathcal{L}(\mathcal{E})$ to be the set of all module homomorphisms $T:\mathcal{E}\longrightarrow\mathcal{E}$ for which there is an adjoint module homomorphism $T^*:\mathcal{E}\longrightarrow\mathcal{E}$ with $\langle T\xi,\eta\rangle = \langle\xi,T^*\eta\rangle$ for all $\xi,\eta\in\mathcal{E}$. Then T is automatically bounded and $\mathcal{L}(\mathcal{E})$ is a C^*-algebra with respect to the operator norm.

Let $\xi,\eta\in\mathcal{E}$. Define a module homomorphism $\theta_{\xi,\eta}$ by $\theta_{\xi,\eta}(\varsigma) = \xi\langle\eta,\varsigma\rangle$ for $\varsigma\in\mathcal{E}$. We denote by $\mathcal{K}(\mathcal{E})$ the closure of the $\mathbb{C}$ linear span of $\{\theta_{\xi,\eta}: \xi,\eta\in\mathcal{E}\}$. An element of $\mathcal{K}(\mathcal{E})$ is called a compact operator on $\mathcal{E}$.

1.4. DEFINITION. *Let* $(\mathcal{E},F)$ *be a pair such that:*

i) $\mathcal{E}$ *is a countably generated Hilbert* A*-module with a given* $\mathbb{Z}_2$*-grading;*

ii) F *is an operator in* $\mathcal{L}(\mathcal{E})$ *of degree* 1 *with* F^2-1 *and* $F-F^*$ *in* $\mathcal{K}(\mathcal{E})$.

Then $(\mathcal{E},F)$ *is called a Kasparov* A*-module.*

1.5. A Kasparov A-module is nothing but a Kasparov $(\mathbb{C},A)$-bimodule and hence it determines an element of $K_0(A) = KK(\mathbb{C},A)$. Then it is known that $(\mathcal{E},F)$ yields the zero element if F is invertible in $\mathcal{L}(\mathcal{E})$. Moreover if F and F' are operators in $\mathcal{L}(\mathcal{E})$ with $F-F'\in\mathcal{K}(\mathcal{E})$, then $(\mathcal{E},F)$ and $(\mathcal{E},F')$

determine the same element in $K_0(A)$. In this paper these two facts are needed in section 4.

2. THE GEOMETRIC K-GROUP AND CONNES' RESULT.

2.1. Let V be a smooth manifold and Γ a countable discrete group. Suppose that Γ acts on V by diffeomorphisms from the right. Then we define a category $C(V,\Gamma)$ in the following way:

1) an object of $C(V,\Gamma)$ is a pair (W,ρ) such that
i) W is a smooth manifold with a proper Γ-action i.e. the map $W \times \Gamma \longrightarrow W \times W$ which takes (x,g) to (x,xg) is proper;
ii) $\rho\colon W \longrightarrow V$ is a Γ-equivariant submersion;
2) a morphism from (W_1,ρ_1) to (W_2,ρ_2) is a Γ-equivariant smooth map $f\colon W_1 \longrightarrow W_2$ with $\rho_1 = \rho_2 f$.

Let $E\Gamma$ be a universal principal Γ-bundle and denote by $V \times_\Gamma E\Gamma$ the orbit space of $V \times E\Gamma$ with the diagonal action. Then we define another category $\mathbb{B}(V,\Gamma)$ as follows:

1) an object of $\mathbb{B}(V,\Gamma)$ is a pair (M,φ) such that
i) M is a smooth manifold;
ii) $\varphi\colon M \longrightarrow V \times_\Gamma E\Gamma$ is a continuous map;
2) a morphism from (M_1,φ_1) to (M_2,φ_2) is a smooth map $h\colon M_1 \longrightarrow M_2$ with $\varphi_1 = \varphi_2 h$.

Moreover let $\mathbb{C}(V,\Gamma)$ denote the full subcategory of $C(V,\Gamma)$ whose objects are all pairs (W,ρ) such that W is a principal Γ-bundle with the base space W/Γ. Then we define a functor T from $\mathbb{C}(V,\Gamma)$ to $\mathbb{B}(V,\Gamma)$ as follows. Let (W,ρ) be an object of $\mathbb{C}(V,\Gamma)$ and f a morphism in $\mathbb{C}(V,\Gamma)$ from (W_1,ρ_1) to (W_2,ρ_2). Let $\psi\colon W \longrightarrow E\Gamma$ be the classifying bundle map for W and $\rho \times_\Gamma \psi\colon W/\Gamma \longrightarrow V \times_\Gamma E\Gamma$ the quotient map of $\rho \times \psi\colon W \longrightarrow V \times E\Gamma$.

Then we put

$$T(W,\rho) = (W/\Gamma, \rho \times_\Gamma \psi)$$

$$T(f) = f/\Gamma\colon W_1/\Gamma \longrightarrow W_2/\Gamma$$

where f/Γ denotes the quotient map of f.

2.2. Let V and W be smooth manifolds with smooth Γ-actions and $f: W \longrightarrow V$ a Γ-equivariant smooth map. Then we call f a Γ-equivariantly K-oriented map if a Γ-equivariant spin^c structure is given for $TW \oplus f^*TV$. If Γ is a trivial group, it is simply called a K-oriented map.

Let f be a Γ-equivariantly K-oriented map from W to V. We assume that the Γ-action on W is proper. Then it is known that f determines an element

$$f_! \rtimes \Gamma \in KK^j[C_0(W) \rtimes \Gamma, C_0(V) \rtimes \Gamma] \cdot$$

and hence induces a Gysin map

$$f_! \rtimes \Gamma\colon K_i[C_0(W) \rtimes \Gamma] \longrightarrow K_{i+j}[C_0(V) \rtimes \Gamma] \qquad (i = 0,1)$$

taking a Kasparov product with the above $f_! \rtimes \Gamma$, where $j = \dim V + \dim W \pmod 2$; see [8]. If Γ is a trivial group, we denote it simply by $f_!$.

2.3. REMARK. A Gysin map $f_! \rtimes \Gamma$ depends only on the homotopy class of f as a Γ-equivariantly K-oriented smooth map. If $f: W_1 \longrightarrow W_2$ and $g: W_2 \longrightarrow W_3$ are as above, then we have $(gf)_! \rtimes \Gamma = (g_! \rtimes \Gamma)(f_! \rtimes \Gamma)$.

2.4. Let (W,ρ) be an object of $C(V,\Gamma)$. Put $\tau = \ker(d\rho\colon TW \longrightarrow TV)$ and $\sigma = \rho^*TV$. We have $TW \cong \tau \oplus \sigma$. Let $\pi\colon \tau \longrightarrow W$ be the projection map. Then we obtain a Γ-equivariantly K-oriented map $\rho\pi\colon \tau \longrightarrow V$. In fact we have

$$\begin{aligned}[T\tau \oplus (\rho\pi)^*\, TV]|_W &\cong TW \oplus \tau \oplus \sigma \\ &\cong TW \oplus TW\end{aligned}$$

and prefer the spin^c structure for $T\tau \oplus (\rho\pi)^*TV$ given by the almost complex structure on $TW \otimes \mathbb{C}$ identifying it with $TW \oplus TW$.

Let (W_i,ρ_i) be an object in $C(V,\Gamma)$ $(i = 1,2)$. We define τ_i and π_i for (W_i,ρ_i) in the same way as above. Let f be a morphism from (W_1,ρ_1) to (W_2,ρ_2) and $f'\colon \tau_1 \longrightarrow \tau_2$ denote the restriction to τ_1 of $df\colon TW_1 \longrightarrow TW_2$. Note $\rho_1\pi_1 = \rho_2\pi_2 f'$. Since $\rho_1\pi_1$ and $\rho_2\pi_2$ are both Γ-equivariantly K-oriented, they uniquely determine a spin^c structure for $T\tau_1 \oplus f'^*T\tau_2$. Therefore f' yields a Gysin map

$$f_! \rtimes \Gamma\colon K_*[C_0(\tau_1) \rtimes \Gamma] \longrightarrow K_*[C_0(\tau_2) \rtimes \Gamma].$$

2.5. DEFINITION. *The geometric* K-*group* $K^*(V,\Gamma)$ *is defined by*

$$K^*(V,\Gamma) = \varinjlim_{C(V,\Gamma)} K_*[C_0(\tau) \rtimes \Gamma]$$

where the direct limit is taken using the Gysin map $f_! \rtimes \Gamma$ *above.*

2.6. Let (W,ρ) be an object of $C(V,\Gamma)$ and ξ an element of $K_*[C_0(\tau) \rtimes \Gamma]$. Let $x \in K^*(V,\Gamma)$ be the element determined by (W,ρ) and ξ. Then we put

$$\mu\colon K^*(V,\Gamma) \longrightarrow K_*[C_0(V) \rtimes \Gamma]$$
$$\mu(x) = [(\rho\pi)_! \rtimes \Gamma](\xi).$$

It is well defined by Remark 2.3.

Here we shall mention the Baum-Connes conjecture. In [2] Baum and Connes conjectured that μ gives an isomorphism for any V and Γ. Actually this is a special case of a more general conjecture proposed in [1]. For geometric and analytic applications of this conjecture we refer the reader to [1].

2.7. Let M be a smooth manifold and E an $\mathbb{R}$ vector bundle on M. Let $E - \{0\}$ denote the space E with the zero section deleted. The K homology of M twisted by E (denoted by $K_*^E(M)$) is by definition the K homology of the pair $(E, E-\{0\})$:

$$K_*^E(M) = K_*(E, E-\{0\}).$$

Let $h\colon M_0 \longrightarrow M$ be a continuous map. Then we note that h induces a natural homomorphism

$$h_*\colon K_*^{h^*E}(M_0) \longrightarrow K_*^E(M).$$

Let F be another $\mathbb{R}$ vector bundle on M. Suppose that $E \oplus F$ is a spin^c manifold i.e. the tangent bundle $T(E \oplus F)$ has a spin^c structure.

Then we have the following Poincaré duality:

$$\cap[E \oplus F]_K \colon K^*(F) \longrightarrow K_*^E(M)$$

where $[E \oplus F]_K$ denotes the fundamental homology class for $E \oplus F$ in topological K homology with proper supports [3].

2.8. Let (M_i, φ_i) be an object of $\mathcal{B}(V,\Gamma)$ $(i = 1,2)$ and h a morphism from (M_1,φ_1) to (M_2,φ_2). We define vector bundles $\tilde{\tau}_i$ and $\tilde{\sigma}_i$ over M_i such as

$$\tilde{\sigma}_i = \varphi_i^*(TV \underset{\Gamma}{\times} E\Gamma) \qquad \tilde{\tau}_i = TM_i/\sigma_i .$$

Note $TM_i \cong \tilde{\tau}_i \oplus \tilde{\sigma}_i$. Then we have the following commutative diagram [2][8][9]:

$$\begin{array}{ccc}
K^*(\tilde{\tau}_1) & \xrightarrow{\;h'_!\;} & K^*(\tilde{\tau}_2) \\
{\scriptstyle \cap[TM_1]_K}\downarrow & & \downarrow{\scriptstyle \cap[TM_2]_K} \\
K_*^{\tilde{\sigma}_1}(M_1) & \xrightarrow[\;h_*\;]{} & K_*^{\tilde{\sigma}_2}(M_2)
\end{array}$$

where $h' \colon \tilde{\tau}_1 \longrightarrow \tilde{\tau}_2$ denotes the map induced by $dh \colon TM_1 \longrightarrow TM_2$. Here we identify the topological K-group $K^*(\tilde{\tau}_i)$ with $K_*[C_0(\tilde{\tau}_i)]$.

2.9. Let $\mathcal{C}(V,\Gamma)$ be the full subcategory of $C(V,\Gamma)$ defined in 2.1 and (W,ρ) an object of $\mathcal{C}(V,\Gamma)$. Define τ for (W,Γ) in the same way as in 2.4. Since τ admits a free Γ-action, there is the Morita equivalence between $C_0(\tau) \rtimes \Gamma$ and $C_0(\tau/\Gamma)$, which yields an isomorphism $K_*[C_0(\tau) \rtimes \Gamma] \longrightarrow K_*[C_0(\tau/\Gamma)]$. Then for a morphism f from (W_1,ρ_1) to (W_2,ρ_2) in $\mathcal{C}(V,\Gamma)$ we have the following commutative diagram:

$$\begin{array}{ccc} K_*[C_0(\tau_1) \rtimes \Gamma] & \xrightarrow{f'_! \rtimes \Gamma} & K_*[C_0(\tau_2) \rtimes \Gamma] \\ \downarrow & & \downarrow \\ K_*[C_0(\tau_1/\Gamma)] & \xrightarrow[(f'/\Gamma)_!]{} & K_*[C_0(\tau_2/\Gamma)] \end{array}$$

where the vertical maps are isomorphisms induced by the Morita equivalence as above.

Let T be the functor from $\mathfrak{A}(V,\Gamma)$ to $\mathfrak{B}(V,\Gamma)$ in 2.1 and put $(M,\varphi) = T(W,\rho)$. We denote by $K_*^{TV}(V \times_\Gamma E\Gamma)$ the K-homology of $V \times_\Gamma E\Gamma$ twisted by the vector bundle $TV \times_\Gamma E\Gamma$. Then we put

$$K^*(V,\Gamma)_0 = \varinjlim_{\mathfrak{A}(V,\Gamma)} K_*[C_0(\tau) \rtimes \Gamma]$$

similarly to 2.5.

2.10. PROPOSITION [2]. *Let* ξ *be an element of* $K_*[C_0(\tau) \rtimes \Gamma]$ and $u_\xi \in K^*(\tau/\Gamma)$ *denote the element corresponding to* ξ *through the Morita equivalence mentioned above. Let* $x \in K^*(V,\Gamma)_0$ *be the element determined by* (W,ρ) *and* ξ. *Then the homomorphism*

$$\varphi\colon K^*(V,\Gamma)_0 \longrightarrow K_*^{TV}(V \times_\Gamma E\Gamma)$$

given by

$$\varphi(x) = \varphi_*(u_\xi \cap [TM]_K)$$

is an isomorphism.

Proof. The well-definedness follows from 2.8 and 2.9 with $\varphi_1 = \varphi_2 h$. Suppose that $E\Gamma$ is a manifold. Then $V \times E\Gamma$ is a final object in $\mathfrak{A}(V,\Gamma)$ with ρ the projection map $V \times E\Gamma \longrightarrow V$ and the corresponding $\varphi\colon V \times_\Gamma E\Gamma \longrightarrow V \times_\Gamma E\Gamma$ to $(V \times E\Gamma,\rho)$ is the identity map. Hence it induces an isomorphism obviously.

For the general case we apply the above argument to the n-skeleton $E\Gamma^{(n)}$ and take the direct limit over n. □

2.11. REMARK. There is a homomorphism

$$\lambda: K^*(V,\Gamma)_0 \longrightarrow K^*(V,\Gamma)$$

induced by the inclusion $\mathfrak{C}(V,\Gamma) \subset C(V,\Gamma)$. If Γ is torsion-free, then these categories coincide to each other and hence λ is an isomoprhism.

2.12. Next we shall discuss the equivariant Chern character ch_Γ from $K^*(V,\Gamma)$ into the equivariant cohomology group $H^*(V,\Gamma)$. First we review the definition of $H^*(V,\Gamma)$. Let $S(\Gamma) = \{g \in \Gamma: g$ is of finite order$\}$ and put

$$\hat{V} = \{(x,g) \in V \times \Gamma: xg = x \text{ and } g \in S(\Gamma)\}.$$

Note that $\hat{V}$ contains a copy of V as the fixed point set of the identity element of Γ. Let Γ act on $\hat{V}$ by $(x,g)h = (xh,h^{-1}gh)$ for $h \in \Gamma$. Then we define

$$H^i(V,\Gamma) = \bigoplus_{q\in\mathbb{N}} H_{2q+i}^{T\hat{V}}(\hat{V} \times_\Gamma E\Gamma) \qquad (i = 0,1).$$

Here $H_j^{T\hat{V}}(\hat{V} \times_\Gamma E\Gamma)$ denotes the jth homology group of $\hat{V} \times_\Gamma E\Gamma$ with coefficients $\mathbb{C}$ twisted by the vector bundle $T\hat{V} \times_\Gamma E\Gamma$ i.e.

$$H_j^{T\hat{V}}(\hat{V} \times_\Gamma E\Gamma) = H_j(T\hat{V} \times_\Gamma E\Gamma, T\hat{V} \times_\Gamma E\Gamma - \{0\}; \mathbb{C}).$$

Then Baum-Connes [2] constructed the equivariant Chern character

$$ch_\Gamma: K^*(V,\Gamma) \longrightarrow H^*(V,\Gamma),$$

which turns out to be $\mathbb{C}$-isomorphism. See also [12]. Note that $H^*(V,\Gamma)$ contains $H_*^{TV}(V \times_\Gamma E\Gamma)$ as a direct summand. We put

$$\mathrm{ch}_0 = P\mathrm{ch}_\Gamma\lambda\colon K^*(V,\Gamma)_0 \longrightarrow H_*^{TV}(V \underset{\Gamma}{\times} E\Gamma)$$

where P denotes the projection map from $H^*(V,\Gamma)$ onto $H_*^{TV}(V \underset{\Gamma}{\times} E\Gamma)$. Then we shall give an explicit formula for ch_0 using the ordinary Chern character $\mathrm{ch}\colon K_*^{TV}(V \underset{\Gamma}{\times} E\Gamma) \longrightarrow H_*^{TV}(V \underset{\Gamma}{\times} E\Gamma)$, which will be needed later.

Let (W,ρ) be an object of $\mathfrak{a}(V,\Gamma)$ and put $(M,\varphi) = T(W,\rho)$. We define τ and $\tilde{\tau}$ for W and M as previously. Note $\tilde{\tau} = \tau/\Gamma$.

2.13. PROPOSITION [2][8]. 1) *We have the following commutative diagram:*

$$\begin{array}{ccccc}
K_*^{TV}(V \underset{\Gamma}{\times} E\Gamma) & \xleftarrow{\ \varphi\ } & K^*(V,\Gamma)_0 & \xrightarrow{\ \lambda\ } & K^*(V,\Gamma) \\
& \searrow^{\mathrm{td}(\sigma_\Gamma\otimes\mathbb{C})^{-1}\cap\mathrm{ch}} & \downarrow{\scriptstyle \mathrm{ch}_0} & & \downarrow{\scriptstyle \mathrm{ch}_\Gamma} \\
& & H_*^{TV}(V \underset{\Gamma}{\times} E\Gamma) & \xleftarrow[\ P\ ]{} & H^*(V,\Gamma)
\end{array}$$

where we denote $TV \underset{\Gamma}{\times} E\Gamma$ by σ_Γ.

2) *Let* $\xi \in K_*[C_0(\tau) \rtimes \Gamma]$ *and* $u_\xi \in K^*(\tilde{\tau})$ *as in* 2.10. *Let* $x \in K^*(V,\Gamma)_0$ *be the element determined by* (W,ρ) *and* ξ. *Then we have*

$$\mathrm{ch}_0(x) = \varphi_*\{\mathrm{ch}(u_\xi)\,\mathrm{td}(\tilde{\tau}\otimes\mathbb{C})\cap[TM]\}$$

where ch *is the ordinary Chern character* $K^*(\tilde{\tau}) \longrightarrow H_c^*(\tilde{\tau};\mathbb{C})$. □

2.14. Next we shall review the secondary characteristic classes for a Γ-manifold V following [4][11]. Let J^kV be the jet bundle of V of order k. Recall that J^kV admits a natural action of the orthogonal group O_n ($n = \dim V$).

Put $\nu^k = J^kV/O_n$. We call ν^k a thickened version of V following [4] in the sense that ν^k is homotopy equivalent to V. Then we denote by $\Omega^*(\nu^k)^{\mathrm{Diff}V}$ the space of all differential forms on ν^k which are invariant under all induced actions of diffeomorphisms of V. Put $\Omega^*(\nu)^{\mathrm{Diff}V}$ to be the direct limit of $\Omega^*(\nu^k)^{\mathrm{Diff}V}$ over k. Then it is known that there is a natural homomorphism $C^*(\mathfrak{A}_n;O_n) \longrightarrow \Omega^*(\nu)^{\mathrm{Diff}V}$, where $C^*(\mathfrak{A}_n;O_n)$ is the space of all O_n-basic cochains over the Lie algebra of formal vector fields on $\mathbb{R}^n$. The cohomology group $H^*(\mathfrak{A}_n;O_n)$ of $C^*(\mathfrak{A}_n;O_n)$ is called the Gel'fand-Fuks cohomology group.

Put $\nu^k_\Gamma = \nu^k \times_\Gamma E\Gamma$. Note that ν^k_Γ is a thickened version of $V \times_\Gamma E\Gamma$. It naturally induces a cochain map from $\Omega^*(\nu^k)^{\mathrm{Diff}V}$ to the de Rham complex $\Omega^*(\nu^k_\Gamma)$ and yields a homomorphism

$$\chi\colon H^*(\mathfrak{A}_n;O_n) \longrightarrow H^*(V \times_\Gamma E\Gamma;\mathbb{R}).$$

Then we call an element in the image of χ a secondary characteristic class for $V \times_\Gamma E\Gamma$.

2.15. Let V be a smooth manifold with a smooth Γ-action as previously. We further assume that V is an oriented manifold and that the Γ-action preserves the orientation of V. Then $TV \times_\Gamma E\Gamma$ is an oriented vector bundle over $V \times_\Gamma E\Gamma$ and induces the Thom isomorphism

$$\Phi\colon H^{TV}_{q+n}(V \times_\Gamma E\Gamma) \longrightarrow H_q(V \times_\Gamma E\Gamma;\mathbb{C})$$

where $n = \dim V$.

Now we can state Connes' result mentioned in the introduction.

2.16. THEOREM [8]. *Let* $\omega \in H^*(V \times_\Gamma E\Gamma;\mathbb{R})$ *be a secondary characteristic class for* $V \times_\Gamma E\Gamma$. *Then there exists an additive map* $\phi_\omega\colon K_*[C_0(V) \rtimes \Gamma] \longrightarrow \mathbb{C}$ *such* that

$$\phi_\omega \mu_0(x) = \langle \omega, \Phi \mathrm{ch}_0(x)\rangle$$

for any $x \in K^*(V,\Gamma)_0$, *where we put* $\mu_0 = \mu\lambda$. *In other words we have the* *following commutative diagram:*

$$
\begin{array}{ccc}
K^*(V,\Gamma)_0 & \xrightarrow{\ \mu_0\ } & K_*[C_0(V) \rtimes \Gamma] \\
\Big\downarrow{\scriptstyle \Phi ch_0} & & \Big\downarrow{\scriptstyle \phi_\omega} \\
H_*(V \times_\Gamma E\Gamma;\mathbb{C}) & \xrightarrow[\langle\omega,\ \rangle]{} & \mathbb{C}
\end{array}
\qquad \square
$$

3. A GEOMETRIC DESCRIPTION OF CERTAIN KASPAROV MODULES OVER $C_0(V) \rtimes \Gamma$.

In this section we shall give a description of certain Kasparov modules over $C_0(V) \rtimes \Gamma$ following the method of [6; section 7].

3.1. Let (W,ρ) be an object of $C(V,\Gamma)$ and E a Γ-equivariant vector bundle on W with a Γ-invariant hermitian metric. Put $\tau = \ker d\rho$ as before. Then we choose a Γ-invariant metric on τ. Put $W_x = \rho^{-1}(x)$ for $x \in V$. Since ρ is a submersion, W_x is a submanifold of W. We note that the metric chosen on τ gives a Riemannian metric on W_x for each $x \in V$.

Let $H_x = L^2(W_x,E)$ be the Hilbert space of all L^2-sections of E over W_x and $\langle\ ,\ \rangle_x$ denote the inner product on H_x (we choose $\langle\ ,\ \rangle_x$ such that it is anti-linear in the first variable and linear in the second). Let H be a family of Hilbert spaces $(H_x)_{x\in V}$. Then we define a unitary operator $\circ g\colon H_x \longrightarrow H_{xg}$ for each $g \in \Gamma$ and $x \in V$ by:

$$(u\circ g)(y) = g[u(yg^{-1})] \qquad (y \in W_x)$$

for $u \in H_x$. Note that this defines a right Γ-action on H: $(u\circ g)\circ h = u\circ gh$ for $g,h \in \Gamma$.

Let $C_c(W,E)$ be the space of all continuous sections of E over W with compact support. We denote by S the space of all sections $\xi = (\xi_x)_{x\in V}$ of H over V such that there is an element $s \in C_c(W,E)$ with $\xi_x = s|W_x$. Note that S admits a left Γ-action such as

$$g(\xi)_x = \xi_{xg}\circ g^{-1} \qquad (x \in V)$$

for $\xi \in S$ and $g \in \Gamma$. Then S is a pre-Hilbert $C_c(V,\Gamma)$-module in the following way:

1) the right $C_c(V,\Gamma)$-action on $\mathcal{S}$ is given by

$$\xi a = \sum_{g\in\Gamma} g^{-1}(\xi)g^{-1}(\rho^* a_g)$$

for $\xi \in \mathcal{S}$ and $a = \sum_{g\in\Gamma} a_g U_g \in C_c(V,\Gamma)$ i.e.

$$(\xi a)_x = \sum_{g\in\Gamma} (\xi_{xg^{-1}} \circ g) a_g(xg^{-1}) \qquad (x \in V);$$

2) the $C_c(V,\Gamma)$-valued inner product is defined by

$$\langle \xi,\eta\rangle = \sum_{g\in\Gamma} \langle \xi,\eta\rangle_g U_g$$

where

$$\langle \xi,\eta\rangle_g(x) = \langle \xi_x, g(\eta)_x\rangle_x = \langle \xi_x \circ g, \eta_{xg}\rangle_{xg} \qquad (x \in V).$$

Since the Γ-action on W is proper, we can easily verify that $\langle \xi,\eta\rangle$ is an element of $C_c(V,\Gamma)$.

3.2. DEFINITION. *We define a Hilbert C*-module* $\mathcal{E} = \mathcal{E}_{W,E}$ *as the completion of* $\mathcal{S}$ *with respect to the norm* $\|\xi\| = \|\langle \xi,\xi\rangle\|^{1/2}$.

3.3. We first recall $G_x = \{(xg^{-1},g) \in V \times \Gamma : g \in \Gamma\}$ $(x \in V)$ and let $C_c(G_x)$ denote the space of all functions on G_x with compact support. Let $\xi = (\xi_x)_{x\in V} \in \mathcal{S}$. Given $f \in C_c(G_x)$, we define $S_{\xi,x}(f) \in H_x$ by

$$S_{\xi,x}(f) = \sum_{g\in\Gamma} (\xi_{xg^{-1}} \circ g) f(xg^{-1},g).$$

For $u \in H_x$ we also define a function $T_{\xi,x}(u)$ on G_x by

$$[T_{\xi,x}(u)](xg^{-1},g) = \langle \xi_{xg^{-1}} \circ g, u\rangle_x \qquad ((xg^{-1},g) \in G_x).$$

3.4. LEMMA. *For* $f \in C_c(G_x)$ *and* $u \in H_x$ *it follows:*

1) $\langle S_{\xi,x}(f), u\rangle_x = \sum_{g\in\Gamma} \overline{f}(xg^{-1},g)[T_{\xi,x}(u)](xg^{-1},g)$;

2) $T_{\xi,x}S_{\xi,x}(f) = \pi_x(\langle\xi,\xi\rangle)f$;

3) $\|S_{\xi,x}(f)\|^2 = \langle f, \pi_x(\langle\xi,\xi\rangle)f\rangle_2$

where $\langle\ ,\ \rangle_2$ *denote the inner product on* $\ell^2(G_x)$.

Proof. 1) A straightforward calculation.
2) Using 1) we obtain

$$[T_{\xi,x}S_{\xi,x}(f)](xh^{-1},x) = \langle \xi_{xh^{-1}}\circ h, S_{\xi,x}(f)\rangle_x$$
$$= \sum_{g\in\Gamma} \langle \xi_{xh^{-1}}\circ h, \xi_{sb^{-1}}\circ g\rangle f(xg^{-1},g).$$

On the other hand we have

$$[\pi_x(\langle\xi,\xi\rangle)f](xh^{-1},h) = \sum_{g\in\Gamma} \langle \xi_{xh^{-1}}\circ g, \xi_{xh^{-1}g}\rangle_{xh^{-1}g} f(xh^{-1}g, g^{-1}h)$$
$$= \sum_{k\in\Gamma} \langle \xi_{xh^{-1}}\circ hk^{-1}, \xi_{xk^{-1}}\rangle_{xk^{-1}} f(xk^{-1},k)$$

where we put $k = g^{-1}h$. Since $\circ h$ is a unitary operator, the above coincide to each other.
3) This follows directly from 1) and 2). □

Then the following proposition is a corollary of the above lemma.

3.5. PROPOSITION. *For a given* $\xi \in \mathcal{S}$ *the formulae in 3.3 yield bounded operators* $S_{\xi,x}\colon \ell^2(G_x) \longrightarrow H_x$ *and* $T_{\xi,x}\colon H_x \longrightarrow \ell^2(G_x)$ *for each* $x \in V$ *satisfying the following properties:*

1) $\langle S_{\xi,x}(f), u\rangle_x = \langle f, T_{\xi,x}(u)\rangle_2$

for $f \in \ell^2(G_x)$ *and* $u \in H_x$;

2) $S_{\xi,x}(\varphi|G_x) = (\xi\varphi)_x$ $\qquad T_{\xi,x}(\eta_x) = \langle\xi,\xi\rangle|_{G_x}$

for $\eta \in \mathcal{S}$ *and* $\varphi \in C_c(V,\Gamma)$ *(here we consider* φ *naturally as a function on* $V \times \Gamma$*);*

3) $T_{\xi,x}S_{\eta,x}(f) = \pi_x(\langle\xi,\eta\rangle)f$ $\qquad S_{\xi,x}T_{\eta,x}(\varsigma_x) = (\xi\langle\eta,\varsigma\rangle)_x$

for $f \in \ell^2(G_x)$ *and* $\eta,\varsigma \in \mathcal{S}$;

4) $\|T_{\xi,x}\| = \|S_{\xi,x}\| = \|\pi_x(\langle\xi,\xi\rangle)\|^{1/2}$,

in particular

$$\sup_{x\in V}\|T_{\xi,x}\| = \sup_{x\in V}\|S_{\xi,x}\| = \|\xi\|. \qquad \square$$

Let $T = (T_x)_{x\in V}$ be a family of bounded operators $T_x\colon H_x \longrightarrow H_x$. Then we define a linear map T and T^* on $\mathcal{S}$ by $T(\xi) = (T_x(\xi_x))_{x\in V}$ and $T^*(\xi) = (T_x^*(\xi_x))_{x\in V}$ for $\xi \in \mathcal{S}$, respectively.

3.6. PROPOSITION. *Suppose that* $T = (T_x)_{x\in V}$ *satisfies the following properties:*

i) $\sup_{x\in V}\|T_x\| < +\infty$;

ii) $T_x(u)\circ g = T_{xg}(u\circ g)$ *for* $u \in H_x$ *and* $g \in \Gamma$;

iii) T *and* T^* *map* $\mathcal{S}$ *into itself.*

Then T *defines an operator in* $\mathfrak{L}(\mathcal{E})$.

Proof. It is easily verified that T is a module homomorphism on $\mathcal{S}$ by ii). For $\xi \in \mathcal{S}$ we note

$$S_{T(\xi),x} = T_x S_{\xi,x}$$

by ii). Then we obtain

$$\begin{aligned}\|T(\xi)\| &= \sup_{x\in V}\|S_{T(\xi),x}\| \\ &\le (\sup_{x\in V}\|T_x\|)(\sup_{x\in V}\|S_{\xi,x}\|) \\ &= (\sup_{x\in V}\|T_x\|)\|\xi\|\end{aligned}$$

by Proposition 3.5 4). Hence the property i) shows that T gives a bounded operator on $\mathcal{E}$. Note that the same argument holds for T^*. Since T^* is the adjoint operator of T, T defines an operator in $\mathcal{L}(\mathcal{E})$. $\square$

3.7. REMARK. If we replace the property iii) above by the following one, the same conclusion holds.

iii′) Let $\mathcal{S}_0$ be a dense subset of $\mathcal{S}$ with $\| \ \|$. Then T and T^* map $\mathcal{S}_0$ into $\mathcal{S}$.

3.8. Let (W,ρ) be an object of $C(V,\Gamma)$ and $\tau = \ker d\rho$ as in 3.1. Suppose that ρ is Γ-equivariantly K-oriented. Then it uniquely determines a Γ-equivariant spin^c structure on τ. Let S be the spinor bundle associated with τ and D_x the Dirac operator along W_x for $x \in V$. Suppose that each W_x is even-dimensional and that W/Γ is compact.

3.9. DEFINITION. *We define* $\rho_! \rtimes \Gamma \in K_0[C_0(V) \rtimes \Gamma]$ *to be the element represented by the following Kasparov module:*

i) *the Hilbert* $C_0(V) \rtimes \Gamma$*-module is* $\mathcal{E}_{W,S}$;

ii) *the operator in* $\mathcal{L}(\mathcal{E}_{W,S})$ *is given by* $F = \{(1+D_x)^{-1/2} D_x\}_{x \in V}$.

We call $\rho_! \rtimes \Gamma$ *the index of* F.

3.10. REMARK. Here we are slightly abusing the notation $\rho_! \rtimes \Gamma$. In fact, if we take the left $C_0(W) \rtimes \Gamma$-action on $\mathcal{E}_{W,S}$ into account, the above Kasparov module yields $\rho_! \rtimes \Gamma \in KK^0[C_0(W) \rtimes \Gamma, C_0(V) \rtimes \Gamma]$ mentioned in 2.2. Let $\sigma_F \in K_0[C_0(\tau) \rtimes \Gamma]$ be the element determined by the principal symbol of F. Then they are related by

$$\rho_! \rtimes \Gamma = [(\rho\pi)_! \rtimes \Gamma](\sigma_F)$$

where $\pi\colon \tau \longrightarrow W$ is the projection map. In other words the index $\rho_! \rtimes \Gamma$ of F is equal to the image of the symbol $\sigma_F \in K_0[C_0(\tau) \rtimes \Gamma]$ by a Gysin map $(\rho\pi)_! \rtimes \Gamma$; see [1][9].

4. A VANISHING THEOREM FOR THE LONGITUDINAL INDEX.

4.1. Let $*$ be the set consisting of a point with a trivial Γ-action and

(W,ρ) an object of $C(*,\Gamma)$. We fix a Γ-invariant Riemannian metric on W as previously. Suppose that W admits a spin^c structure invariant for the Γ-action. Then the spinor bundle S on W admits a lifted Γ-action. We also fix a Γ-invariant hermitian metric on S.

We further assume that there is a smooth S^1-action on W from the left which preserves the spin^c structure on W and commutes with the Γ-action. Taking an average over S^1 if necessary, we may assume that the metrics chosen above are also invariant for the S^1-action. Note that S admits a lifted S^1-action. Then we choose a metric connection ∇ on S invariant under both Γ and S^1.

Let $C_c^\infty(W,S)$ be the space of all smooth sections of S over W with compact support. Note that $C_c^\infty(W,S)$ is naturally endowed with an inner product invariant under Γ and S^1. Let g_t denote the action of $t \in S^1 = \mathbb{R}/\mathbb{Z}$ and X the Killing vector field generated by $\{g_t\}_{t\in S^1}$. Then we put

$$(L_X u)(y) = \lim_{t\to 0} \frac{1}{t}\{u(y)-g_t[u(g_{-t}y)]\} \qquad (y \in W)$$

for $u \in C_c^\infty(W,S)$. Note that L_X is a skew self-adjoint operator on $C_c^\infty(W,S)$.

4.2. LEMMA. *Suppose that* W/Γ *is compact.*
1) *Let* D *be the Dirac operator and* $c(X)$ *denote the Clifford multiplication by* X. *Then we have*

$$[L_X,D] = 0 = [L_X,c(X)].$$

2) *Put* $A = L_X - \nabla_X$. *Then* A *is an endomorphism of the vector bundle* S *such that*

$$|A| = \sup_{y\in W}||A_y|| < +\infty .$$

3) *For a given* $\epsilon > 0$, *there exists* $\delta > 0$ *such that we have* $||L_X u|| \le \delta||u||$

for any $u \in C_c^\infty(W,S)$ *with* $\|Du\| \le \varepsilon\|u\|$.

Proof. 1) Since g_t commutes with D, L_X commutes with D. Moreover L_X commutes with $c(X)$ since X is invariant under S^1.

2) Suppose that A is an endomorphism of S. Then it is obvious that $|A| < +\infty$ since W/Γ is compact. To prove that A is an endomorphism it suffices to show

$$A(fu) = f(Au)$$

for any $f \in C_c^\infty(W)$ and $u \in C_c^\infty(W,S)$. In fact we have

$$\begin{aligned} A(fu) &= L_X(fu) - \nabla_X(fu) \\ &= (Xf)u + f(L_Xu) - (Xf)u - f(\nabla_Xu) \\ &= f(Au). \end{aligned}$$

3) Let $\nabla\colon C_c^\infty(W,S) \longrightarrow C_c^\infty(W,S\otimes T^*W)$ be the connection of S chosen before. We recall that the Weitzenböck formula for D yields

$$\int_W (Du,Du)dv = \int_W (\nabla u,\nabla u)dv + \int_W (Ru,u)dv$$

for $u \in C_c^\infty(W,S)$, where R is an endomorphism of S which can be expressed in terms of the curvature tensors of S and TW; see [10]. Let $\{e_\alpha\}$ be an orthonormal basis for T_yW. Then we recall $(\nabla u,\nabla u) = \sum_\alpha(\nabla_{e_\alpha}u,\nabla_{e_\alpha}u)$ at $y \in W$. We also note $|R| = \sup_{y\in W}\|R_y\| < +\infty$.

Without loss of generality we may assume $\|X\| \le 1$. Then the above observation shows

$$\|Du\|^2 \ge \|\nabla u\|^2 - |R|\|u\|^2$$

and

$$\|\nabla u\| \ge \|\nabla_X u\|$$

since $(\nabla u,\nabla u) \ge (\nabla_X u,\nabla_X u)$. On the other hand we have

$$\|\nabla_X u\| \geq \|L_X u\| - |A|\|u\|$$

by 2). Hence putting them together we obtain

$$\|Du\| + |R|^{1/2}\|u\| + |A|\|u\| \geq \|L_X u\|$$

which proves 3). □

Consider the same situation as in 4.1. Let $\mathcal{S} = C_c(W,S)$ and H denote the L^2-completion of $\mathcal{S}$.

4.3. LEMMA. *Let* P *be the spectral projection of* $\sqrt{-1}\, L_X$ *on* H *corresponding to some interval* $I \subset \mathbb{R}$. *Then we have:*

1) P *is a self-adjoint bounded operator on* H;
2) $(Pu)\circ g = P(u\circ g)$ *for* $u \in H$ *and* $g \in \Gamma$;
3) P *maps* $C_c^\infty(W,S)$ *into* $\mathcal{S}$.

Proof. The first claim is a well known fact. The second follows from $[L_X, g] = 0$. To prove the third we take a continuous function f on W with compact support which is S^1-invariant i.e. $f(g_t y) = f(y)$ for any $t \in S^1$ and $y \in W$. Let M_f denote the multiplication by f. Since $[L_X, M_f] = 0$, we then obtain $[P, M_f] = 0$. For a given $u \in C_c^\infty(W,S)$ we choose f such as $f(y) = 1$ for $y \in \operatorname{supp} u$. Then we have

$$Pu = P(M_f u) = M_f(Pu)$$

which shows $\operatorname{supp} Pu \subset \operatorname{supp} f$.

The remaining part is to show that Pu is a continuous section. Let σ be the spectrum of L_X. Since σ is a discrete set, the characteristic function χ on I is continuous over σ. Then the functional calculus for χ yields that Pu is continuous since $(zI - L_X)^{-1}u \in C_c^\infty(W,S)$ for $z \in \mathbb{C}$. □

4.4. THEOREM. *Let* (W,ρ) *be as in* 4.1. *We assume that* W *is even-dimensional and that* W/Γ *is compact. Let* $\rho_! \rtimes \Gamma$ *be the element of*

$K_0(C^*\Gamma)$ *defined by the Kasparov module* $(\mathcal{E}_{W,S}, F)$ *in* 3.9. *Suppose that the* S^1*-action on* W *is locally free. Then we have* $\rho_! \rtimes \Gamma = 0$.

Proof. For $t \in \mathbb{R}$ we define an operator $D(t)$ on $C_c^\infty(W,S)$ by

$$D(t) = D + \sqrt{-1}t\ c(X).$$

Note that $D(t)$ is self-adjoint on $C_c^\infty(W,S)$ and anti-commutes with the involution on $C_c^\infty(W,S)$. Put

$$F(t) = |1+D(t)^2|^{-1/2}D(t).$$

Then $F(t)$ yields an operator in $\mathcal{L}(\mathcal{E}_{W,S})$. Here we note that $F(t)$ is a compact perturbation of $F = (1+D^2)^{-1/2}D$ since the principal symbols are equal; see [6][9].

Let L_X be the operator defined in 4.1. For $\epsilon = 1$ we fix $\delta > 0$ given in Lemma 4.2 3). Let P_1, P_2 and P_3 denote the spectral projections of $\sqrt{-1}\ L_X$ on H corresponding to the intervals $(-\infty,-\delta]$, $(-\delta,\delta]$ and $(\delta,+\infty)$, respectively. Then by Lemma 4.3 with Proposition 3.6, each P_i $(i = 1,2,3)$ yields an operator in $\mathcal{L}(\mathcal{E}_{W,S})$ with $P_i = P_i^2$. Hence $\mathcal{E}_{W,S}$ splits into a direct sum of Hilbert $C^*\Gamma$-modules $\mathcal{E}_{W,S} = \mathcal{E}_1 \oplus \mathcal{E}_2 \oplus \mathcal{E}_3$ where $\mathcal{E}_i$ denote the image of $\mathcal{E}_{W,S}$ by P_i. Note that $F(t)$ commutes with each P_i since $D(t)$ commutes with L_X by Lemma 4.2 1). Hence $F(t)$ also splits into a direct sum $F_1(t) \oplus F_2(t) \oplus F_3(t)$ correspondingly and it yields

$$(\mathcal{E}_{W,S}, F(t)) = (\mathcal{E}_1, F_1(t)) \oplus (\mathcal{E}_2, F_2(t)) \oplus (\mathcal{E}_3, F_3(t)).$$

Note that $F_i(t)$ is a compact perturbation of $F_i = P_iFP_i$ since the restriction of a compact operator is compact. Then we shall prove that each Kasparov module $(\mathcal{E}_i, F_i(t))$ defines the zero element in $K_0(C^*\Gamma)$ by showing that $F_i(t)$ is invertible for some $t \in \mathbb{R}$.

For $i = 1$ and 3 Lemma 4.2 3) implies that D is bounded from the below on $P_i(H)$ by $\epsilon = 1$. Then F_i is invertible on $\mathcal{E}_i$ and hence

$(\mathcal{E}_i, F_i)$ defines the zero element.

For $i = 2$ we need the compact perturbation $F_2(t)$ of F_2. First we note

$$|D(t)|^2 u = D^2 u + \sqrt{-1}t(2A-2L_X+B)u + t^2\|X\|^2 u$$

for $u \in C_c^\infty(W,S)$, where B is an endomorphism of S defined by

$$B_y = \sum_\alpha c(e_\alpha)c(\nabla_{e_\alpha} X) \qquad (y \in W).$$

The relevant property is that $\sqrt{-1}t(2A-2L_X+B)$ yields a bounded operator on $P_2(H)$ and that the operator $\sqrt{-1}t(2A-2L_X+B) + t^2\|t^2\|X\|^2$ is bounded from below by a strictly positive number for t sufficiently large. In fact we notice that L_X is bounded on $P_2(H)$ by the choice of P_2 and that X is nowhere vanishing on W since the S^1-action is locally free. Thus thinking of the order of t, we can verify that $D(t)^2$ is bounded from the below on $P_2(H)$ by a strictly positive number for a large $t > 0$. This implies that $F_2(t)$ is invertible for a large $t > 0$ and hence $(\mathcal{E}_2, F_2)$ is zero in the abelian group $K_0(C^*\Gamma)$.

Since $(\mathcal{E}_{W,s}, F)$ is a direct sum of $(\mathcal{E}_i, F)$, we finally obtain $\rho_! \rtimes \Gamma = 0$. □

4.5. Next we shall prove a similar vanishing theorem for the index of a family of the Dirac operators in the following situation. Let V be a closed manifold and Γ act on V by orientation preserving diffeomorphisms. Let M be a closed spin^c manifold and $\tilde{M}$ a principal Γ-bundle on M. Suppose that there is a smooth S^1-action on $\tilde{M}$ which preserves the spin^c structure on $\tilde{M}$ induced from M and commutes with the Γ-action. Put $W = V \times \tilde{M}$ and view W as a free Γ-space with the diagonal Γ-action. Let $p: V \times \tilde{M} \longrightarrow V$ be the projection map. Then (W,p) defines an object of $\mathfrak{a}(V,\Gamma)$.

4.6. THEOREM. *Let* $p_! \rtimes \Gamma \in K_0[C_0(V) \rtimes \Gamma]$ *be the element defined in* 3.9

from (W,p). *Suppose that the* S^1*-action on* $\tilde{M}$ *is locally free. Then we have* $p_! \rtimes \Gamma = 0$.

Proof. Since W is a product space, we can apply the same argument as in 4.4 to a family of Hilbert spaces $(L^2(W_x,S))_{x\in V}$. □

4.7. COROLLARY. *Let* (W,p) *be as in 4.5 and* $N = V \times_\Gamma \tilde{M}$ *the foliated bundle over* M *with* $\pi: N \longrightarrow M$ *the projection map. Suppose that the* S^1*-action on* $\tilde{M}$ *is locally free. Then for any secondary characteristic class* ω *of the foliated bundle* N *we obtain*

$$\langle \omega\pi^*\hat{\mathfrak{A}}(M), [N]\rangle = 0$$

where $\hat{\mathfrak{A}}(M)$ *is the total* $\hat{A}$*-class of the* $spin^c$ *manifold* M.

Proof. We first assume that M is even-dimensional. Let $(\mathcal{E}_{W,S},F)$ be the Kasparov module defined in 3.9 from (W,p) and $\xi \in K_0[C_0(\tau) \rtimes \Gamma]$ the element defined by the principal symbol of F as in 3.10. Let $x \in K^0(V,\Gamma)_0$ denote the element defined by (W,P) and ξ. By Remark 3.10 we have

$$\mu_0(x) = p_! \rtimes \Gamma.$$

Let ω_0 be the universal secondary characteristic class for $V \times_\Gamma E\Gamma$ whose pull back to N by φ is a given ω. We apply Theorem 2.6 to the above. Then it follows

$$\begin{aligned} 0 &= \phi_\omega \mu_0(x) \\ &= \langle \omega_0, \Phi ch_0(x)\rangle \\ &= \langle \omega_0, \Phi\varphi_*\{ch(u_\xi)td(\tilde{\tau}\otimes\mathbb{C})\cap[TN]\}\rangle \\ &= \langle \omega td(\tilde{\tau})^{-1}td(\tilde{\tau}\otimes\mathbb{C}), [N]\rangle \\ &= \langle \omega\pi^*\hat{\mathfrak{A}}(M), [N]\rangle \end{aligned}$$

by Proposition 2.13 2) and Theorem 4.6.

When M is odd-dimensional, we take $M \times M$ instead of M. Applying the above argument to this, we obtain

$$\begin{aligned} 0 &= \langle (\omega\times\omega)(\pi\times\pi)^*\hat{\mathfrak{A}}(M\times M), [N\times N]\rangle \\ &= \langle \omega\pi^*\hat{\mathfrak{A}}(M), [N]\rangle^2. \end{aligned}$$ □

Then Theorem A in the introduction is a direct corollary of the above since the assumption in Theorem A implies that there is a lifted locally free S^1-action on $\tilde{M}$.

If we use the signature operator instead of the Dirac operator, we also obtain a similar results to the above. We conclude this section by stating this result since the proof is completely same as in 4.7; see also [11].

4.8. THEOREM. *Let* V *be a closed manifold diffeomorphisms. Let* M *be a closed manifold and* $\tilde{M}$ *a principal* Γ*-bundle over* M. *Let* $N = V \times_{\Gamma} \tilde{M}$ *with* $\pi\colon N \longrightarrow M$ *the projection map as above. Suppose that there is a locally free* S^1*-action on* $\tilde{M}$ *commuting with* Γ. *Then for any secondary characteristic class* ω *of the foliated bundle* N *we obtain*

$$\langle \omega\pi^*\mathfrak{L}(M), [N]\rangle = 0$$

where $\mathfrak{L}(M)$ *is the total* L*-class of* M *corresponding to the formal power series* $x/\tanh \frac{x}{2}$. □

REFERENCES

1. P. Baum and A. Connes, "Geometric K-theory for Lie groups and foliations," preprint (1982).

2. P. Baum and A. Connes, "Chern character for discrete groups," A Fête of Topology, Academic Press, 1988.

3. P. Baum and R.G. Douglas, "Relative K homology and C*-algebras," preprint (1988).

4. R. Bott, "On some formulas for the characteristic classes of group actions," Lecture Note in Math. vol. 652, Springer-Verlag, Berlin-Heidelberg-New York, 1978, 25-61.

5. B. Blackadar, K theory for operator algebras, MSRI Publications 5, Springer-Verlag, New York, 1986.

6. A. Connes, "A survey of foliations and operator algebras," Operator algebras and applications, Proc. Symp. Pure Math. 38 (1982), Part I, 521-628.

7. A. Connes, "Non-commutative differential geometry," Chapter I and II, Publ. Math. IHES 62 (1986), 257-360.

8. A. Connes, "Cyclic cohomology and the transversal fundamental class of a foliation," Geometric methods in operator algebras, Pitman research notes in mathematics series 123 (1986), 52-144.

9. A. Connes and G. Skandalis, "The longitudinal index theorem for foliations," Publ. Res. Inst. Math. Sci. Kyoto Univ. 20 (1984), no. 6, 1139-1183.

10. M. Gromov and H.B. Lawson Jr., "Positive scalar curvature and the Dirac operator on complete riemannian manifolds," Publ. Math. IHES 58 (1983), 83-196.

11. A. Haefliger, "Differentiable cohomology," CIME (1976), 21-70.

12. H. Moriyoshi, "Chern character for proper Γ-manifolds," preprint (1988).

13. E. Witten, "Supersymmetry and Morse theory," J. Diff. Geom. 17 (1982), 661-692.

DEPARTMENT OF MATHEMATICS
FACULTY OF SCIENCE
UNIVERSITY OF TOKYO
HONGO, TOKYO, 113 JAPAN

Current Address:
Department of Mathematics
The Pennsylvania State University
University Park, PA 16802

Contemporary Mathematics
Volume 105, 1990

L^2-INDEX THEORY, ETA INVARIANTS AND VALUES OF L-FUNCTIONS

Werner Müller

0. INTRODUCTION. The purpose of this paper is to discuss some aspects of L^2-index theory on noncompact manifolds which are related to locally symmetric spaces of finite volume.

Many analysts have produced generalizations of the Atiyah-Singer index theorem to noncompact manifolds of various sorts, and these have been successfully employed in diverse fields such as number theory, representation theory, geometry of foliations, study of scalar curvature and complex analysis (see [Co], [C-M], [C-S], [A-D-S], [A-S] , [G-L], [M2], [R]). We do not intend to give a survey of all these aspects of the theory. Our main concern is that part which is related to number theory. This brings together the three subjects mentioned in the title and it has its origin in the work of Hirzebruch [Hi] and Atiyah, Patodi and Singer [A-P-S].

Hirzebruch [Hi] made an extensive study of the "cusps" of Hilbert modular surfaces for real quadratic fields. He attached to each cusp α a signature defect $\delta(\alpha)$ which is the correction term due to a cusp in the Hirzebruch signature formula of a 4-manifold. By explicit computations he was able to show that

$$\delta(\alpha) = L(\mathbf{M}, \mathbf{V}, 0) \tag{0.1}$$

where $L(\mathbf{M},\mathbf{V},0)$ is the special value of a certain L-function $L(\mathbf{M},\mathbf{V},s)$ attached to a real quadratic field F. Here $\mathbf{M}$ is an abelian subgroup of rank 2 in F and $\mathbf{V}$ is a group of totally positive units acting on $\mathbf{M}$. The L-function is defined as

$$L(\mathbf{M}, \mathbf{V}, s) = \sum_{(\mathbf{M}-0)/\mathbf{V}} \frac{\operatorname{sign} N(\mu)}{|N(\mu)|^s}, \quad \operatorname{Re}(s) > 1, \tag{0.2}$$

1980 Mathematics Subject Classification (1985 Revision): 58G10, 11M41

where $N(\mu) = \mu\mu'$ denotes the norm of μ. Hirzebruch then conjectured that (0.1) might continue to hold for totally real number fields of any degree. Motivated in part by this conjecture Atiyah, Patodi and Singer studied in [A-P-S] Hirzebruch's result (0.1) in the wider context of Riemannian geometry. They introduced a differential geometric analogue of the L-function (0.2) which is defined in terms of eigenvalues λ of a certain first order selfadjoint differential operator A on an odd-dimensional manifold by the series

$$\eta(s) = \sum_{\lambda\neq 0} (\operatorname{sign}\lambda)|\lambda|^{-s} . \tag{0.3}$$

This is the so-called eta function of the differential operator A. One of the main results of [A-P-S] is an extension of Hirzebruch's signature formula to a $4k$-dimensional compact oriented Riemannian manifold X with boundary Y. Under the assumption that the metric of X is a product near the boundary one has

$$\operatorname{Sign}(X) = \int_X L_k(p) - \eta(0) \tag{0.4}$$

where $L_k(p)$ is the k-th Hirzebruch L-polynomial in the Pontrjagin forms of the Riemannian metric of X and $\eta(0)$ is the special value of the eta function $\eta(s)$ of Y. This formula was obtained from an index formula for a certain boundary value problem for the signature operator on X. It can be also interpreted as L^2-index formula for the signature operator on the non-compact manifold $\hat{X} = X \cup (\mathbb{R}^+ \times Y)$. This formula has been rederived in the L^2-framework by many authors [C1], [M1], [Br](cf. §1). It appears as natural signature formula for manifolds with ends which are warped products.

Formula (0.4) is analogous to (0.1) and the strategy of Atiyah, Patodi and Singer to prove Hirzebruch's conjecture was to apply (0.4) to manifolds which occur as boundaries of neighborhoods of the cusps of Hilbert modular varieties. A proof of Hirzebruch's conjecture along these lines was given by Atiyah, Donnelly and Singer in [A-D-S1], [A-D-S2]. An independent proof was given by the author in [M2,Ch.XII]. Partial results were obtained in [M1]. Our starting point was a formula of Shimizu [Sh] for the dimension of the space of cusp forms of Hilbert modular varieties. One contribution in Shimizu's formula is given by the same special values of L-functions which occur in Hirzebruch's conjecture (cf. Theorem 11 in [Sh]). On the other hand, it is known that the dimension of the space of cusp forms of a given weight for a Hilbert modular variety $\Gamma\backslash H^n$ is the same as the multiplicity of a certain discrete series representation occurring in $L^2(\Gamma\backslash SL(2,\mathbb{R})^n)$ and, for sufficiently high weight, this multiplicity coincides with the L^2-index of some twisted Dirac operator on $\Gamma\backslash H^n$ (cf. [Mo, Theorem 3.2]). Thus, for sufficiently high weight, the formula of Shimizu can be interpreted as L^2-index formula for twisted Dirac operators. This suggests that a similar formula might be true for the signature operator. Using Selberg's trace formula we established in [M1] such a formula which gives the following signature theorem for a Hilbert modular variety $\Gamma\backslash H^n$

$$\text{Sign}(\Gamma\backslash H^n) = \sum_{j=1}^{r} \delta(z_j) + \sum_{i=1}^{p} L(\mathbf{M}_i^*, \mathbf{V}_i, 0) . \tag{0.5}$$

Here $z_1, \ldots, z_r$ are the quotient singularities of $\Gamma\backslash H^n$ and $\delta(z_i)$ is the signature defect of z_i , p is the number of cusps of $\Gamma\backslash H^n$ and $L(\mathbf{M}_i^*, \mathbf{V}_i, 0)$ is the special value of an L-function attached to the i-th cusp α_i (we have used the functional equation to replace the special values at s=1 by those at s=0). If we compare (0.5) with the signature formula of Hirzebruch [Hi, p.228] then we obtain a weaker form of (0.1)

$$\sum_{i=1}^{p} \delta(\alpha_i) = \sum_{i=1}^{p} L(\mathbf{M}_i^*, \mathbf{V}_i, 0).$$

Now the obvious thing to do is to chop off a single cusp of $\Gamma\backslash H^n$, glue it to a compact manifold with the same boundary (this is always possible because the boundary is a compact solvemanifold and therefore its tangent bundle admits a framing) and prove an L^2-index formula for the signature operator on this manifold. This idea has been developed in [M2] in a wider context and a proof of Hirzebruch's conjecture based on this approach is given in Chapter XII of [M2].

In [M2] we consider complete manifolds with a finite number of ends which are generalizations of "cusps" of $\mathbb{Q}$-rank one locally symmetric spaces. On such manifolds X we consider Dirac type operators. It turns out that each Dirac type operator $D_+ : C^\infty(X, S_+) \longrightarrow C^\infty(X, S_-)$ has a finite L^2-index defined by

$$L^2\text{-Ind}(D_+) = \dim(\ker(D_+) \cap L^2) - \dim(\ker(D_+^*) \cap L^2)$$

and we prove an index formula. In section 2,3 and 4 we recall the main results. Actually, the index formula obtained in [M2] is only a preliminary version in the general case because it contains a contribution (the unipotent term) which is not yet in a satisfactory form. We treat this problem in section 4 and, under a mild restriction on the cusp, we obtain a final index formula. The main contribution of the cusp to the index formula is given by a special value $L(0)$ of some L-function attached to the cusp. In the Hilbert modular case this is an L-function attached to a totally real number field similar to (0.2). In section 5 we investigate these L-functions more closely and it turns out that they are L-functions attached to prehomogeneous vector spaces in the sense of Sato and

Shintani [Sa], [S-S].

We observe that all $\mathbb{Q}$-rank one locally symmetric spaces are among the manifolds we are considering. Thus our index formula contains in particular an index formula for twisted Dirac operators on $\mathbb{Q}$-rank one locally symmetric spaces. Furthermore, it is known that the index of any locally invariant elliptic differential operator on such a manifold can be expressed through indices of certain twisted Dirac operators [B-M]. We also note that M.Stern [St] has obtained an L^2-index formula for the signature operator on $\mathbb{Q}$-irreducible Hermitian locally symmetric spaces of any rank. In his formula the contribution of the cusps is not given in terms of L-functions. The expressions he obtains seem to be related to the formulas appearing in the adiabatic limits of eta invariants [B-C1], but they can be certainly converted into Sato-Shintani L-functions.

In section 6 we specialize to the signature operator and in section 7 we assume in addition that the cusp of our manifold X is a cusp of a Hermitian locally symmetric space of $\mathbb{Q}$-rank one. Then we can employ the Baily-Borel compactification of the locally symmetric space to obtain a compactification $\bar{X}$ of our manifold X. The cross section N of the cusp is a fibration

$$(0.6) \qquad \Gamma \cap U\backslash U \longrightarrow N \longrightarrow \Gamma_M\backslash X_M$$

over a compact locally symmetric space with fibre a compact nilmanifold. For simplicity we assume that the fibre $\Gamma \cap U\backslash U$ is a torus. Then we get the following signature formula:

$$(0.7) \qquad \mathrm{Sign}_{IH^*}(\bar{X}) = \int_X L_k(p) + \mathrm{Vol}(\Gamma \cap U\backslash U)L(0) .$$

Here $\mathrm{Sign}_{IH^*}(\bar{X})$ means the signature of $\bar{X}$ with respect to middle intersection cohomology and $L(s)$ is an L-function attached to the torus bundle (0.6). The difference

$$\mathrm{Sign}_{IH^*}(\bar{X}) - \int_X L_k(p)$$

can be interpreted as differential geometric signature defect so that (0.7) generalizes Hirzebruch's result to some extent and puts it in a larger context. In particular, if we are dealing with tube domains then the singular stratum of $\bar{X}$ is a single point and in this case we have $\mathrm{Sign}_{IH^*}(\bar{X}) = \mathrm{Sign}(X)$. The L-functions can also be described explicitly which provides us with examples different from the Hilbert modular case.

In the final section 8 we discuss the relation with the work of Bismut and Cheeger on adiabatic limits of eta invariants [C2],[B-C1]. The metric on the cusp provides us with a family of metrics $\{g_t\}$, $t \geq 1$, on the torus bundle (0.6). Then the limit

$$\lim_{t \to \infty} \eta(N, g_t) \tag{0.8}$$

of the eta invariants of N with respect to the metric g_t exists and coincides with the adiabatic limit considered in [B-C1]. Furthermore, the special value of the L-function occurring in (0.7) is closely related to the adiabatic limit (0.8). In the special case of a cusp comming from an arithmetic quotient of a tube domain we obtain equality

$$\lim_{t \to \infty} \eta(N, g_t) = \mathrm{Vol}(\Gamma \cap U \backslash U) L(0). \tag{0.9}$$

For 2-torus bundles over the circle which arise as cross sections of cusps of Hilbert modular surfaces we compute the left hand side of (0.9) by using the formula obtained by Cheeger in [C2]. Then (0.9) reduces to a formula which is due to Hecke [He] and has also other interesting interpretations (cf. [A]).

1. ETA INVARIANTS. In this section we briefly recall the index theorem of Atiyah, Patodi and Singer [A-P-S] and we restate it in different ways.

Let W be a compact C^∞ manifold with smooth boundary $\partial W = N$ and let

$$D\colon C^\infty(W, E) \longrightarrow C^\infty(W, F)$$

be a first order elliptic differential operator on W which, in a neighborhood $[0, \varepsilon) \times N$ of the boundary, takes the special form

$$D = \sigma\left(\frac{\partial}{\partial u} + A\right) \tag{1.1}$$

where u is the inward normal coordinate, $\sigma\colon E|N \longrightarrow F|N$ a bundle isomorphism and A a selfadjoint operator N with respect to a given hermitian inner product on $E|N$ and a Riemannian metric on N. Let P_+ (resp. P_-) denote the orthogonal projection operator onto the direct sum of eigenspaces of A corresponding to non-negative (resp. negative) eigenvalues of A. Let $C^\infty(W, E; P_+)$ denote the space of C^∞ sections Φ of E such that

$$P_+(\Phi|N) = 0. \tag{1.2}$$

Then $D\colon C^\infty(W, E; P_+) \longrightarrow C^\infty(W, F)$ has a finite index given by

$$\mathrm{Ind}(D) = \int_W \omega - \bar{\eta}(0) \tag{1.3}$$

where ω is the local index density [A-B-P] and $\bar{\eta}(0)$ is the reduced eta invariant of A. It is related to the usual eta invariant $\eta(0)$ by

$$\bar{\eta}(0) = \frac{1}{2}(\eta(0) + \dim\ker A)$$

and $\eta(0)$ is the special value of the eta function (0.3) where λ runs over the eigenvalues of A. In the case of the signature operator

$$D_S\colon \Lambda_+^*(W) \longrightarrow \Lambda_-^*(W)$$

on a 4k-dimensional oriented Riemannian manifold W which is isometric to a product near the boundary the index formula (1.3) reduces to the signature formula (0.4). The operator A occurring in (1.1) acts on $\Lambda^*(N)$ and is given by

$$A\Phi = (-1)^{k+p+1}(\varepsilon * d - d *)\Phi$$

where Φ is either a 2p-form ($\varepsilon=1$) or a (2p-1)-form ($\varepsilon=-1$). Moreover, A preserves the parity of forms and commutes with $\Phi \longrightarrow (-1)^p * \Phi$ so that A splits as $A = A^{ev} \oplus A^{odd}$ and A^{ev} is isomorphic to A^{odd}. The eta invariant in (0.4) is then the eta invariant of A^{ev}.

It was observed in [A-P-S] that the index formula (1.3) has an alternative description as L^2-index formula on the noncompact manifold

$$X = W \cup (\mathbb{R}^+ \times N).$$

We extend the Riemannian metric on $[0,\infty) \times N$ to X so that $\mathbb{R}^+ \times N$ is equipped with the product metric. The vector bundles E and F extend in the obvious way to vector bundles $\mathcal{E}$ and $\mathcal{F}$ over X and we equip these vector bundles with inner products which on $\mathbb{R}^+ \times N$ are the natural inner products induced from $E|N$ and $F|N$. Because of (1.1) we can also extend D to an elliptic differential operator

$$(1.4) \qquad \mathcal{D}\colon C^\infty(X,\mathcal{E}) \longrightarrow C^\infty(X,\mathcal{F}).$$

One defines the L^2-index of $\mathcal{D}$ by

$$(1.5) \qquad L^2\text{-Ind}(\mathcal{D}) = \dim(\ker(\mathcal{D}) \cap L^2) - \dim(\ker(\mathcal{D}^*) \cap L^2)\ .$$

Equivalently, we may take $\bar{\mathcal{D}}$, the closure in L^2 of $\mathcal{D}$ restricted to compactly supported C^∞ sections. Then

$$L^2\text{-Ind}(\mathcal{D}) = \dim(\ker(\bar{\mathcal{D}})) - \dim(\ker(\bar{\mathcal{D}}^*)).$$

where $\bar{\mathcal{D}}^*$ is the adjoint of $\bar{\mathcal{D}}$ in the Hilbert space sense which is the same as the closure in L^2 of $\mathcal{D}^*$. It was proved in [A-P-S, Corollary 3.14] that

(1.6) $$\operatorname{Ind}(D) = L^2\text{-}\operatorname{Ind}(\mathcal{D}) - h_\infty(F)$$

where $h_\infty(F)$ is the dimension of the subspace of $\ker(A)$ consisting of limiting values of extended L^2 sections φ of F satisfying $\mathcal{D}^*\varphi = 0$ (cf. [A-P-S] p.58). In particular, if $\ker(A) = 0$, we have

$$\operatorname{Ind}(D) = L^2\text{-}\operatorname{Ind}(\mathcal{D}) .$$

Let $h_\infty(E)$ be the dimension of the subspace of $\ker(A)$ consisting of limiting values of extended L^2 sections φ of E satisfying $\mathcal{D}\varphi = 0$. Then we can rewrite (1.3) as

(1.7) $$L^2\text{-}\operatorname{Ind}(\mathcal{D}) = \int_X \omega - \frac{\eta(0)}{2} - \frac{1}{2}(h_\infty(E) - h_\infty(F))$$

where $\eta(0)$ is the eta invariant of A. In [M4] we have given an interpretation of $h_\infty(E) - h_\infty(F)$ in terms of scattering theory. Let H be the closure in L^2 of $\mathcal{D}^*\mathcal{D}$ acting in $C_c^\infty(X,E)$. Furthermore, let H_o be the selfadjoint extension of $-\partial^2/\partial u^2 + A^2$, acting in $C_c^\infty(\mathbb{R}^+ \times N, E)$, obtained by imposing Dirichlet boundary conditions at $\{0\} \times N$. Associated to (H, H_o) there is a scattering operator S with scattering matrix $S(\lambda)$. The properties of the scattering matrix imply that $S(0)$ is a linear operator in $\ker(A)$ and

$$\operatorname{Tr} S(0) = h_\infty(E) - h_\infty(F).$$

Unfortunately, one does not know how to compute this term in general. However, if $\ker(A) = 0$, it vanishes automatically and the index formula (1.7) simplifies to

(1.8) $$L^2\text{-}\operatorname{Ind}(\mathcal{D}) = \int_X \omega - \frac{\eta(0)}{2} .$$

In [C1], Cheeger gave an alternative proof of the signature formula for a manifold with boundary using the heat expansion on spaces with conical singularities. He attaches a cone $C(N)$ to the boundary N of W, so that $W' = W \cup C(N)$ is a manifold with a conical singularity. Let r be the radial coordinate on $C(N)$ and g^N the metric on N. Then the cone $C(N)$ is equipped with the metric

$$dr^2 + r^2 g^N .$$

This metric is then extended to a smooth metric on W'. Assume that W' is of dimension $4k$. Applying the heat equation method to the calculation of the signature of W', Cheeger obtains the following signature formula

$$\text{Sign}(W') = \int_{W'} L_k(p) - \eta(0) \tag{1.9}$$

which is equivalent to (0.4) because $\text{Sign}(W') = \text{Sign}(W)$ and $L_k(p) \equiv 0$ on the cone $C(N)$. By the same approach one can derive the index formula (1.3) for twisted Dirac operators (cf. [B-C2]).

In [M3] it was shown that one can also work on manifolds with cusps. In this case we equip the cylinder $\mathbb{R}^+ \times N$ with the metric

$$dr^2 + e^{-2r} g^N$$

and glue it to W. The resulting manifold W' is a manifold with a cusp. Again we can apply the heat equation method to calculate the signature and we obtain a formula similar to (1.9).

J.Brüning [Br] has treated general warped products. Consider a complete Riemannian manifold X and assume that there exists an open subset U of X such that $W = X - U$ is a compact manifold with boundary N and U is isometric to $\mathbb{R}^+ \times N$ with metric

$$dr^2 + f(r)^2 g^N \tag{1.10}$$

where g^N is again the metric on N. The metric (1.10) is a warped product metric. On such manifolds Brüning considers a class of first order elliptic differential operators which includes all operators of "geometric origin". Under certain assumptions on f and the elliptic operator he proves an L^2-index formula. The main contribution of the noncompact part is again given by an eta invariant.

Thus the index theorem of Atiyah, Patodi and Singer has various interpretations as L^2-index theorem for elliptic operators on certain noncompact manifolds which have a finite number of ends and each end is equipped with a warped product metric.

2. MANIFOLDS WITH CUSPS OF RANK ONE. In this section we consider complete Riemannian manifolds with a more complicated structure of the metric on the ends. Each manifold has a finite number of ends, but they are now bundles of cusps. All locally symmetric spaces of $\mathbb{Q}$-rank one have this structure. Manifolds with cusps of rank one have been considered in [M2] and we recall some facts from this paper. For all details we refer to [M2].

Let G be a connected noncompact semisimple Lie group with finite center and fix a maximal compact subgroup K of G. The Lie algebras of G and K will be denoted by $\mathfrak{g}$ and $\mathfrak{k}$ respectively. Let $\mathfrak{g} = \mathfrak{k} \oplus \mathfrak{p}$ be the Cartan decomposition, Θ the Cartan involution of $\mathfrak{g}$ (or of G) with respect to K, $B(.,.)$ the Cartan-Killing form of $\mathfrak{g}$ and $(.,.)_\Theta$ the inner product on $\mathfrak{g}$ defined by $(X_1,X_2)_\Theta = -B(X_1,\Theta(X_2))$, $X_1,X_2 \in \mathfrak{g}$. Put

$$\tilde{Y} = G/K.$$

Since the inner product $(.,.)_\Theta$ is Ad(K) invariant, by translation we obtain a G-invariant Riemannian metric on $\tilde{Y}$. From now on we shall assume that $\tilde{Y}$ is equipped with this metric. Let P be a split rank one parabolic subgroup of G with split component A and corresponding Langlands decomposition

$$P = UAM. \tag{2.1}$$

U is the unipotent radical of P, A is a $\mathbb{R}$-split torus of dimension one and M centralizes A. We always assume that the split component A is Θ-stable. By $\mathfrak{a}$, $\mathfrak{m}$ and $\mathfrak{u}$ we shall denote the Lie algebras of A, M and U respectively. Set $S = UM$ and let Γ be a discrete uniform torsion free subgroup of S. Then the manifold

$$Y = \Gamma\backslash\tilde{Y} = \Gamma\backslash G/K \tag{2.2}$$

is a (complete) cusp of rank one. Let

$$K_M = M \cap K .$$

K_M is a maximal compact subgroup of M. Put

$$X_M = M/K_M . \tag{2.3}$$

There is a canonical diffeomorphism

$$\tilde{\xi}: \mathbb{R}^+ \times U \times X_M \xrightarrow{\ \simeq\ } \tilde{Y} \tag{2.4}$$

Given $t \geq 0$, set

$$\tilde{Y}_t = \tilde{\xi}([t,\infty) \times U \times X_M) . \tag{2.5}$$

$\tilde{Y}_t$ is invariant under Γ and we set

$$Y_t = \Gamma\backslash\tilde{Y}_t . \tag{2.6}$$

Each manifold of this type will be called a cusp of rank one. We describe its structure more explicitly. Let

$$\Gamma_M = M \cap (U\Gamma). \tag{2.7}$$

Γ_M is a discrete subgroup of M, $\Gamma_M\backslash M$ and $\Gamma\cap U\backslash U$ are compact and the sequence $1 \longrightarrow \Gamma\cap U \longrightarrow \Gamma \to \Gamma_M \longrightarrow 1$ is exact. Set $Z = S/S\cap K$. There is a canonical fibration

$$p\colon \Gamma\backslash Z \longrightarrow \Gamma_M\backslash X_M \tag{2.8}$$

with fibre $\Gamma\cap U\backslash U$ which is a compact nilmanifold. Furthermore, there is a canonical diffeomorphism

$$\xi\colon [t,\infty)\times \Gamma\backslash Z \xrightarrow{\sim} Y_t \ . \tag{2.9}$$

Next we describe the metric on the cusp Y. Let λ be the unique simple root of (P,A). Then $\mathfrak{u}$ admits a direct sum decomposition

$$\mathfrak{u} = \mathfrak{u}_\lambda \oplus \mathfrak{u}_{2\lambda}$$

where $\mathfrak{u}_\alpha = \{X\in\mathfrak{u} \mid [H,X] = \alpha(H)X,\ H\in\mathfrak{a}\}$. Let h_λ (resp. $h_{2\lambda}$) be the right invariant inner product on $\mathfrak{u}$ which is zero on $\mathfrak{u}_{2\lambda}$ (resp. $\mathfrak{u}_\lambda$) and equal to $(.,.)_\Theta$ on $\mathfrak{u}_\lambda$ (resp. $\mathfrak{u}_{2\lambda}$). For $x = mK_M$, $m\in M^o$, and $\alpha\in\{\lambda,2\lambda\}$ set $du_\alpha^2(x)=\frac{1}{2}(\mathrm{Int}(m))^*h_\alpha$. If g denotes the G-invariant metric on $\tilde{Y}$ defined above then

$$\xi^*g = dr^2 + dx^2 + e^{-2br}du_\lambda^2(x) + e^{-4br}du_{2\lambda}^2(x) \tag{2.10}$$

where $b = |\lambda|$, dx^2 is the invariant metric on X_M induced by the restriction of the Killing form and ξ is the diffeomorphism (2.4). This is also locally the description of the metric on the cusp. The volume element is given by

$$d\mathrm{Vol} = r^{-(m+1)}dr\, d\mathrm{Vol}(x)\wedge d\mathrm{Vol}(u) \tag{2.11}$$

where $m = \dim(\mathfrak{u}_\lambda)|\lambda| + 2\dim(\mathfrak{u}_{2\lambda})|\lambda|$. Note that $d\mathrm{Vol}(u)$ is independent of x.

<u>REMARKS</u>. 1) Let $x\in\Gamma_M\backslash X_M$ and let $p^{-1}(x)$ be the fibre over x of (2.8). The restriction of the metric (2.10) to $\mathbb{R}^+\times p^{-1}(x)$ is the metric on a cusp of a $\mathbb{R}$-rank one locally symmetric space of finite volume. In this sense we may regard the cusp $Y = \Gamma\backslash G/K$ as a bundle of $\mathbb{R}$-rank one cusps over the compact locally symmetric space $\Gamma_M\backslash X_M$.

2) Assume that N is commutative, that is, $\mathfrak{u}_{2\lambda} = 0$. Then $p\colon \Gamma\backslash Z \longrightarrow \Gamma_M\backslash X_M$ is a torus bundle. Fix $r\in\mathbb{R}^+$ and let g_r be the metric on the slice $\{r\}\times\Gamma\backslash Z$ induced by (2.10). Equipped with g_r, $p\colon \Gamma\backslash Z \longrightarrow \Gamma_M\backslash X_M$ is a Riemannian submersion and g_r, $r\in\mathbb{R}^+$, is a family of metrics on $\Gamma\backslash Z$ similar to the families considered in the adiabatic approximation (cf. [C2]). We shall return to this viewpoint in section 8.

DEFINITION. *A complete Riemannian manifold* X *is called a manifold with cusps of rank one if* X *admits a decomposition* $X = X_o \cup X_1 \cup \ldots \cup X_h$ *satisfying:*
(i) X_o *is a compact manifold with boundary. If* $i,j \geq 1$ *and* $i \neq j$ *, then* $X_i \cap X_j = \emptyset$*. Moreover* $X_o \cap X_i = \partial X_i$ *for* $i \geq 1$.
(ii) For each i *,* $1 \leq i \leq h$ *,* X_i *is a cusp of rank one.*

Note that each locally symmetric space of $\mathbb{Q}$-rank one admits such a decomposition. For simplicity we shall assume that the manifolds we are considering have a single cusp. Such a manifold X has a decomposition

$$X = W \cup Y_1$$

where W is a compact manifold with boundary and Y_1 is a cusp of rank one.

3. DIRAC-TYPE OPERATORS. Let X be a manifold with a cusp of rank one. In this section we consider Dirac-type operators on X which are locally invariant on the cusp. All differential operators of geometric origin such as the signature operator or twisted Dirac operators are contained in this class. We shall review some facts about analysis of these operators. More details can be found in [G-L], [M2].

A vector bundle $\mathcal{E}$ over $Y = \Gamma\backslash G/K$ is called locally homogeneous if there exists a homogeneous vector bundle $\tilde{\mathcal{E}}$ over G/K such that $\mathcal{E} = \Gamma\backslash\tilde{\mathcal{E}}$. A Hermitian metric $h^{\mathcal{E}}$ in $\mathcal{E}$ is called locally invariant if $h^{\mathcal{E}}$ is obtained by pushing down a G-invariant Hermitian metric in $\tilde{\mathcal{E}}$. The same applies to connections in $\mathcal{E}$. We shall assume that all vector bundles E over X are locally homogeneous on the cusp, that is, there exists a locally homogeneous vector bundle $\mathcal{E}$ over Y such that $E|Y_1$ and $\mathcal{E}|Y_1$ are isomorphic through a bundle map which covers the identity of Y_1. We also assume that all Hermitian metrics and connections in E are such that their restriction to $E|Y_1$ are obtained from locally homogeneous metrics or connections in $\mathcal{E}$.

Let $\mathrm{Cliff}^c(X)$ be the complexified Clifford algebra bundle on X. There is a canonical embedding $T(X) \subset \mathrm{Cliff}^c(X)$. Furthermore, $\mathrm{Cliff}^c(X)$ is equipped with a natural connection ∇ and Hermitian metric extending the Riemannian connection and metric on $T(X)$. ∇ preserves the metric and satisfies

$$\nabla(\varphi\cdot\psi) = (\nabla\varphi)\cdot\psi + \varphi\cdot\nabla\psi$$

for all sections $\varphi, \psi \in C^\infty(\mathrm{Cliff}^c(X))$. Now suppose that $S \longrightarrow X$ is a vector bundle which is a bundle of left modules over the bundle of algebras $\mathrm{Cliff}^c(X)$. S will be called a <u>Clifford bundle</u> over X if it is equipped with a Hermitian metric and compatible connection such that

(i) For each unit vector $e \in T_x(X)$, the module multiplication $e: S_x \longrightarrow S_x$ is an isometry for each $x \in X$.

(ii) For all $\varphi \in C^\infty(\mathrm{Cliff}^c(X))$ and $s \in C^\infty(S)$, we have $\nabla(\varphi s) = \varphi \nabla s + (\nabla\varphi)s$.

On a Clifford bundle S there is a natural first-order differential operator $D: C^\infty(S) \longrightarrow C^\infty(S)$, called the Dirac operator of S. D is the composition

$$C^\infty(S) \xrightarrow{\nabla} C^\infty(T^*(X) \otimes S) \longrightarrow C^\infty(T(X) \otimes S) \longrightarrow C^\infty(S).$$

Here the second arrow is defined by the metric and the third one by Clifford multiplication. If $x \in X$ and $\{e_1, \ldots e_n\}$ is an orthonormal basis for $T_x(X)$ then Ds is given by

$$(3.1) \qquad (Ds)(x) = \sum_{k=1}^{n} e_k \cdot (\nabla_k s)(x).$$

A (generalized) Dirac operator $D: C^\infty(S) \longrightarrow C^\infty(S)$ has the following properties

(1) If we regard D as an unbounded operator in $L^2(S)$ with domain $C_c^\infty(S)$, then D is essentially selfadjoint.

(2) Let $\bar{D}$ be the closure of D in $L^2(S)$. Since D is elliptic we have by elliptic regularity that

$$\ker(\bar{D}) = \{\Phi \in C^\infty(S) \mid D\Phi = 0,\ \Phi \in L^2(S)\}.$$

Moreover

$$(3.2) \qquad \dim(\ker(\bar{D})) < \infty$$

(3) The spectrum of $H = \bar{D}^2$ consists of a point spectrum and an absolutely continuous spectrum. Let $L^2(S) = L_d^2(S) \oplus L_c^2(S)$ be the decomposition into the subspace $L_d^2(S)$ which is spanned by the eigenvectors of H and the absolutely continuous subspace $L_c^2(S)$. Let $H_d = H | L_d^2(S)$. Then $\exp(-tH_d)$ is of the trace class for each $t>0$. In particular, all eigenvalues of H have finite multiplicity and infinity is the only possible point of accumulation of the eigenvalue sequence.

Assume that X is of dimension 21 and let $\omega \in C^\infty(\mathrm{Cliff}^c(X))$ be defined by

$$\omega = i^l e_1 \cdots e_{2l}$$

where $e_1, \ldots, e_{2l}$ is a local tangent frame field. ω has the following properties: (i) $\omega^2 = 1$, (ii) $\nabla\omega = 0$, (iii) $\omega e = -e\omega$, $e \in T(X)$. This implies that S admits a parallel orthogonal spliting

$$S = S_+ \oplus S_-$$

where $S_\pm$ are the ± 1-eigenbundles for left multiplication by ω. By restriction we obtain a pair of elliptic operators

$$D_\pm : C^\infty(S_\pm) \longrightarrow C^\infty(S_\mp)$$

which are formal adjoint of one another. Let $\bar{D}_\pm$ be the closure of $D_\pm$ in L^2. Then $\bar{D} = \bar{D}_+ \oplus \bar{D}_-$. In view of property (3.2), $\ker(\bar{D}_+)$ and $\ker(\bar{D}_-)$ are finite-dimensional and we may define the L^2-index of D_+ by

(3.3) $$L^2\text{-Ind}(D_+) = \dim(\ker(\bar{D}_+)) - \dim(\ker(\bar{D}_-)).$$

Note that

(3.4) $$\ker(\bar{D}_\pm) = \ker(D_\pm) \cap L^2(S_\pm) .$$

4. <u>THE L^2-INDEX OF DIRAC OPERATORS</u>. In [M2, Theorem 10.32] we have derived a preliminary version of an index formula for Dirac operators as above. This index formula contains one term, the so-called unipotent term $\mathcal{U}$, which is not yet in a satisfactory form in the general case. In the present section we shall continue with the investigation of the unipotent term and bring the index formula to its final form. First we recall the index formula derived in [M2]. Let

$$D_+ : C^\infty(X, S_+) \longrightarrow C^\infty(X, S_-)$$

be a Dirac operator as in section 3. According to our assumption, there exist locally homogeneous vector bundles $\mathcal{S}_\pm$ over Y such that $S_\pm|Y_1 = \mathcal{S}_\pm|Y_1$, Y_1 being the cusp of X. $\mathcal{S}_+$ and $\mathcal{S}_-$ are obtained from homogeneous vector bundles $\tilde{\mathcal{S}}_+$ and $\tilde{\mathcal{S}}_-$ over G/K by $\mathcal{S}_\pm = \Gamma\backslash\tilde{\mathcal{S}}_\pm$. The isotropy representations of $\tilde{\mathcal{S}}_\pm$ will be denoted by $\sigma^\pm : K \longrightarrow GL(V_\pm)$. There is also a locally invariant elliptic differential operator

(4.1) $$\mathcal{D}_+ : C^\infty(Y, \mathcal{S}_+) \longrightarrow C^\infty(Y, \mathcal{S}_-)$$

which coincides with D_+ on Y_1. Note that the lift $\tilde{\mathcal{D}}_+ : C^\infty(\tilde{\mathcal{S}}_+) \rightarrow C^\infty(\tilde{\mathcal{S}}_-)$ of $\mathcal{D}_+$ commutes with the action of G. Next recall that a section $\varphi \in C^\infty(\mathcal{S}_\pm)$ may be regarded as a C^∞ function $\varphi : G \longrightarrow V_\pm$ satisfying

$$\varphi(\gamma g k) = \sigma^\pm(k^{-1})\varphi(g)$$

for all $\gamma \in \Gamma$, $k \in K$, $g \in G$. The constant term φ_0 of φ is then defined as

$$\varphi_0(g) = \int_{\Gamma \cap U\backslash U} \varphi(ug)\,du$$

where the measure on U is normalized by the condition $\mathrm{Vol}(\Gamma\cap U\backslash U) = 1$. The constant term φ_o is contained in the space $(C^\infty((\Gamma U)\backslash G)\otimes V_\pm)^K$. Let $\sigma_M^\pm\colon K_M \to GL(V_\pm)$ be the restriction of $\sigma^\pm$ to $K_M = K\cap M$ and let $S_M^\pm$ be the associated locally homogeneous vector bundles over $\Gamma_M\backslash X_M$. Using (2.4), it follows that

$$(C^\infty((\Gamma U)\backslash G)\otimes V_\pm)^K \simeq C^\infty(\mathbb{R}^+\times\Gamma_M\backslash X_M, S_M^\pm) \tag{4.2}$$

The locally invariant operator $\mathcal{D}_+$ maps $(C^\infty((\Gamma U)\backslash G)\otimes V_+)^K$ to $(C^\infty((\Gamma U)\backslash G)\otimes V_-)^K$ and, in view of (4.2), induces an elliptic differential operator

$$\mathcal{D}_{+,o}\colon C^\infty(\mathbb{R}^+\times\Gamma_M\backslash X_M, S_M^+)\longrightarrow C^\infty(\mathbb{R}^+\times\Gamma_M\backslash X_M, S_M^-)$$

It is easy to see that there exists a selfadjoint differential operator

$$D_M\colon C^\infty(\Gamma_M\backslash X_M, S_M^+)\longrightarrow C^\infty(\Gamma_M\backslash X_M, S_M^-)$$

and a bundle isomorphism $\beta\colon S_M^+\to S_M^-$ such that

$$\mathcal{D}_{+,o} = \beta\left(r\frac{\partial}{\partial r} + D_M\right). \tag{4.3}$$

Put

$$\tilde{D}_M = D_M + \frac{m}{2}\mathrm{Id} \tag{4.4}$$

where m has the same meaning as in (2.11). Assume that $\ker(\tilde{D}_M) = 0$. Then the index formula proved in [M2] is the following

$$L^2\text{-}\mathrm{Ind}(D_+) = \int_X \omega + \mathcal{U} + \frac{1}{2}\eta(0) \tag{4.5}$$

where ω is the local index density of D_+ [A-B-P], $\eta(0)$ is the eta invariant of $\tilde{D}_M$ and $\mathcal{U}$ is the unipotent term which we shall discuss below. If $\ker(\tilde{D}_M) \neq 0$, there is an additional contribution to the index formula which is similar to the term $1/2(h_\infty(E) - h_\infty(F))$ in (1.7). It is given by the scattering matrix for zero-energy. This is a linear map

$$C_+(0)\colon \ker(\tilde{D}_M)\longrightarrow\ker(\tilde{D}_M).$$

The additional term in the index formula is then $-1/2\mathrm{Tr}(C_+(0))$. For details we refer to [M2].

Now we turn to the unipotent term. In Chapter XI of [M2] its calculation has been reduced to a problem of harmonic analysis. First we describe the result obtained in [M2,Ch.XI]. Let

$$\mathcal{D}_-: C^\infty(Y,S_-) \longrightarrow C^\infty(Y,S_+)$$

be the formal adjoint operator of $\mathcal{D}_+$. Let $\tilde{\mathcal{D}}_\pm: C^\infty(Y,\tilde{S}_\pm) \longrightarrow C^\infty(Y,\tilde{S}_\mp)$ be the lift of $\mathcal{D}_\pm$. The operator $\tilde{\mathcal{D}}_\mp\tilde{\mathcal{D}}_\pm$ is a G-invariant elliptic second order differential operator which is essentially selfadjoint in $L^2(\tilde{S}_\pm)$ with domain the smooth compactly supported sections. Denote by $\tilde{\Delta}_\pm$ its unique selfadjoint extension. Then, for $t>0$, $\exp(-t\tilde{\Delta}_\pm)$ is a G-invariant integral operator with a smooth kernel. The kernel of $\exp(-t\tilde{\Delta}_\pm)$ is given by a C^∞ function

$$h_t^\pm: G \longrightarrow \mathrm{End}(V_\pm)$$

which satisfies

$$h_t^\pm(k_1gk_2) = \sigma^\pm(k_1)\circ h_t^\pm(g)\circ\sigma^\pm(k_2),$$

$g\in G$, $k_1,k_2\in K$. That $h_t^\pm$ is the kernel of the heat operator $\exp(-t\tilde{\Delta}_\pm)$ means that

$$(\exp(-t\tilde{\Delta}_\pm)\varphi)(g) = \int_G h_t^\pm(g^{-1}g')\varphi(g')dg'$$

for $\varphi\in L^2(\tilde{S}^\pm)$, $g\in G$. Set

$$f_t(g) = \mathrm{tr}\, h_t^+(g) - \mathrm{tr}\, h_t^-(g),\ g\in G,\ t>0. \tag{4.6}$$

f_t is a K-finite C^∞ function on G and, for all $p>0$, it belongs to Harish-Chandra's L^p-Schwartz space $\mathcal{C}^p(G)$ of p-integrable rapidly decreasing functions.

The structure of the parabolic group P whose Langlands decomposition is given by (2.1) enters also into the description of $\mathcal{U}$. For simplicity, we shall assume that the unipotent radical U of P is commutative. Via $\exp: \mathfrak{u} \longrightarrow U$ we may identify U with $\mathbb{R}^k$. Put

$$L = MA.$$

Then L is a Levi component of P and it normalizes U. Let

$$\mathrm{Ad}_L: L \longrightarrow GL(\mathbb{R}^k)$$

be the adjoint representation of L on $\mathfrak{u} = \mathbb{R}^k$. There is an L-invariant $\mathbb{Q}$-irreducible polynomial $p\in\mathbb{Q}[x_1,\dots,x_k]$ such that

$$\mathring{U} = \{u\in\mathbb{R}^k \mid p(u) \neq 0\}$$

is the union of finitely many open L-orbits $C_L(u)$. Moreover $\Gamma\cap U\subset\mathring{U}$. Put

$$N(u) = p(u)^{1/\deg(P)}, \quad u \in U.$$

Furthermore, given $\gamma \in \Gamma \cap U$, we denote by M_γ the centralizer of γ in M and by $(\Gamma_M)_\gamma$ the centralizer of γ in Γ_M. Let $s \in \mathbb{C}$ with $\mathrm{Re}(s)>0$ and $u \in \overset{\circ}{U}$. Set

$$\zeta_u(s) = \sum_{\gamma \in (C_L(u) \cap \Gamma)/\Gamma_M} \frac{\mathrm{Vol}((\Gamma_M)_\gamma \backslash M_\gamma)}{|N(\gamma)|^{s+k}} \tag{4.7}$$

This series is absolutely and uniformly convergent on compacta of $\mathrm{Re}(s)>0$ and it admits a meromorphic continuation to the entire complex plane whose only possible singularities are simple poles at $s=0$ and $s=-1$. Moreover, one has

$$\operatorname*{Res}_{s=0} \zeta_u(s) = \frac{\mathrm{Vol}(\Gamma_M \backslash M)}{|\lambda| \mathrm{Vol}(\Gamma \cap U \backslash U)} \tag{4.8}$$

Let $u_1, \ldots, u_q$ be a set of representatives for the L-orbits in $\overset{\circ}{U}$ such that $N(u_j) = 1$, $j=1,\ldots,q$. Denote by C_j the constant term of the Laurent expansion of $\zeta_{u_j}(s)$ at $s=0$. Then the unipotent term $\mathcal{U}$ is the constant term in the asymptotic expansion as $t \to 0$ of

$$\begin{aligned} &\mathrm{Vol}(\Gamma \cap U \backslash U) \sum_{j=1}^{q} C_j \int_{G_{u_j} \backslash G} f_t(g^{-1} u_j g)\, dg \, + \\ &+ \frac{\mathrm{Vol}(\Gamma_M \backslash M)}{|\lambda|} \int_U f_t(u) \log N(u)\, du . \end{aligned} \tag{4.9}$$

To determine $\mathcal{U}$ we have to investigate the orbital integrals occurring in (4.9). First we deal with the invariant orbital integrals. Let $u \in G$ be unipotent and set

$$I_G(u,f) = \int_{G_u \backslash G} f(g^{-1} u g)\, dg, \quad f \in \mathcal{C}(G).$$

This is a unipotent orbital integral. The assignment $f \longmapsto I_G(u,f)$, $f \in \mathcal{C}(G)$, is a distribution on G [Ra]. By an unpublished result of Harish-Chandra, unipotent orbital integrals can be expressed in terms of semisimple orbital integrals. A proof of this fact is given in the appendix of [Ar]. Before stating this formula we have to introduce some notation. Let J be a Cartan subgroup of G and denote by J_o the center of J. For each $f \in \mathcal{C}(G)$, the invariant integral Φ_f^J relative to J is a function on J' (= regular elements on J) which is defined by

$$\Phi_f^J(j) = \varepsilon_R(j)\Delta(j) \int_{J_o\backslash G} f(g^{-1}jg)dg \ , \quad j\in J' \ .$$

For the definition of ε_R and Δ see [W1,8.1.1]. Given a unipotent element $u\in G$ there exist Cartan subgroups $J_1,\dots,J_q$ of G and constant coefficient differential operators D_i on J_i such that

$$I_G(u,f) = \sum_i \lim_{\substack{j\to 1\\ j\in J_i'}} (D_i\Phi_f^{J_i})(j) \ , \quad f\in \mathcal{C}(G). \tag{4.10}$$

We shall apply this formula to the function f_t defined by (4.6). First we recall one important property satisfied by f_t. Let $\hat{G}$ denote the unitary dual of G, i.e., the set of equivalence classes of irreducible unitary representations of G. Let π be an irreducible unitary representation and let $f\in C_c^\infty(G)$. Set

$$\pi(f) = \int_G f(g)\pi(g)dg$$

Then $\pi(f)$ is of the trace class and the assignment $f\in C_c^\infty(G)\longmapsto \operatorname{Tr}\pi(f)$ is a distribution on G. This distribution is the character of π, denoted by Θ_π. The function f_t does not belong to $C_c^\infty(G)$, but $\pi(f_t)$ is also of the trace class for every irreducible unitary representation π and we have

PROPOSITION 4.1. *There are only finitely many* $\pi\in\hat{G}$ *such that*

$$\operatorname{Tr}\pi(f_t) \neq 0 \ .$$

For the proof see Proposition 11.26 in [M2].

COROLLARY 4.2. *Assume that* J *is a noncompact Cartan subgroup of* G. *Then*

$$\Phi_{f_t}^J = 0 \ .$$

PROOF. To prove the corollary we employ the Fourier inversion formula for the invariant integral established by R.Herb [Hr]. To describe her result we introduce some notation. For details we refer to [Hr]. Let J be a Θ-stable Cartan subgroup of G with Lie algebra $\mathfrak{j}$. As usually, set $J_K = J\cap K$, $\mathfrak{j}_\mathfrak{p} = \mathfrak{j}\cap\mathfrak{p}$ and $J_\mathfrak{p} = \exp(\mathfrak{j}_\mathfrak{p})$. Let $C_G(J_\mathfrak{p})$ be the centralizer of $J_\mathfrak{p}$ in G. Then one has $C_G(J_\mathfrak{p}) = M_1J_\mathfrak{p}$, where M_1 is reductive with compact Cartan subgroup J_K. Let $\operatorname{Car}(M_1)$ denote a full set Θ-stable representatives of M_1-conjugacy classes of Cartan subgroups of M_1. These representatives can be chosen so that, for each

$B \in Car(M_1)$, one has $B_K^o \subset J_K$. If $B \in Car(M_1)$, then $\tilde{B} = BJ_p$ is a Cartan subgroup of G. To each pair (j^*,ν) with $j^* \in \hat{J}_K$ and $\nu \in j_p^*$ there corresponds a certain tempered invariant distribution $\Theta(J,j^*,\nu)$ on G. If j^* is regular, then $\Theta(J,j^*,\nu)$ is, up to sign, the character of a tempered unitary representation of G induced from a parabolic subgroup of G with split component J_p. Otherwise, $\Theta(J,j^*,\nu)$ is a linear combination of characters of representations which can be embedded in a unitary principal series representation associated to a different class of cuspidal parabolics. Now we can state Theorem 1 of [Hr]: Let $j = j_K j_p$ be a regular element of J with $j_K \in J_K$ and $j_p \in J_p$. Then

$$\Phi_{f_t}^J(j) = \sum_{B\in Car(M_1)} \sum_{j^*\in\hat{B}_K} \int_{j_p^*} j_p^{-i\nu'} \int_{b_p^*} C(M_1,B,b^*,\nu,j_K)\cdot \tag{4.11}$$

$$\cdot\, \Theta(BJ_p,b^*,\nu'\times\nu)(f_t)d\nu d\nu' ,$$

where the coefficients $C(M_1,B,b^*,\nu,j_K)$ are explicitly computable. If J is noncompact we have $\dim(j_p^*) \geq 1$. Combined with Proposition 4.1, this implies the Corollary. Q.E.D.

COROLLARY 4.3. *Assume that* rank G > rank K . *Then, for each unipotent element* $u \in G$ *, we have*

$$\int_{G_u\backslash G} f_t(g^{-1}ug)dg = 0.$$

Thus we can assume that rank G = rank K . To compute the unipotent orbital integrals in this case we employ (4.10) and the Fourier inversion formula (4.11) for the invariant integrals. In view of Corollary 4.2, we may assume that each J_i occurring in (4.10) is compact. After conjugation we may assume that J_i is a maximal torus $T \subset K$ with Lie algebra $\mathfrak{t}$. Let $L_T \subset \sqrt{-1}\mathfrak{t}^*$ be the lattice which corresponds to the unitary character group $\hat{T}$ of T. Recall that to each $\tau \in L_T$ there is associated a central eigendistribution Θ_τ on G characterized uniquely by certain properties [H]. Employing (4.10) and (4.11) we obtain

PROPOSITION 4.4. *Assume that* rank G = rank K . *Let* $u \in G$ *be unipotent. There exist complex numbers* $a_u(\tau)$, $\tau \in L_T$, *such that*

$$\int_{G_u\backslash G} f_t(g^{-1}ug)dg = \sum_{\tau\in L_T} a_u(\tau)\,\Theta_\tau(f_t) .$$

We observe that the series on the right hand side is finite. In fact, this follows from the calculation of $\Theta_\tau(f_t)$ in [M2]. We recall the final result. Let $k_{\mathbb{C}}$ and $t_{\mathbb{C}}$ denote the complexifications of the Lie algebras k and t respectively. Let $S_\pm$ be the half-spin $k_{\mathbb{C}}$-modules (cf. [B-M,p.157]). Since $\tilde{D}_+$ is elliptic, it follos from the results of Miatello [Mi] that there exists a unique virtual $k_{\mathbb{C}}$-module V such that, in the representation ring $R(k_{\mathbb{C}})$ of $k_{\mathbb{C}}$, one has

$$[V_+]-[V_-]=[V]\otimes([S_+]-[S_-]).$$

Assume that

$$V=\sum_\mu n_\mu E_\mu \tag{4.12}$$

where μ ranges over a finite subset of L_T, E_μ is an irreducible $k_{\mathbb{C}}$-module with highest weight μ and $n_\mu \in \mathbb{Z}$. Let W_K be the Weyl group of $(k_{\mathbb{C}}, t_{\mathbb{C}})$ and let ρ_c be the half-sum of the positive roots of $(k_{\mathbb{C}}, t_{\mathbb{C}})$. Then

$$\Theta_\tau(f_t)=(-1)^{1/2\dim(G/K)}\sum_{w\in W_K}\det(w)n_{-w\tau-\rho_c} \tag{4.13}$$

(see p.130 in [M2] for the proof). In particular, $\Theta_\tau(f_t)\neq 0$ implies that there exists $w\in W_K$ such that $\tau=-w(\mu+\rho_c)$ for some μ occurring in (4.12). Furthermore, (4.13) shows that, for any $\tau\in L_T$, $\Theta_\tau(f_t)$ is independent of t. Therefore, by Proposition 4.4, we get

COROLLARY 4.5. *Let* $u\in G$ *be unipotent. Then*

$$\int_{G_u\backslash G} f_t(g^{-1}ug)\,dg$$

is a constant independent of t.

Let $u_1,\ldots,u_q$ be as in (4.9) and set

$$a_j=\int_{G_{u_j}\backslash G} f_t(g^{-1}u_jg)\,dg\ ,\quad j=1,\ldots,q. \tag{4.14}$$

LEMMA 4.6. *The numbers* a_j *satisfy* $\sum_{j=1}^{q} a_j=0$.

PROOF. According to (11.55) in [M2] we have

$$\int_{G_{u_j}\backslash G} f_t(g^{-1}u_jg)\,dg=\int_{C_L(u_j)} f_t(u)\,du\ .$$

Note that the number $\Delta(u_j)$ occurring in this formula equals 1 in our case. This follows from the assumption that U is commutative. Hence

$$\sum_{j=1}^{q} a_j = \sum_{j=1}^{q} \int_{C_L(u_j)} f_t(u)du = \int_U f_t(u)$$

which vanishes by Lemma 11.28 in [M2]. Q.E.D.

Now introduce the following analytic function

$$L(s) = \sum_{j=1}^{q} a_j \zeta_{u_j}(s) , \quad s\in\mathbb{C}, \tag{4.15}$$

where $\zeta_{u_j}(s)$ is defined by (4.7). By (4.8), the residue of $\zeta_{u_j}(s)$ at s=0 is independent of j. Using Lemma 4.6, it follows that L(s) is a meromorphic function of $s\in\mathbb{C}$ whose only possible singularity is a simple pole at s=-1. In particular, L(s) is holomorphic at s=0. Summarizing we have proved that the contribution to $\mathcal{U}$ obtained from the first part in (4.9) is given by

$$\mathrm{Vol}(\Gamma\cap U\backslash U)L(0). \tag{4.16}$$

The L-function L(s) will be further investigated in the next section. It turns out that L(s) is an example of an L-function associated to a prehomogeneous vector space.

We add some remarks about the nature of L(s). The zeta functions $\zeta_{u_j}(s)$ are completely determined by the discrete group Γ and the parabolic group P. In particular, they do not depend on the differential operator $\mathcal{D}_+$. The constants a_j are independent of Γ. They may depend on the unipotent orbit $G_{u_j}\backslash G \simeq C_L(u_j)$ and they depend on the differential operator $\mathcal{D}_+$ through (4.12) and (4.13).

It remains to investigate the weighted orbital integral

$$\int_U f_t(u) \log(N(u))du \tag{4.17}$$

occurring in (4.9). The calculation of this integral is more difficult. If the $\mathbb{R}$-rank of G is greater than 1, a Fourier inversion formula is not yet available. For $SL(2,\mathbb{R})$ such a formula was obtained by Arthur, Herb and Sally [A-H-S] and G.Warner [W2] has treated the general $\mathbb{R}$-rank one case. The contribution of the weighted orbital integrals to the L^2-index of a twisted Dirac operator on an $\mathbb{R}$-rank one locally symmetric space has been determined by Barbasch and Moscovici [B-M]. Their results show that (4.17) does not vanish in general. In place of computing (4.17) in general we establish conditions which imply that this term vanishes. We make the following assumption:

(A) G *is nonsimple. Let* $p: \tilde{G} \longrightarrow G$ *be the universal covering of* G *and* $\tilde{G} = G_1 \times \cdots \times G_r$, $r > 1$, *its decomposition into simple factors. Let* $p_i: \tilde{G} \to G_i$ *denote the projection onto the i-th component. Then the parabolic subgroup* P *is such that* $p_i(p^{-1}(P)) \neq G_i$ *for each* i, $1 \leq i \leq r$.

PROPOSITION 4.7. *Suppose that* G *and* P *satisfy assumption* (A). *Furthermore, assume that* D_+ *is one of the following operators:*
(1) The Gauß-Bonnet operator, (2) The signature operator or (3) A twisted Dirac operator.
Then the weighted unipotent orbital integral (4.17) vanishes.

PROOF. First note that the unipotent radical U of P is simply connected. Let $p: \tilde{G} \to G$ be the universal covering and let $\tilde{P} = p^{-1}(P)$. Since A and U in (2.1) are simply connected, $\tilde{P}$ has the same split component A and the Langlands decomposition $\tilde{P} = UA\tilde{M}$ where $\tilde{M} = p^{-1}(M)$. Furthermore, the adjoint action of $\tilde{L} = \tilde{M}A$ on $U \simeq \mathbb{R}^k$ factors through the adjoint action of MA. Therefore, the weighted orbital integrals (4.17) coincide and we may assume that G is simply connected. Now we can proceed as in the proof of Lemma 11.70 in [M2]. Q.E.D.

We can now state the final index formula.

THEOREM 4.8. *Let the assumptions be the same as in Proposition 4.7. Then the* L^2*-index of* D_+ *is given by*

$$L^2\text{-Ind}(D_+) = \int_X \omega + \mathrm{Vol}(\Gamma \cap U \backslash U) L(0) + \frac{1}{2}\eta(0) - \frac{1}{2}\mathrm{Tr}(C_+(0)) .$$

Here ω *and* $\eta(0)$ *have the same meaning as in (4.5),* $L(s)$ *is the L-function (4.15) and* $C_+(0): \ker(\tilde{D}_M) \longrightarrow \ker(\tilde{D}_M)$ *is the scattering matrix for zero energy.*

REMARK. Throughout our discussion of the unipotent term $\mathcal{U}$ we have worked under the assumption that U is commutative. In general one has to work with the decomposition $U = U_1U_2$ where $U_1 = \exp(\mathfrak{u}_\lambda)$ and $U_2 = \exp(\mathfrak{u}_{2\lambda})$. Put $\Gamma_2 = \Gamma \cap U_2$ and $\Gamma_1 = \Gamma \cap U/\Gamma_2$. Then $\Gamma_1 \subset U_1$, $\Gamma_2 \subset U_2$ and we can proceed essentially in the same way as above. We end up with two L-functions associated to (U_1, Γ_1) and (U_2, Γ_2) whose values at s=0 are the corresponding contribution to the unipotent term.

5. ZETA FUNCTIONS AND L-FUNCTIONS ASSOCIATED TO PREHOMOGENEOUS VECTOR SPACES.

In this section we shall investigate the L-functions which occur in the index formula of Theorem 4.8. more closely. They are related to the zeta functions studied by Sato and Shintani [Sa], [S-S]. Let us recall the definition of these zeta functions.

Let L be a reductive complex linear algebraic group and σ a rational representation of L on a finite dimensional complex vector space $\underline{V}$. The triple $(L,\sigma,\underline{V})$ is called a <u>prehomogeneous vector space</u> if there exists a proper algebraic subset S of $\underline{V}$ such that $\underline{V}-S$ is a single L-orbit. The set S is called the singular set of $(L,\sigma,\underline{V})$. We shall assume that L and $\underline{V}$ have $\mathbb{Q}$-structures such that σ is defined over $\mathbb{Q}$ and that S is a $\mathbb{Q}$-irreducible hypersurface. Then there exists a $\mathbb{Q}$-irreducible homogeneous polynomial $p\in\mathbb{Q}[\underline{V}]$ and a $\mathbb{Q}$-rational character χ of L such that

$$S = \{ x\in\underline{V} \mid p(x) = 0 \}$$

and

$$p(\sigma(l)x) = \chi(l)p(x) \ , \ l\in L \ , \ x\in\underline{V}.$$

The polynomial p is a relative invariant of $(L,\sigma,\underline{V})$ and any other relative invariant is of the form cp^m for some $c\in\mathbb{Q}$ and $m\in\mathbb{N}$.

Put $M = \ker \chi$. Since L is reductive there exists a 1-dimensional torus A defined over $\mathbb{Q}$ in L such that $L = MA$. Let $L = L(\mathbb{R})^o$ (the connected component of the identity in $L(\mathbb{R})$), $M = M(\mathbb{R})^o$ and $V = \underline{V}(\mathbb{R})$. Then $V - S\cap V$ is the union of finitely many L-orbits $V_1,\dots,V_r$. Each V_i is a connected open subset of V.

Let Γ_M be an arithmetic subgroup of M and Λ a lattice in $V(\mathbb{Q})$ such that $\Gamma_M \Lambda = \Lambda$. Given $v\in V$, we denote by M_v the stabilizer of v in M and by $(\Gamma_M)_v$ the stabilizer of v in Γ_M. There is a natural normalization of the Haar measure on M_v and we denote by $\mathrm{Vol}((\Gamma_M)_v\backslash M_v)$ the volume with respect to the normalized Haar measure. Then for each L-orbit V_j we define the following zeta function

$$\zeta_j(s;\Lambda,\Gamma_M) = \sum_{\mu\in(V_j\cap\Lambda)/\Gamma_M} \frac{\mathrm{Vol}((\Gamma_M)_v\backslash M_v)}{|p(\mu)|^{-s}} \tag{5.1}$$

The summation means that we sum over a complete set of representatives of the Γ_M-orbits in $V_j\cap\Lambda$. It is proved in [Sa], [S-S] that the series (5.1) is absolutely and uniformly convergent on compacta of $\mathrm{Re}(s) > \dim(V)$. Furthermore, the zeta functions $\zeta_j(s;\Lambda,\Gamma_M)$, $j=1,\dots,r$, admit meromorphic continuations to the whole complex plane and they satisfy a system of functional equations.

In the case our dicrete group Γ arises from an arithmetic situation the zeta function (4.7) is of this type. In general we get slight modifications of the zeta functions described above, but the examples we are considering are all of the type (5.1).

EXAMPLES. 1) Let F be a totally real number field of degree n. Set $\underline{V} = F \otimes_{\mathbb{Q}} \mathbb{C} \simeq \mathbb{C}^n$, $L = R_{F/\mathbb{Q}}GL(1)$ (restriction of scalars á la Weil) and let σ be the natural representation of L on $\underline{V}$. L can be identified with $(\mathbb{C}^*)^n$ and σ is given by $\sigma(l)z = (l_1z_1,\dots,l_nz_n)$, $z\in\mathbb{C}^n$, $l\in(\mathbb{C}^*)^n$. L and $\underline{V}$ have natural $\mathbb{Q}$-structures with $L(\mathbb{Q}) = F^*$ and $\underline{V}(\mathbb{Q}) = F$. Then $(L,\sigma,\underline{V})$ is a prehomogeneous vector space. The $\mathbb{Q}$-irreducible polynomial p is given by $p(z) = z_1\cdots z_n$. Furthermore, $L = (\mathbb{R}^+)^n$ and $M = \{ y\in(\mathbb{R}^+)^n \mid y_1\cdots y_n = 1 \}$. The orbits of the action of L on $V - S\cap V$ are parametrized by $\{\pm1\}^n$ and the orbit corresponding to $\varepsilon\in\{\pm1\}^n$ is given by $V_\varepsilon = \{x\in\mathbb{R}^n \mid \varepsilon_ix_i > 0\}$. Denote by $x\in F \longrightarrow x^{(i)}\in\mathbb{R}$ the n different embeddings of F into $\mathbb{R}$. Let $\mathcal{O}_F$ be the ring of intergers of F. The mapping $\mu \longrightarrow (\mu^{(1)},\dots,\mu^{(n)})$ identifies $\mathcal{O}_F$ with a lattice in $\mathbb{R}^n$ which is contained in $\underline{V}(\mathbb{Q})$. This mapping also identifies the group U_F^+ of totally positive units in F with an arithmetic subgroup of M. Our zeta function $\zeta_\varepsilon(s)$ associated to V_ε is then given by

$$\zeta_\varepsilon(s) = \sum_{\substack{\mu\in(\mathcal{O}_F-0)/U_F^+\\ \varepsilon_i\mu^{(i)}>0}} |N(\mu)|^{-s}, \quad Re(s) > 1, \tag{5.2}$$

where $N(\mu) = \mu^{(1)}\cdots\mu^{(n)}$ is the norm of the algebraic number μ. This is a partial Dedekind zeta function of the number field F corresponding to the "ray class" $\mathcal{O}_F$. These zeta functions arise for Hilbert modular cusps (cf. [M2, Ch. XII]).

2) Other examples are zeta functions associated to self-dual homogeneous cones [S-O]. Let V be a real vector space of dimension n and let V^* be its dual. If C is a convex cone in V its dual C^* is defined by

$$C^* = \{ x^*\in V^* \mid \langle x,x^*\rangle > 0 \text{ for all } x\in\bar{C}-0 \}.$$

C is called self-dual if there exists a linear isomorphism $T\colon (V,C)\longrightarrow(V^*,C^*)$ and it is called homogeneous if the automorphism group

$$L = \mathrm{Aut}(V,C)^o = \{g\in GL(V) \mid gC = C\}^o$$

is transitive on C. If C is a self-dual homogeneous cone, we fix a positive definite inner product $\langle.,.\rangle$ on V defined by the isomorphism T. Then the automorphism group L is the connected component of the identity of the set of real points of a reductive algebraic group and for any $c_o\in C$

$$K = \{ g\in L \mid gc_o = c_o \}$$

is a maximal compact subgroup of L. Thus $C \simeq L/K$ is a Riemannian symmetric space. As explained in [S-O], to each self-dual homogeneous cone (V,C) with a base point $c_o \in C$ there is associated a polynomial function

$$N: V \longrightarrow \mathbb{R}$$

called <u>norm</u>. If C is irreducible the norm is given as follows: Let

$$\Phi_C(x) = \int_{C^*} e^{-\langle x,x^*\rangle} dx^*$$

be the characteristic function of C. Then

$$N(x) = (\Phi_C(c_o)^{-1}\Phi_C(x))^{-r/n}, \quad x \in C,$$

where r is the $\mathbb{R}$-rank of L. Now assume that (V,C) has a $\mathbb{Q}$-simple $\mathbb{Q}$-structure. This means that there is a $\mathbb{Q}$-vector space $V_{\mathbb{Q}}$ such that $V = V_{\mathbb{Q}} \otimes_{\mathbb{Q}} \mathbb{R}$, for which L is the connected component of the identity of the group of real points of an algebraic group defined over $\mathbb{Q}$ and, if $(V,C) = \prod_i (V_i,C_i)$ is the decomposition into irreducible cones, no partial product is defined over $\mathbb{Q}$. Assume that $c_o \in V_{\mathbb{Q}}$. Then the norm $N: V \longrightarrow \mathbb{R}$ is a polynomial function with coefficients in $\mathbb{Q}$. Let $\Lambda \subset V_{\mathbb{Q}}$ be a lattice and Γ an arithmetic subgroup of L such that $\Gamma\Lambda = \Lambda$. Then the zeta function associated to C, Λ and Γ is defined as

$$(5.3) \qquad \zeta_{C,c_o}(s) = \sum_{\mu \in C \cap \Lambda/\Gamma} |\Gamma_\mu|^{-1} N(\mu)^{-s}, \quad s \in \mathbb{C},$$

where $\Gamma_\mu = \{\gamma \in \Gamma \mid \gamma\mu = \mu\}$. This is a zeta function in the previous sense. The zeta function (5.2) is subsummed under these examples. The self-dual homogeneous cone in this case is $(\mathbb{R}^+)^n$. For the further investigation of the zeta functions (5.3) we refer to [S-O].

If we take appropriate linear combinations of the zeta functions (5.1) we get analytic functions which in some cases may be considered as L-functions associated to prehomogeneous vector spaces. In general it is not clear which linear combinations give rise to reasonable L-functions. In example 1) we may take the character $\chi: \{\pm 1\}^n \longrightarrow \{\pm 1\}$ defined by $\chi(\varepsilon) = \prod \varepsilon_j$. Then

$$(5.4) \qquad L_F(s) = \sum_{\varepsilon \in \{\pm 1\}^n} \chi(\varepsilon)\zeta_\varepsilon(s) = \sum_{\mu \in (\mathcal{O}_F - 0)/\mathcal{O}_F^*} \frac{\operatorname{sign} N(\mu)}{|N(\mu)|^s}$$

is a well-known Shimizu L-function that arises in different situations related to cusps of Hilbert modular varieties (cf. [Sh], [Hi], [A], [A-D-S], [M2] and [C2]). In particular, its value at s=0 appears in the index formula for the signature operator on Hilbert modular varieties [M1]. In the case of self-dual homogeneous cones appropriate L-functions have been investigated in [S-O].

6. THE L^2-INDEX OF THE SIGNATURE OPERATOR. Let X be an oriented $2l$-dimensional Riemannian manifold with a cusp of rank one. As above we assume that the cusp $Y = \Gamma\backslash G/K$ of X is defined by a parabolic subgroup P of G whose unipotent radical U is commutative. Recall that this means that Y is diffeomorphic to the half-cylinder $\mathbb{R}^+ \times \Gamma\backslash Z$ and $\Gamma\backslash Z$ is a torus bundle over the compact locally symmetric space $\Gamma_M\backslash X_M$. Examples will be discussed below.

Let τ be the involution on the space of C^∞ differential forms $\Lambda^*(X)$ defined by $\tau\Phi = i^{p(p-1)+l} * \Phi$ for $\Phi\in\Lambda^p(X)$. Denote by $\Lambda^*_\pm(X)$ the ± 1-eigenspaces of τ. Then $d + d^*$ interchanges $\Lambda^*_+(X)$ and $\Lambda^*_-(X)$, and hence defines by restriction an operator

$$D_+ : \Lambda^*_+(X) \longrightarrow \Lambda^*_-(X)$$

which is called the signature operator. D_+ is a first order elliptic differential operator and one verifies that D_+ is a generalized Dirac operator in the sense of section 3. Besides of the usual properties satisfied by Dirac type operators D_+ has to be locally invariant on the cusp Y_1. In fact, $D_+|Y_1$ coincides with the restriction to Y_1 of the signature operator

$$\mathcal{D}_+ : \Lambda^*_+(Y) \longrightarrow \Lambda^*_-(Y)$$

on Y and it is obvious that $\mathcal{D}_+$ is locally invariant, i.e., its lift to $\tilde{Y}$ is a G-invariant differential operator. Thus we can apply Theorem 4.8 to compute the L^2-index of the signature operator D_+. To this end, we have to compute the various terms occurring in the index formula. First we relate the L^2-index of D_+ to cohomology. Recall that the L^2 norm in $\Lambda^*(X)$ is given by

$$\|\Phi\|^2 = \int_X \Phi\wedge *\bar{\Phi} , \quad \Phi\in\Lambda^*(X) .$$

Let $L^2\Lambda^*(X)$ be the completion of the space $\Lambda^*_c(X)$ of compactly supported C^∞ forms on X with respect to this norm. Let $\Delta = dd^* + d^*d$ be the Laplacian acting on $\Lambda^*(X)$ and set

$$H^*_{(2)}(X) = \{\Phi\in\Lambda^*(X) \mid \Delta\Phi = 0, \|\Phi\| < \infty\}.$$

This is the space of square integrable harmonic forms on X. The involution τ maps $H^*_{(2)}(X)$ to itself. Denote by $H^*_{(2),\pm}(X)$ the ± 1-eigenspaces of τ in $H^*_{(2)}(X)$. Then it follows from (3.3) and (3.4) that

$$\text{(6.1)} \qquad L^2\text{-Ind}(D_+) = \dim H^*_{(2),+}(X) - \dim H^*_{(2),-}(X) .$$

Since τ maps $H^k_{(2)}(X)$ to $H^{2l-k}_{(2)}(X)$, $0 \leq k \leq 2l$, it follows that

$$H^k_{(2)}(X) \oplus H^{2l-k}_{(2)}(X) \ , \ 0 \le k < l, \text{ and } \ H^l_{(2)}(X)$$

are invariant under τ. Let $H^k_{(2),\pm}(X)$, $0 \le k \le l$, denote the ±1-eigenspaces of τ in these spaces. Then we have

$$(6.2) \qquad H^*_{(2),\pm}(X) = \bigoplus_{0 \le k \le l} H^k_{(2),\pm}(X) \ .$$

Moreover, if $0 \le k < l$, then one has

$$H^k_{(2),\pm}(X) = \{ \Phi \pm \tau\Phi \mid \Phi \in H^k_{(2)}(X) \} \ .$$

Hence $\dim H^k_{(2),+}(X) = \dim H^k_{(2),-}(X)$ for $k < l$, and using (6.1) and (6.2), we obtain

$$(6.3) \qquad L^2\text{-Ind}(D_+) = \dim H^l_{(2),+}(X) - \dim H^l_{(2),-}(X) \ .$$

Assume that $l=2m+1$. In this case, the mapping $\tau : H^l_{(2)}(X) \longrightarrow H^l_{(2)}(X)$ coincides with $i*$. Since $*$ is a real operator one has

$$\tau(\bar{\Phi}) = -\tau(\Phi) \ , \quad \Phi \in H^l_{(2)}(X) \ .$$

This implies that the map $\Phi \longrightarrow \bar{\Phi}$ induces an isomorphism of $H^l_{(2),+}(X)$ and $H^l_{(2),-}(X)$. Therefore, by (6.3), we get

$$L^2\text{-Ind}(D_+) = 0 \ , \text{ if } \ l=2m+1.$$

Hence we can assume that $n=4k$, $k \in \mathbb{N}$. The local signature theorem of [A-B-P] implies that the local index density ω of D_+ is the Hirzebruch L-polynomial $L_k(p_1,\dots,p_k)$ in the Pontrjagin forms p_i of the Riemannian metric of X. This is the first term in the index formula of Theorem 4.8.

Next we shall compute the eta invariant $\eta(0)$. For this purpose we have to determine the operator D_M occurring in (4.3). Recall that the cross section of the cusp Y is the torus bundle (2.8) with typical fibre $\Gamma \cap U \backslash U$. Therefore the harmonic forms on the fibres of (2.8) can be identified with the Lie algebra cohomology $H^*(\mathfrak{u};\mathbb{C}) \simeq \Lambda^*(\mathfrak{u}_{\mathbb{C}})$. Let $\mathbf{H}^*(\mathfrak{u})$ be the locally constant sheaf over $\Gamma_M \backslash X_M$ associated to the canonical representation of Γ_M on $H^*(\mathfrak{u};\mathbb{C})$. Then the space of U-invariant differential forms on the cusp can be identified with the space

$$\Lambda^*(\mathbb{R}^+ \times \Gamma_M \backslash X_M ; \mathbf{H}^*(\mathfrak{u}))$$

- the space of differential forms on $\mathbb{R}^+ \times \Gamma_M \backslash X_M$ with values in the local system $\mathbf{H}^*(\mathfrak{u})$. The involution τ induces an involution τ_o on this space and we denote the ±1-eigenspaces of τ_o by

$$\Lambda^*_{\pm}(\mathbb{R}^+ \times \Gamma_M \backslash X_M; \mathbf{H}^*(u),$$

which in turn can be identified with the space $C^\infty(\mathbb{R}^+) \otimes \Lambda^*(\Gamma_M \backslash X_M; \mathbf{H}^*(u))$. Let d_M and $*_M$ be the exterior derivative and the Hodge operator in the space $\Lambda^*(\Gamma_M \backslash X_M; \mathbf{H}^*(u))$. Let $*_U$ be the Hodge operator in $\Lambda^*(u_{\mathbb{C}})$ and τ_U the associated involution which we also regard as endomorphism of the local system $\mathbf{H}^*(u)$. Furthermore, let $C: \Lambda^*(u_{\mathbb{C}}) \longrightarrow \Lambda^*(u_{\mathbb{C}})$ be the operator defined by $C\Psi = -p\Psi$ for $\Psi \in \Lambda^p(u_{\mathbb{C}})$. C induces an operator in $\Lambda^*(\Gamma_M \backslash X_M; \mathbf{H}^*(u))$ which we denote by C_M. Let B_M be the first order differential operator in the space $\Lambda^*(\Gamma_M \backslash X_M; \mathbf{H}^*(u))$ defined by

$$B_M \Phi = (-1)^{k+p+1}(\varepsilon *_M d_M - d_M *_M)\Phi$$

where Φ is either a 2p-form ($\varepsilon = 1$) or a (2p-1)-form ($\varepsilon = -1$). Then a little computation shows that the restriction of D_+ to the space of U-invariant differential forms $\Lambda^*_+(\mathbb{R}^+ \times \Gamma_M \backslash X_M; \mathbf{H}^*(u))$ is equivalent to

$$\mathcal{D}_{+,o} = *_M\left(r \frac{\partial}{\partial r} + B_M \circ \tau_U + C_M\right) \tag{6.4}$$

(this corrcts a misstatement on p.142 in [M2]). Thus the operator $\tilde{D}_M$ defined by (4.4) is given by

$$\tilde{D}_M = B_M \circ \tau_U + \tilde{C}_M$$

where $\tilde{C}_M = C_M + \frac{m}{2}\mathrm{Id}$ and $m = \dim(U)$. Note that $\mathrm{Tr}(\tilde{C}_M) = 0$. Let $e(x,x,t)$ be the kernel of $\exp(-t\tilde{D}_M^{\,2})$. Then it is easy to see that $\mathrm{tr}\, \tilde{D}_M e(x,x,t) = 0$. Thus the eta invariant $\eta(0)$ of $\tilde{D}_M$ vanishes.

To deal with the fourth term $\mathrm{Tr}(C_+(0))$ we have to recall some facts about the scattering matrix established in [M2]. By definition, $C_+(0)$ is a linear operator in $\ker(\tilde{D}_M)$. Assume that $\ker(\tilde{D}_M) \neq 0$. To each $\Phi \in \ker(\tilde{D}_M)$ and $\lambda \in \mathbb{C} - [0,\infty)$ one can associate a C^∞ differential form $E(\Phi,\lambda)$ on X which satisfies

$$\Delta E(\Phi,\lambda) = \lambda\, E(\Phi,\lambda)$$

(cf. [M2,VIII]). If X is a locally symmetric space then $E(\Phi,\lambda)$ is an Eisenstein series. Let $0 = \mu_1 < \mu_2 < \cdots$ be the eigenvalues of $\tilde{D}_M^{\,2}$. Then $E(\Phi,.)$ can be continued to a meromorphic function on the Riemann surface Σ which is associated to the square roots $z \longrightarrow \sqrt{z-\mu_i}$, $i \in \mathbb{N}$. The functional equation satisfied by $E(\Phi,\lambda)$ implies that $E(\Phi,\lambda)$ is regular at $s=0$ and $dE(\Phi,0) = $ $= d^*E(\Phi,0) = 0$. Moreover, in the cusp, we may write

$$E(\Phi,0) = \Psi_o + \frac{dr}{r}\wedge\Psi_1 + \varphi$$

where Ψ_o, $\Psi_1 \in \ker(\tilde{D}_M)$ and φ is in L^2. Set

$$E^{\pm}(\Phi,0) = E(\Phi,0) \pm \tau E(\Phi,0).$$

Then $E^{\pm}(\Phi,0)\in\Lambda^*(X)$ and we have

$$dE^{\pm}(\Phi,0) = d^*E^{\pm}(\Phi,0) = 0$$

so that $E^{\pm}(\Phi,0)$ defines a cohomology class $[E^{\pm}(\Phi,0)]\in H^*(X)$. It follows from the definition of the scattering matrix that in the cusp we have

$$E^{\pm}(\Phi,0) = \Phi + C_{\pm}(0)\Phi + \frac{dr}{r}\wedge\tau_o(\Phi + C_{\pm}(0)\Phi) + \Psi^{\pm} \tag{6.5}$$

where $\Psi^{\pm}$ is in L^2. Here τ_o is the involution in $\Lambda^*(\Gamma\backslash Z)$ defined by the Hodge operator and $C_{\pm}(0)\colon \ker(\tilde{D}_M) \longrightarrow \ker(\tilde{D}_M)$ are linear operators (cf. [M2, VIII]). The following facts were established in [M2,pp.97-98] : The operators $C_{\pm}(0)$ satisfy

$$C_{\pm}(0)^2 = \mathrm{Id}\ ,\ C_+(0) = -{*}C_-(0){*}\ . \tag{6.6}$$

In our case the kernel of $\tilde{D}_M$ has a description in terms of cohomology groups. Let

$$\Delta_M = d_M d_M^* + d_M^* d_M\ .$$

A computation shows that

$$\tilde{D}_M^{\,2} = \Delta_M + (p - m/2)^2\mathrm{Id}$$

on $\Lambda^*(\Gamma_M\backslash X_M;H^p(u))$. Since $\Delta_M \geq 0$ it follows that

$$\ker(D_M) = \begin{cases} 0\ , & m \text{ odd} \\ H^*(\Gamma_M\backslash X_M;H^{m/2}(u)), & m \text{ even} \end{cases} \tag{6.7}$$

where we have identified the De Rham cohomology group $H^*(\Gamma_M\backslash X_M;H^{m/2}(u))$ with the space of $H^{m/2}(u)$-valued harmonic forms on $\Gamma_M\backslash X_M$. There is a canonical embedding

$$i\colon \Lambda^p(\Gamma_M\backslash X_M;H^q(u)) \longrightarrow \Lambda^{p+q}(\Gamma\backslash Z)$$

which induces an isomorphism in cohomology

$$i^*: \bigoplus_{p+q=d} H^p(\Gamma_M\backslash X_M; H^q(u)) \xrightarrow{\simeq} H^d(\Gamma\backslash Z)$$

(cf. [Ha, Theorem 2.8]). Hence we may identify $\ker(\tilde{D}_M)$ with a subspace of $H^*(\Gamma\backslash Z)$. Also note that the split component A of P acts on $H^*(\Gamma\backslash Z)$ by algebraic characters χ which gives a decomposition

(6.8) $$H^*(\Gamma\backslash Z) = \bigoplus_{\chi} H^*_{\chi}(\Gamma\backslash Z) .$$

Then

$$\ker(\tilde{D}_M) \simeq H^*_{-\rho}(\Gamma\backslash Z)$$

and we may regard $C_{\pm}(0)$ as linear operators in $H^*_{-\rho}(\Gamma\backslash Z)$. In view of (6.6) we have

$$H^*_{-\rho}(\Gamma\backslash Z) = H^*_{-\rho}(\Gamma\backslash Z)_+ \oplus H^*_{-\rho}(\Gamma\backslash Z)_-$$

where $H^*_{-\rho}(\Gamma\backslash Z)_{\pm}$ are the ± 1-eigenspaces of $C_+(0)$. Now recall that $X = W \cup Y_1$ where W is a compact manifold with boundary $\Gamma\backslash Z$ and $Y_1 \simeq [1,\infty) \times \Gamma\backslash Z$. Therefore we get a map

$$b: H^*(X) \longrightarrow H^*(W) \longrightarrow H^*(\Gamma\backslash Z) .$$

LEMMA 6.1. *Let* $p: H^*(\Gamma\backslash Z) \longrightarrow H^*_{-\rho}(\Gamma\backslash Z)$ *denote the orthogonal projection. The image of the map*

$$p \circ b: H^*(X) \longrightarrow H^*_{-\rho}(\Gamma\backslash Z)$$

is $H^*_{-\rho}(\Gamma\backslash Z)_+$.

PROOF. We may essentially follow the proof of Theorem 4.6.3 in [Ha]. Let R^q be the subspace in $H^q_{-\rho}(\Gamma\backslash Z)$ consisting of all cohomology classes which are restrictions of cohomology classes on X. Then R^q and R^{n-q-1} are orthogonal with respect to the nondegenerate pairing

$$<[\varphi], [\psi]> = \int_{\Gamma\backslash Z} \varphi\wedge\psi$$

Hence

(6.9) $$\dim(R^q) + \dim(R^{n-q-1}) \leq \dim(H^q_{-\rho}(\Gamma\backslash Z)).$$

Let $\Phi\in H^*_{-\rho}(\Gamma\backslash Z)$ and consider $E^{\pm}(\Phi,0)$. The form $\Psi^{\pm}$ in (6.5) is an L^2 form on Y_1 which satisfies $d\Psi^{\pm} = d^*\Psi^{\pm} = 0$. We may decompose $\Psi^{\pm}$ as $\Psi^{\pm} = \Psi^{\pm}_o + \Psi^{\pm}_1$ where $\Psi^{\pm}_o$ is the constant term of $\Psi^{\pm}$. Then $\Psi^{\pm}_1$ is a "cusp form" and therefore it is rapidly decreasing on Y_1. On the other hand, $\Psi_o^{\pm}$ is an L^2 form in $\Lambda^*([1,\infty)\times\Gamma_M\backslash X_M;H^*(\mathcal{U}))$ with $d\Psi_o^{\pm} = d^*\Psi_o^{\pm} = 0$. If we expand $\Psi_o^{\pm}$ in terms of the eigenfunctions of $\tilde{D}_M^2$ it is easy to see that $|\Psi_o^{\pm}(r,x)| \leq Cr^{-c}$ for some constants $C > 0$, $c > 0$ and $r \to \infty$. Thus $|\Psi^{\pm}(r,x)| \leq Cr^{-c}$ as $r \to \infty$ and, using (6.5), we obtain

$$b([E^{\pm}(\Phi,0)]) = [\Phi] + [C_{\pm}(0)\Phi] .$$

This implies $H^*_{-\rho}(\Gamma\backslash Z)_+ \subset \mathrm{Im}(p\circ b)$. Now assume that $\Phi\in H^*_{-\rho}(\Gamma\backslash Z)_-$, i.e., $C_+(0)\Phi = -\Phi$. By (6.6) we get $C_-(0)*\Phi = *\Phi$ and therefore $*(H^*_{-\rho}(\Gamma\backslash Z)_-) \subset \mathrm{Im}(p\circ b)$. Hence

$$\dim(R^q) + \dim(R^{n-q-1}) \geq \dim(H^q_{-\rho}(\Gamma\backslash Z)_+) + \dim(*H^q_{-\rho}(\Gamma\backslash Z)_-) = \dim(H^q_{-\rho}(\Gamma\backslash Z)) .$$

Combined with (6.9) we get $H^q_{-\rho}(\Gamma\backslash Z)_+ = R^q$. Q.E.D.

Now Poincaré duality in the exact sequence

$$H^*(W) \xrightarrow{j} H^*(\partial W) \longrightarrow H^*(W,\partial W)$$

shows that $\mathrm{Im}(j)$ is dual to its orthogonal complement. Therefore

$$(\mathrm{Im}\, b)^{\perp} = *\mathrm{Im}\, b$$

Since $*$ is compatible with the decomposition (6.8), it follows from Lemma 6.1 that

$$H^*_{-\rho}(\Gamma\backslash Z)_- = \mathrm{Im}(p\circ b)^{\perp} = *(H^*_{-\rho}(\Gamma\backslash Z)_+) .$$

Hence $\dim(H^*_{-\rho}(\Gamma\backslash Z)_+) = \dim(H^*_{-\rho}(\Gamma\backslash Z)_-)$. But this is equivalent to

$$\mathrm{Tr}(C_+(0)) = 0 .$$

Summarizing our results we get the following formula for the L^2-index of the signature operator:

$$L^2\text{-}\mathrm{Ind}(D_+) = \int_X L_k(p_1,\dots,p_k) + \mathrm{Vol}(\Gamma\cap U\backslash U)L(0) \tag{6.10}$$

where $L(s)$ is the L-function defined by (4.15).

7. RELATION WITH INTERSECTION COHOMOLOGY. We would like to give the L^2 index of the signature operator D_+ a topological interpretation similar to the compact case. For this purpose we need additional assumptions on X (actually we expect that these assumptions can be removed).

In this section we suppose that the cusp Y of X is a cusp of a Hermitian locally symmetric space of $\mathbb{Q}$-rank one. This means the following: We assume that G is the group of real points of a connected semisimple algebraic group G defined over $\mathbb{Q}$ whose $\mathbb{Q}$-rank equals one and which is such that the symmetric space $\tilde{Y} = G/K$ is Hermitian. Let $\tilde{\Gamma}$ be a neat arithmetic subgroup of G. Here "neat" means that, for each $\gamma \in \tilde{\Gamma}$, γ^p is unipotent for some positive integer p implies that γ itself is unipotent; in particular $\tilde{\Gamma}$ is torsion-free. Then $V = \tilde{\Gamma}\backslash\tilde{Y}$ is a Hermitian locally symmetric space of $\mathbb{Q}$-rank one. It is also a manifold with cusps of rank one in our sense. The cusps of V correspond to the $\tilde{\Gamma}$-conjugacy classes of minimal $\mathbb{Q}$-rational parabolic subgroups of G. Let P be such a subgroup and set

$$\Gamma = \tilde{\Gamma} \cap P .$$

Then $Y = \Gamma\backslash\tilde{Y}$ is a cusp of V and we assume that the cusp of X is of this type.

The complex manifold V can be compactified to a normal projective variety $\bar{V}$ - the Baily-Borel-Satake compactification of V [B-B]. Since G has $\mathbb{Q}$-rank one, the singular stratum $S = \bar{V} - V$ is itself a smooth projective variety. The connected components of S are smooth compact arithmetic quotients of lower-dimensional Hermitian symmetric spaces., the "rational boundary components" of $\tilde{Y}$. To each cusp of V we may add the corresponding component of S to get a partial compactification at this cusp, and thus a compactification $\bar{X}$ of X. The singular stratum $\bar{X} - X$ is a compact manifold of even dimension.

Now we return to (6.1) and try to give a topological interpretation to the right hand side. First we relate the L^2 harmonic forms to L^2-cohomology. Let

$$\Lambda^*_{(2)}(X) = \{\, \Phi \in \Lambda^*(X) \mid \Phi,\ d\Phi \in L^2 \,\}.$$

This space is stable under exterior differentiation and the L^2-cohomology group $H^*_{(2)}(X;\mathbb{C})$ is by definition the cohomology of the complex $(\Lambda^*_{(2)}(X), d)$. It is known that for Hermitian locally symmetric spaces of finite volume the L^2-cohomology is finite dimensional [B-C]. The same is true in our case. To prove this fact recall that $H^*_{(2)}(X;\mathbb{C})$ is finite-dimensional iff $d\,\Lambda^*_{(2)}(X)$ is a closed subspace of $\Lambda^*_{(2)}(X)$. This can be restated as follows: Let Δ be the Laplacian in $\Lambda^*(X)$ and let $\bar{\Delta}$ be its closure in L^2. Then d has closed range as operator in $\Lambda^*_{(2)}(X)$ iff the essential spectrum of $\bar{\Delta}$ has a positive lower bound. Now the continuous spectrum of $\bar{\Delta}$ can be determined using results of [M2]. Let μ_1 be the lowest eigenvalue of $\tilde{D}_M^{\,2}$. Then it follows from Theorem 4.38 and Theorem 6.17 in [M2] that the essential spectrum of $\bar{\Delta}$ is the interval $[\mu_1, \infty)$. Therefore the essential spectrum of $\bar{\Delta}$ has a positive lower bound iff

$\ker(\tilde{D}_M) = 0$. Using (6.7), we obtain

LEMMA 7.1 *There exists* $c > 0$ *such that* $\text{ess spec}(\bar{\Delta}) = [c,\infty)$ *iff* $H^*_{-\rho}(\Gamma\backslash Z)=0$.

This is a condition which depends only on the cusp. Using the results of [B-C] and the observations above, it follows that $H^*_{(2)}(X;\mathbb{C})$ is finite-dimensional. Moreover the canonical map of $\mathcal{H}^*_{(2)}(X)$ to $H^*_{(2)}(X;\mathbb{C})$ induces an isomorphism

$$\alpha : \mathcal{H}^*_{(2)}(X) \xrightarrow{\sim} H^*_{(2)}(X;\mathbb{C}) \tag{7.1}$$

Now consider the compactification $\bar{X}$ defined above. Since the singular stratum $\bar{X} - X$ is of even dimension the middle intersection cohomology $IH^*(\bar{X})$ can be defined by

$$IH^p(\bar{X}) = IH_{n-p}(\bar{X}) , \quad p\in\mathbb{N},$$

where $IH_*(\bar{X})$ denotes the middle intersection homology of $\bar{X}$ [G-M1].

PROPOSITION 7.2 *There is a natural isomorphism*

$$H^*_{(2)}(X;\mathbb{C}) \simeq IH^*(\bar{X}) .$$

The proof of this result for arithmetic quotients of bounded symmetric domains of $\mathbb{Q}$-rank one given by Borel [Bo] is purely local in the sense that one has to verify the vanishing of certain cohomology groups which are completely determined by the cusps. Therefore Proposition 7.2 follows immediately from the results of [Bo]. Combinig (7.1) and Proposition 7.2 we get an isomorphism

$$\mathcal{H}^*_{(2)}(X) \simeq IH^*(\bar{X}) . \tag{7.2}$$

Now middle intersection cohomology satisfies Poincaré duality, i.e., there is a perfect pairing

$$IH^i(\bar{X}) \times IH^{n-i}(\bar{X}) \longrightarrow \mathbb{C} , \quad i\in\mathbb{N} . \tag{7.3}$$

The functorial construction of the isomorphism (7.2) implies that this pairing corresponds to the pairing

$$\mathcal{H}^i_{(2)}(X) \times \mathcal{H}^{n-i}_{(2)}(X) \longrightarrow \mathbb{C}$$

given by

$$\langle\Phi,\Psi\rangle = \int_X \Phi\wedge\Psi$$

(cf. [G-M2]). Let $\mathrm{Sign}_{IH*}(\bar{X})$ be the signature of the intersection form (7.3). Then by (6.1) we get

$$L^2\text{-Ind}(D_+) = \mathrm{Sign}_{IH*}(\bar{X}) \quad . \tag{7.4}$$

Now assume that G is not simple. Then we can apply the index formula (6.10) which together with (7.4) gives the following signature theorem for Riemannian manifolds X as above

$$\mathrm{Sign}_{IH*}(\bar{X}) = \int_X L_k(p_1,\ldots,p_k) + \mathrm{Vol}(\Gamma \cap U\backslash U)L(0). \tag{7.5}$$

We proceed now to discuss examples. Let V, C and $L = \mathrm{Aut}(V,C)^{\circ}$ have the same meaning as in Example 2) of section 5. Then $\mathcal{D} = V + \sqrt{-1}C$ is a tube domain in $V_{\mathbb{C}} = \mathbb{C}^n$. Let $G = (\mathrm{Hol}\,\mathcal{D})^{\circ}$ denote the connected component of the identity of the group of holomorphic automorphisms of $\mathcal{D}$. $\mathcal{D}$ is a Hermitian symmetric space of noncompact type and G is a semisimple Lie group of Hermitian type with center reduced to the identity. Moreover, the semi-direct product

$$P = L\cdot V \tag{7.6}$$

is a parabolic subgroup of G corresponding to a point boundary component of $\mathcal{D}$ which we denote symbolically as $\sqrt{-1}\,\infty$. P can be identified with the group of affine automorphisms of $\mathcal{D}$. Note that the unipotent radical of P coincides with V which is commutative. A given $\mathbb{Q}$-structure on V determines uniquely a $\mathbb{Q}$-structure on G such that P and L are subgroups defined over $\mathbb{Q}$. The tube domain $\mathcal{D}$ with a $\mathbb{Q}$-structure on G determined in this manner is called a "rational symmetric tube domain". We shall assume that the $\mathbb{Q}$-rank of G is one. It follows from the classification theory that a rational symmetric tube domain $\mathcal{D}$ with $\mathbb{Q}\text{-rank}\,G = 1$ is of the form

$$\mathcal{D} = \mathcal{D}_1 \times \cdots \times \mathcal{D}_1$$

(m-fold product), where $\mathcal{D}_1$ is an irreducible symmetric domain of type (III_1), (III_2) or $(\mathrm{I}_{p,p})$. We write symbolically that $\mathcal{D}$ is of type $(\mathrm{III}_1)^m$, $(\mathrm{III}_2)^m$ or $(\mathrm{I}_{p,p})^m$. For $E = \mathbb{R}, \mathbb{C}$ denote by $\mathcal{P}_r(E)$ the cone of positive definite Hermitian matrices of size r with entries in E. Then C and $G(\mathbb{Q})$ (matrices with rational entries) are given by the following list

	C	$G(\mathbb{Q})$
$(\mathrm{III}_1)^m$	$(\mathbb{R}^+)^m$	$SL_2(F)/\{\pm 1\}$
$(\mathrm{III}_2)^m$	$\mathcal{P}_2(\mathbb{R})^m$	$SU_2(\mathbf{D}/F)/\{\pm 1\}$
$(\mathrm{I}_{p,p})^m$	$\mathcal{P}_p(\mathbb{C})^m$	$SU_2(\mathbf{D}'/F'/F)/\{\pm 1\}$

Here F is a totally real number field of degree m, $\mathbf{D}$ is a totally indefinite quarternion algebra over F, F' is a totally imaginary quadratic extension of F, $\mathbf{D}'$ is a central division algebra over F' with involution of the second kind relative to F'/F, and SU_2 denotes the special unitary group for the Hermitian form

$$\langle x,y\rangle = x_1y_2{*} + x_2y_1{*}\ , \quad x,y \in \mathbf{D}^2 \text{ (or } \mathbf{D}'^2)$$

where $x \to x^*$ denotes the involution of $\mathbf{D}$ or $\mathbf{D}'$. Let $\tilde{\Gamma}$ be a neat arithmetic subgroup of G. Then the Baily-Borel compactification of $\tilde{\Gamma}\backslash\mathcal{D}$ is obtained by adding finitely many points $\alpha_1,\dots,\alpha_r$:

$$\overline{\tilde{\Gamma}\backslash\mathcal{D}} = (\tilde{\Gamma}\backslash\mathcal{D}) \cup \{\alpha_1,\dots,\alpha_r\}.$$

Each α_j corresponds to a $\tilde{\Gamma}$-conjugacy class of $\mathbb{Q}$-rational parabolic subgroups of G. Put

$$\Gamma = \tilde{\Gamma} \cap P$$

with P given by (7.6) and consider the cusp $Y = \Gamma\backslash\mathcal{D}$. It can be partially compactified as analytic space by adding the point $\alpha_1 = \sqrt{-1}\,\infty$:

$$\bar{Y} = Y \cup \{\alpha_1\}\ .$$

A fundamental system of neighborhoods of α_1 in $\bar{Y}$ is given by $\bar{Y}_t = Y_t \cup \{\alpha_1\}$, $t > 0$. If we resolve the singular point α_1 of $\bar{Y}_1$ we get a smooth compact manifold W with boundary $\partial Y_1 = \Gamma\backslash Z$. Let X be the manifold obtained by gluing W and Y_1 along their common boundary and extend the given Riemannian metric on Y_1 to a smooth Riemannian metric on X. The compactification $\bar{X}$ of X considered above is the one-point compactification of X:

$$\bar{X} = X \cup \{\alpha_1\}.$$

As above, assume that $\dim(X) = 4k$. Since the singular stratum of $\bar{X}$ is an isolated singular point its middle intersection cohomology is given by

$$IH^p(\bar{X}) = \begin{cases} H^p(X), & p < 2k \\ \operatorname{Im}(H_c^p(X) \to H^p(X)), & p=2k \\ H_c^p(X), & p > 2k \end{cases}$$

where $H_c^p(X)$ denotes cohomology with compact supports (cf. [G-M1]). Put

$$H_!^{2k}(X) = \operatorname{Im}(H_c^{2k}(X) \longrightarrow H^{2k}(X))\ .$$

The cup product defines a non-degenerate qudratic form on $H_!^{2k}(X)$ with signature denoted by $\mathrm{Sign}(X)$. This quadratic form coincides with the intersection pairing on $IH^{2k}(\bar{X})$. Thus

$$\mathrm{Sign}_{IH*}(\bar{X}) = \mathrm{Sign}(X) \quad .$$

It remains to determine the L-function in (7.5). According to the list above we have to distinguish three cases:

$(\mathrm{III}_1)^m$: This is the Hilbert modular case which has been considered in Chapter XII of [M2] and we refer to it for details. We suppose that $m=2k$. The cross section $\Gamma\backslash Z$ of the cusp is a torus bundle

$$T^{2k} \longrightarrow \Gamma\backslash Z \longrightarrow T^{2k-1} \quad , \tag{7.7}$$

where T^l denotes a torus of dimension l. This torus bundle is determined by the following data: A totally real number field F of degree m, a complete $\mathbb{Z}$-module $\mathbf{M}$ in F and a group $\mathbf{V}$ of totally positive units ε satisfying $\varepsilon\mathbf{M} = \mathbf{M}$. Let $L(\mathbf{M},\mathbf{V},s)$ be the L-function (0.2). Then the L-function in (7.5) is given by

$$L(s) = \frac{(-1)^k}{(2\pi)^m} L(\mathbf{M},\mathbf{V},\frac{s}{m}+1) \quad .$$

Set $d(\mathbf{M}) = \mathrm{Vol}(\mathbb{R}^m/\mathbf{M})$. Then the signature formula is

$$\mathrm{Sign}(X) = \int_X L_k(p_1,\ldots,p_k) + \frac{(-1)^k}{(2\pi)^m} d(\mathbf{M})L(\mathbf{M},\mathbf{V},1) \quad . \tag{7.8}$$

By the functional equation of $L(\mathbf{M},\mathbf{V},s)$ the second term on the right hand side equals $L(\mathbf{M}^*,\mathbf{V},0)$ where $\mathbf{M}^*$ is the dual lattice which is also $\mathbf{V}$-invariant. We observe that a proof of Hirzebruch's conjecture is an immediate consequence of (7.8). After Hirzebruch [Hi,§3], the signature defect $\delta(\alpha_1)$ of the cusp α_1 is defined as follows: Since $N = \Gamma\backslash Z$ is a compact solvmanifold, its tangent bundle admits a framing which can be obtained by pushing down a left invariant framing on the solvable group Z. Let W be a compact manifold with boundary N. Using the framing of $T(W)|N$, we can push down $T(W)$ to some $SO(n)$-bundle ξ over W/N. Let $p_j \in H^{4j}(W,N;\mathbb{Z})$ be the Pontrjagin classes of ξ. Then Hirzebruch defines

$$\delta(\alpha_1) = L_k(p_1,\ldots,p_k)[W,N] - \mathrm{Sign}(W) \quad .$$

It is not difficult to show that

$$\delta(\alpha_1) = \mathrm{Sign}(X) - \int_X L_k(p_1,\ldots,p_k) \quad .$$

Together with (7.8) this implies $\delta(\alpha_1) = L(\mathbf{M}^*,\mathbf{V},0)$ which is Hirzebruch's conjecture.

$(\mathrm{III}_2)^m$: In this case we have $C = P_2(\mathbb{R})^m$. Therefore the tube domain is given by

$$\mathcal{D} = H_2 \times \cdots \times H_2$$

(m copies) where $H_2 = \{ Z \in M_2(\mathbb{C}) \mid {}^tZ = Z,\ \mathrm{Im}(Z) > 0 \}$ is the Siegel upper half-plane of degree two. Then $G \simeq Sp(2,\mathbb{R})^m$ and the parabolic group P is given by $P \simeq P_o{}^m$ where

$$P_o = \left\{ \begin{pmatrix} A & B \\ 0 & {}^tA^{-1} \end{pmatrix} \;\middle|\; A \in GL(2,\mathbb{R}),\ B \in M_2(\mathbb{R}),\ A\,{}^tB = B\,{}^tA \right\} .$$

Here $M_2(\mathbb{R})$ is the matrix algebra of degree two over $\mathbb{R}$. Note that P is the semi-direct product of

(7.9) $$L = GL(2,\mathbb{R})^m \quad \text{and} \quad V = \mathrm{Sym}_2(\mathbb{R})^m$$

where $\mathrm{Sym}_2(\mathbb{R})$ is the space of real symmetric matrices of degree two and $GL(2,\mathbb{R})$ acts on $\mathrm{Sym}_2(\mathbb{R})$ by $g\cdot x = g\,x\,{}^tg$, $g \in GL(2,\mathbb{R})$, $x \in \mathrm{Sym}_2(\mathbb{R})$.

Next we describe the $\mathbb{Q}$-structures on G. Let F be a totally real number field of degree m and $\mathbf{D}$ a totally indefinite quarternion algebra over F. "Totally indefinite" means that, for each infinite place v in F, $\mathbf{D}_v = \mathbf{D} \otimes_F F_v$ is naturally isomorphic to $M_2(F_v)$. In other words, there is a natural isomorphism

$$\mathbf{D}_{\mathbb{R}} = \mathbf{D} \otimes_{\mathbb{Q}} \mathbb{R} \simeq M_2(\mathbb{R})^m \quad .$$

Therefore $\mathbf{D}$ has m different embeddings into $M_2(\mathbb{R})$ denoted by $x \longrightarrow x^{(i)}$, $x \in \mathbf{D}$. Let $u = \begin{pmatrix} 0 & 1 \\ -1 & 0 \end{pmatrix}$. $\mathbf{D}$ has a canonical involution $x \longrightarrow x^*$ such that

$$(x^*)^{(i)} = u\,{}^t(x^{(i)})u^{-1}, \quad x \in \mathbf{D}\ ,\ i=1,\ldots,m\ .$$

The norm and the trace in $\mathbf{D}$ are given by $N_{\mathbf{D}/F}(x) = xx^*$ and $tr(x) = x + x^*$. Now observe that the vector space V can be identified with

$$V' = \{ x \in M_2(\mathbb{R})^m \mid tr(x_i) = 0,\ i=1,\ldots,m \} .$$

Set

$$V'(\mathbb{Q}) = \{ y \in \mathbf{D} \mid tr(y) = 0 \}$$

which we regard as a subspace of V' by sending $y \longrightarrow (y^{(1)},\ldots,y^{(m)})$. This determines the $\mathbb{Q}$-structure in V and therefore the $\mathbb{Q}$-structure in G. The group of $\mathbb{Q}$-rational points is then given by

$$G(\mathbb{Q}) = \left\{ S \in M_2(\mathbf{D}) \mid S \begin{pmatrix} 0 & 1 \\ 1 & 0 \end{pmatrix} S^* = \begin{pmatrix} 0 & 1 \\ 1 & 0 \end{pmatrix} \right\}$$

where $S^* = \begin{pmatrix} a^* & b^* \\ c^* & d^* \end{pmatrix}$ for $S = \begin{pmatrix} a & b \\ c & d \end{pmatrix} \in M_2(\mathbf{D})$.

Let $\mathcal{O}$ be a maximal order in $\mathbf{D}$. $\mathcal{O}$ is an additive subgroup of finite rank in $\mathbf{D}$. Put

$$\tilde{\Gamma} = G(\mathbb{Q}) \cap M_2(\mathcal{O}) .$$

Then $\tilde{\Gamma}$ is an arithmetic subgroup of G. $\tilde{\Gamma}$ may have torsion, but for our purpose it is sufficient to work with $\tilde{\Gamma}$. Set

$$\Gamma = \tilde{\Gamma} \cap P .$$

This is the discrete group which defines the cusp Y. The cross section $\Gamma\backslash Z$ is again a torus bundle $\Gamma \cap V\backslash V \longrightarrow \Gamma\backslash Z \longrightarrow \Gamma_M\backslash X_M$ which we describe now more explicitly. Let

$$M = \left\{ g \in GL(2,\mathbb{R})^m \mid \prod_{i=1}^{m} \det(g_i) = \pm 1 \right\}$$

$$A = \{ \lambda I \mid \lambda \in \mathbb{R}^+ \} .$$

Then A is a $\mathbb{Q}$-split component of P and $P = MAV$ the Langlands decomposition over $\mathbb{Q}$ with respect to A. Furthermore, let $\mathcal{O}^*$ be the group of units in $\mathcal{O}$ and $N = N_{F/\mathbb{Q}} \circ N_{\mathbf{D}/F}$. Note that

(7.10)
$$N(x) = \prod_{i=1}^{m} \det(x^{(i)}) , \quad x \in \mathbf{D} .$$

Set

$$\Lambda = \{ x \in \mathcal{O} \mid tr(x) = 0 \}$$

$$\mathcal{O}_1^* = \{ x \in \mathcal{O}^* \mid N(x) = 1 \} .$$

By sending $x \in \Lambda \longrightarrow (x^{(1)},\ldots,x^{(m)}) \in M_2(\mathbb{R})^m$ we may identify Λ with a lattice in V', and therefore with a lattice in V. The same map identifies $\mathcal{O}_1^*$ with a discrete subgroup of M. Then

$$\Gamma \cap V = \Lambda , \quad \Gamma_M = \mathcal{O}_1^*$$

and the action of Γ_M on $\Gamma \cap V$ corresponds to the natural action of $\mathcal{O}_1^*$ on Λ. We emphasize that Γ_M is now noncommutative. For $m=1$, Γ_M is the fundamental group of a compact Riemann surface Σ of genus > 1 and $\Gamma\backslash Z$ is a 3-torus bundle over the Riemann surface Σ.

To describe the L-function we have to determine the orbit structure of V with respect to the action of L and to calculate the orbital integrals occurring in (4.9). The orbits are the connected components of $\{x \in V \mid N(x) \neq 1\}$ and, using (7.9) and (7.10), it is easy to describe the orbits. Employing the orbit structure and the structure of f_t the computation of the orbital integrals can be reduced to the computation of orbital integrals for $Sp(2,\mathbb{R})$. We are not going into details here. For $\mu \in \Lambda$ let $v(\mu) = \mathrm{Vol}((\Gamma_M)_\mu \backslash M_\mu)$. Denote by Λ^* the dual lattice and set

$$L(\Lambda^*;s) = \sum_{\mu \in (\Lambda^*-0)/(\mathcal{O}_1^*)} v(\mu) \frac{\mathrm{Sign}\, N(\mu)}{|N(\mu)|^s}, \quad \mathrm{Re}(s) > 3/2.$$

Then the signature formula is

$$\mathrm{Sign}(X) = \int_X L_k(p_1,\ldots,p_k) + CL(\Lambda^*;0) \tag{7.11}$$

where C is a constant independent of $\mathbf{D}$ and $\mathcal{O}$. Now

$$\mathrm{Sign}(X) - \int_X L_k(p)$$

coincides again with the signature defect $\delta(\alpha_1)$ of the singular point α_1 so that (7.11) implies

$$\delta(\alpha_1) = CL(\Lambda^*;0).$$

This gives a proof of a generalization of Hirzebruch's conjecture in this case (cf. [S-O]).

The case $(I_{p,p})^m$ is similar.

8. CONNECTIONS WITH THE ADIABATIC LIMIT. Let X be a 4k-dimensional Riemannian manifold with a rank one cusp $Y \simeq \mathbb{R}^+ \times \Gamma\backslash Z$. For $t \geq 1$ put $X_t = X - Y_t$ where $Y_t \simeq (t,\infty) \times \Gamma\backslash Z$. Then each X_t is a compact manifold with boundary $N = \Gamma\backslash Z$ and we can apply the signature formula of [A-P-S]. Since the metric near the boundary is not a product we get an additional boundary integral [G]. Let g_t be the metric on N induced by (2.10) on the slice $\{t\} \times N$. Then

$$\mathrm{Sign}(X) = \int_{X_t} L_k + \int_{\partial X_t} TG(L_k) + \eta(N,g_t) \tag{8.1}$$

where TG is the transgression and $\eta(N,g_t)$ is the eta invariant of N with respect to the metric g_t. Now we pass to the limit $t \to \infty$. It is an easy exercise to show that the boundary term disappears in the limit and we get the following signature formula

(8.2) $$\mathrm{Sign}(X) = \int_X L_k(p) + \lim_{t\to\infty} \eta(N,g_t) .$$

Now assume that $N \xrightarrow{\pi} \Gamma_M\backslash X_M$ is a torus bundle. Let $g(x)$ be the metric on the fibre $\pi^{-1}(x)$, $x\in\Gamma_M\backslash X_M$, and denote by dx^2 the invariant metric on X_M. Then the metric g_t on N can be written as

$$g_t = dx^2 + e^{-2bt}g(x)$$

By rescaling we obtain a family metrics $\{g_\delta\}$ on N as considered by Bismut and Cheeger [B-C1],[C2]. Since rescaling leaves the eta invariant unchanged the limit in (8.2) coincides with the adiabatic limit of the eta invariant (cf. [B-C1],[C2]). If we compare (8.2) and (7.5) we get

$$\lim_{t\to\infty} \eta(N,g_t) = \mathrm{Vol}(\Gamma\cap U\backslash U)L(0) + \mathrm{Sign}(X) - \mathrm{Sign}_{IH*}(\bar{X}) .$$

for manifolds X as considered in section 7. Of particular interest are the cases when $\mathrm{Sign}(X) = \mathrm{Sign}_{IH*}(\bar{X})$. In the previous section we have seen that this is so for all manifolds X which can be compactified (in the analytic sense) by one point. If X is a manifold with $\mathrm{Sign}(X) = \mathrm{Sign}_{IH*}(\bar{X})$ we get

(8.3) $$\lim_{t\to\infty} \eta(N,g_t) = \mathrm{Vol}(\Gamma\cap U\backslash U)L(0)$$

which expresses the adiabatic limit as special value of an L-function. The adiabatic limits of eta invariants have been studied in a broader context by Bismut and Cheeger in [B-C1].In the special case of torus bundles as above we can compare their formula with (8.3). The simplest case is that of a 2-torus bundle over the circle which arises in the context of Hilbert modular surfaces.

Such a torus bundle is defined by a hyperbolic matrix $B\in SL(2,\mathbb{Z})$. Thus

$$B = \begin{pmatrix} a & b \\ c & d \end{pmatrix} , \quad |a+d| > 2.$$

Then B acts on the standard torus $T^2 = \mathbb{R}^2/\mathbb{Z}^2$ and we set

$$N = [0,1]\times T^2/\sim$$

where $(0,x) \sim (1,B(x))$, $x\in T^2$. The projection on the first factor induces a map $p\colon N \longrightarrow S^1$ which is the projection of a flat torus bundle

(8.4) $$T^2 \longrightarrow N \longrightarrow S^1$$

with holonomy B. If we regard B as a linear fractional transformation of the upper half-plane H then B has two real fixed points $w > w'$. Set $\delta = (a+d)^2-4$. Then

$$w = \frac{1}{2c}(a-d+\sqrt{\delta}), \quad w' = \frac{1}{2c}(a-d-\sqrt{\delta}).$$

Let

$$\mathbf{M} = \mathbb{Z}w \oplus \mathbb{Z} .$$

$\mathbf{M}$ is a complete $\mathbb{Z}$-module in the real quadratic field $F = \mathbb{Q}(\sqrt{\delta})$. Let $\varepsilon > 1 > \varepsilon' > 0$ be the eigenvalues of B and set

$$\mathbf{V} = \{ \varepsilon^n \mid n \in \mathbb{Z} \} .$$

Then $\mathbf{V}$ is the group of totally positive units in $\mathcal{O}_F$ with $\varepsilon\mathbf{M} = \mathbf{M}$. Put

$$\Gamma = \left\{ \begin{pmatrix} \varepsilon & \mu \\ 0 & \varepsilon^{-1} \end{pmatrix} \middle| \ \varepsilon \in \mathbf{V}, \ \mu \in \mathbf{M} \right\} .$$

Let $x \in F \longrightarrow x' \in F$ be the nontrivial automorphism of F. Then we may regard Γ as a discrete subgroup of $SL(2,\mathbb{R})^2$ by sending $\gamma \in \Gamma \longrightarrow (\gamma,\gamma') \in SL(2,\mathbb{R})^2$ where $\gamma' = \begin{pmatrix} a' & b' \\ c' & d' \end{pmatrix}$ for $\gamma = \begin{pmatrix} a & b \\ c & d \end{pmatrix}$. Γ leaves the submanifold

$$Z_t = \{ z \in H^2 \mid \operatorname{Im}(z_1)\operatorname{Im}(z_2) = t \}, \quad t > 0,$$

of H^2 invariant and $N \simeq \Gamma \backslash Z_t$. Therefore the L-function associated to the torus bundle (8.4) is the L-function $L(\mathbf{M},\mathbf{V},s)$ defined by (0.2). Moreover, if we equip Z_t with the metric induced from H^2 then $\Gamma\backslash Z_t$ is the torus bundle N equipped with the metric g_t considered above. Now we apply the formula for the adibatic limit obtained by Cheeger [C2]. By formula (A3.21) in [C2] we have

$$\lim_{t\to\infty} \eta(N,g_t) = 8\pi \int_0^{\log\varepsilon} \sum_{\mu\in\mathbf{M}-0} \frac{\mu\mu'}{|e^v\mu + ie^{-v}\mu'|^{2(s+1)}} \Bigg|_{s=0} dv =$$

$$= -4\pi i \int_0^{\log\varepsilon} \sum_{\mu\in\mathbf{M}-0} \frac{e^v\mu + ie^{-v}\mu'}{(e^v\mu - ie^{-v}\mu')\,|e^v\mu + ie^{-v}\mu'|^{2s}} \Bigg|_{s=0} dv .$$

By definition of $\mathbf{M}$ we have $\mu = mw+n$, $\mu' = mw'+n$ for $m,n \in \mathbb{Z}$. If we change variables by

$$z = \frac{e^v w + ie^{-v} w'}{e^v + ie^{-v}}$$

we obtain

$$(8.5) \qquad 2\pi \int_\gamma \sum_{(m,n)}{}' \frac{(m\bar{z}+n)\, y^{s-1}}{(mz+n)\,|mz+n|^{2s}} \Bigg|_{s=0} dz \quad , \quad y = \operatorname{Im}(z).$$

Here γ is an arc on the semi-circle in H with endpoints w,w'. The endpoints of γ are $z_o = \frac{w+iw'}{1+i}$ and $B(z_o)$. We observe that the series in (8.5) is the Eisenstein series for (1,0)-forms on the modular variety $SL(2,\mathbb{Z})\backslash H$. In fact, consider the usual Eisenstein series for $SL(2,\mathbb{Z})$:

$$E(z,s) = \sum_{m,n} \frac{y^s}{|mz+n|^{2s}} ,$$

where $z\in H$, $y = \mathrm{Im}(z)$, $\mathrm{Re}(s) > 1$ and (m,n) runs through $\mathbb{Z}^2 - 0$. It is known that this series is absolutely convergent in $\mathrm{Re}(s) > 1$ and admits a meromorphic continuation to the whole complex plane with a simple pole at $s=1$ and no other poles. The residue of $E(z,s)$ at $s=1$ equals π and therefore is independent of z. Moreover $E(z,s)$ satisfies the following functional equation:

$$(8.6)\qquad \begin{aligned} E(z,s) &= c(s)E(z,1-s) \\ c(s) &= \sqrt{\pi}\, \frac{\Gamma(s-1/2)}{\Gamma(s)} \frac{\zeta(2s-1)}{\zeta(2s)} \end{aligned}$$

where $\zeta(s)$ denotes the Riemann zeta function. A direct computation shows that the series in (8.5) equals $2is^{-1}\, \partial/\partial z E(z,s)$. Using the functional equation (8.6) and the fact that the residue of $E(z,s)$ at $s=1$ is independent of z it follows that $E(z,0)$ is independent of z. Hence $s^{-1}\partial/\partial z E(z,s)$ is holomorphic at $s=0$. Combining (8.3) with the computations above we obtain

$$(8.7)\qquad 4\pi i \lim_{s\to 0} \frac{1}{s} \int_\gamma \frac{\partial}{\partial z} E(z,s)dz = L(\mathbf{M^*},\mathbf{V},0) .$$

Of course, this equation can be also verified by direct computation (cf. [C2, p.205]). Formula (8.7) was first proved by Hecke (cf. [He,p.415]) by a method which is sometimes called the "Hecke trick".

The integral over γ has the following interpretation. The semi-circle in H with endpoints w and w' is the unique B-invariant geodesic in H and it descends to a closed geodesic in the quotient H_B of H by the infinite cyclic group generated by B. This closed geodesic can be identified with γ. We may also regard it as closed geodesic in $SL(2,\mathbb{Z})\backslash H$. However the differential form

$$(8.8)\qquad \omega = \lim_{s\to 0} \frac{1}{s} \frac{\partial}{\partial z} E(z,s)\, dz$$

is not closed so that the left hand side of (8.7) can not be interpreted as period. But Hecke has shown that certain differences of special values of L-functions as above arise as periods of integrals of the third kind [He]. We will discuss this result from our viewpoint elsewhere.

There are other interesting facts connected with (8.7). First we point out that the eta invariants $\eta(N,g_t)$ are actually independent of t. This follows from (8.1) and the fact that the first Pontrjagin form vanishes on the cusp $\Gamma\backslash H^2$. Therefore we have also the following formula

$$\eta(N,g_1) = 4\pi i \int_\gamma \omega \tag{8.9}$$

where ω is the form (8.8). This formula has the following interpretation. The torus bundle $p\colon N \longrightarrow S^1$ equipped with the metric g_1 can be considered as a family of flat 2-dimensional tori parametrized by $t\in S^1$. The flat torus $p^{-1}(t)$ determines a unique point $\gamma(t)\in SL(2,\mathbb{Z})\backslash H$. Therefore we get a smooth map $\gamma\colon S^1 \longrightarrow SL(2,\mathbb{Z})\backslash H$ and the eta invariant $\eta(N,g_1)$ is then given as $\int_{S^1}\gamma^*\omega$.

We proceed now to investigate the differential form ω. To this end, recall that $SL(2,\mathbb{Z})\backslash H$ parametrizes 2-dimensional flat tori with area 1. Such a torus is of the form $\mathbb{C}/\Lambda$, $\Lambda = \mathbb{Z}\,\omega_1 \oplus \mathbb{Z}\,\omega_2$, and we can always choose a basis ω_1, ω_2 of Λ such that

$$z = \omega_1/\omega_2 \in H .$$

Since the area is normalized to be one, z determines the torus $\mathbb{C}/\Lambda$ uniquely. Now consider the Laplacian Δ_z on the torus $\mathbb{C}/\Lambda$. Let Λ^* denote the dual lattice which is also of co-area 1 and let ω_1^*, ω_2^* be a basis of Λ^*. Since Λ is of co-area 1 we have

$$\omega_1^* = -i\nu\omega_2, \quad \omega_2^* = i\nu\omega_1 \tag{8.10}$$

with $\nu\in\{\pm 1\}$. The eigenvalues of Δ_z are $4\pi^2|\mu|^2$, $\mu\in\Lambda^*$. In view of (8.10), the eigenvalues are also given by $4\pi^2|\mu|^2$, $\mu\in\Lambda$. Hence the zeta function of Δ_z is given by

$$\zeta(s;z) = \sum_{\mu\in\Lambda-0} \frac{1}{(2\pi|\mu|)^{2s}} = (2\pi)^{-2s} \sum_{m,n}{}' \frac{1}{|m\omega_1 + n\omega_2|^{2s}} =$$

$$= (2\pi)^{-2s} \sum_{m,n}{}' \frac{\operatorname{Im}(z)^s}{|mz+n|^{2s}} = (2\pi)^{-2s} E(z,s) .$$

This implies

$$\lim_{s\to 0} \frac{1}{s}\frac{\partial}{\partial z} E(z,s) = \frac{\partial}{\partial z}\zeta'(0;z)$$

and

$$\omega = \partial_z\zeta'(0).$$

Now recall that the regularized determinant of Δ_z is defined as

$$\det{}' \Delta_z = e^{-\zeta'(0;z)} .$$

By (8.9) we get the following interesting relation between eta invariants of torus bundles $N \longrightarrow S^1$, values of L-functions and the determinant of Δ_z:

$$L(\mathbf{M^*}, \mathbf{V}, 0) = \eta(N, g_1) = -4\pi i \int_\gamma \partial_z \log \det{}' \Delta_z .$$

For the further disussion of this formula in terms of the Quillen determinant line bundle of H we refer to [A]. It will be very interesting to have similar results in the higher dimensional cases.

REFERENCES

[A] M.F.Atiyah, The logarithm of the Dedekind η-function, Math. Annalen **278**(1987), 335-380.

[A-B-P] M.F.Atiyah, R.Bott, V.K.Patodi, On the heat equation and the Index Theorem, Invent. Math. **19** (1973), 279-330.

[A-D-S1] M.F.Atiyah, H.Donnelly, I.M.Singer, Eta invariants, signature defects of cusps, and values of L-functions, Ann. of Math. **118**(1983), 131-177.

[A-D-S2] M.F.Atiyah, H.Donnelly, I.M.Singer, Signature defects of cusps and values of L-functions: The nonsplit case, Ann. of Math. **119**(1984), 635-637.

[A-P-S] M.F.Atiyah, V.K.Patodi, I.M.Singer, Spectral asymmetry and Riemannian geometry I, Math. Proc. Cambridge Philos. Soc. **77**(1975), 43-69.

[A-S] M.F.Atiyah, W.Schmid, A geometric construction of the discrete series for semisimple Lie groups, Invent. Math. **42**(1977), 1-62.

[A-H-S] J.Arthur, R.A.Herb, P.J.Sally, Jr., The Fourier transform of weighted orbital integrals on SL(2,ℝ), in:"The Selberg trace formula and related topics", Contemp. Math. **53**, AMS, Providence, 1985, 17-37.

[Ar] J.Arthur, The L^2-Lefschetz numbers of Hecke operators, Preprint.

[B-B] W.L.Baily, Jr., A.Borel, Compactification of arithmetic quotients of bounded symmetric domains, Ann. of Math. **84**(1966), 442-528.

[B-M] D.Barbasch, H.Moscovici, L^2-index and the Selberg trace formula, J. Funct. Analysis **53**(1983), 151-201.

[B-C1] J.M.Bismut, J.Cheeger, Eta invariants and their adiabatic limits, Preprint.

[B-C2] J.M.Bismut, J.Cheeger, Families index for manifolds with boundary, superconnections and cones, Preprint IHES/M/88/26, 1988.

[Bo] A.Borel, L^2-cohomology and intersection cohomology of certain arithmetic varieties, Proc. of E.Noether Symposium at Bryn Mawr, Berlin-Heidelberg-New York, Springer-Verlag (1983), 119-131.

[B-C] A.Borel, W.Casselman, L^2-cohomology of locally symmetric manifolds of finite volume, Duke Math. J. **50**(1983), 625-647.

[Br] J.Brüning, L^2-index theorems on certain complete manifolds, Preprint Universität Augsburg, Nr. 173, 1988.

[C1] J.Cheeger, Spectral geometry of singular Riemannian spaces, J. Diff. Geometry **18**(1983), 575-657.

[C2] J.Cheeger, η-invariants, the adiabatic approximation and conical singularities, J. Diff. Geometry **26**(1987), 175-221.

[Co] A.Connes, A survey of foliations and operator algebras, Proc. Sympos. Pure Math., No. **38**, Amer. Math. Soc., Providence, RI, 1982, 521-628.

[C-M] A.Connes, H.Moscovici, The L^2-index theorem for homogeneous spaces of of Lie groups, Ann. of Math. **115**(1982), 291-330.

[C-S] A.Connes, G.Skandalis, The longitudinal index theorem for foliations, Publ. Res. Inst. Math. Soc. **20**(1984), 1139-1183.

[G] P.B.Gilkey, The boundary integrand in the formula for the signature and Euler characteristic of a Riemannian manifold with boundary, Advances in Math. **15**(1975), 334-360.

[G-M1] M.Goresky, R.MacPherson, Intersection homology theory, Topology **19** (1980), 135-162.

[G-M2] M.Goresky, R.MacPherson, Intersection homology theory II, Inventiones Math. **72**(1983), 77-129.

[G-L] M.Gromov, H.B.Lawson, Positive scalar curvature and the Dirac operator Publ. Math. I.H.E.S. **58**(1983), 83-196.

[Ha] G.Harder, On the cohomology of discrete arithmetically defined groups, In: Proc. Intern. Colloquium on Discrete Subgroups of Lie Groups and Applications to Moduli, Bombay, 1973, pp. 129-160, Oxford Univ. Press 1975.

[H] Harish-Chandra, Discrete series for semisimple Lie groups, I, Acta Math. **113**(1965), 241-318, II, Acta Math. **116**(1966), 1-111.

[He] E.Hecke, Darstellung von Klassenzahlen als Perioden von Integralen 3. Gattung aus dem Gebiet der elliptischen Modulfunktionen, Mathematische Werke, Göttingen, Vandenhoeck & Ruprecht, 1959, pp.405-417.

[Hr] R.A.Herb, Discrete series characters and Fourier inversion on semisimple real Lie groups, Trans. Amer. Math. Soc. **277**(1983), 241-262.

[Hi] F.Hirzebruch, Hilbert modular surfaces, L'Enseign. Math. **19**(1973), 183-281.

[Mi] R.J.Miatello, Alternating sum formulas for multiplicities in $L^2(\Gamma\backslash G)$, II, Math. Zeitschrift **182**(1983), 35-44.

[Mo] H.Moscovici, L^2-index of elliptic operators on locally symmetric spaces of finite volume, In: Operator Algebras and K-Theory, Contemp. Math. **10**(1982), Amer. Math. Soc., Providence, R.I., pp.129-138.

[M1] W.Müller, Signature defects of cusps of Hilbert modular varieties and values of L-series at s=1, J. Diff. Geometry **20**(1984), 55-119.

[M2] W.Müller, Manifolds with cusps of rank one: Spectral theory and L^2-index theorem, Lect. Notes in Math. **1244**, Springer-Verlag, Berlin-Heidelberg-New York, 1987.

[M3] W.Müller, Spectral theory of noncompact Riemannian manifolds with cusps and a related trace formula, Preprint IHES/M/80/46, Paris 1980.

[M4] W.Müller, L^2-index and resonances, In: Geometry and Analysis on Manifolds, Lect. Notes in Math. **1339**, Springer-Verlag, Berlin-Heidelberg-New York, 1988, pp.203-211.

[Ra] R.R.Rao, Orbital integrals in reductive groups, Ann. of Math. **96**(1972) 505-510.

[R] J.Roe, An index theorem on open manifolds, I, J. Diff. Geometry **27** (1988), 87-113, II, J. Diff. Geometry **27**(1988), 115-136.

[S-O] I.Satake, S.Ogata, Zeta functions associated to cones and their special values, Preprint MSRI 01520-87, Berkeley 1987.

[Sa] F.Sato, Zeta functions in several variables associated with prehomogeneous vector spaces I: Functional equations, Tôhoku Math. J. **34**(1982), 437-483.

[S-S] M.Sato, T.Shintani, On zeta functions associated with prehomogeneous vector spaces, Ann. of Math. **100**(1974), 131-170.

[Sh] H.Shimizu, On discontinuous groups operating on the product of upper half planes, Ann. of Math. **77**(1963), 33-71.

[St] M.Stern, L^2-index theorems on locally symmetric spaces, Preprint.

[W1] G.Warner, Harmonic Analysis on Semi-Simple Lie Groups, II, Springer-Verlag, Berlin-Heidelberg-New York, 1972.

[W2] G.Warner, Noninvariant integrals on semisimple Lie groups of $\mathbb{R}$-rank one, J. Functional Analysis **64**(1985), 19-111.

AKADEMIE DER WISSENSCHAFTEN DER DDR
KARL-WEIERSTRASS-INSTITUT FÜR MATHEMATIK
DDR-1086 BERLIN, MOHRENSTRASSE 39
GERM. DEM. REPUBLIC

Contemporary Mathematics
Volume **105**, 1990

NON-COMMUTATIVE TORI — A CASE STUDY OF NON-COMMUTATIVE DIFFERENTIABLE MANIFOLDS

Marc A. Rieffel*

ABSTRACT. The non-commutative tori are perhaps the most accessible and best studied interesting examples of non-commutative differentiable manifolds. We give a survey of many of the results which have been obtained about them.

Led by Alain Connes [Cn1, Cn3, Cn4], we have during this decade gotten a glimpse of a new kind of mathematical object, namely non-commutative differentiable manifolds. So far no one has given a satisfactory definition for these objects. But by now a number of naturally arising examples are known which will surely be included when a good definition is found. There is every indication that large parts of the mathematics which one does on ordinary manifolds will be extended to non-commutative manifolds. Substantial motivation for doing this comes from the contributions to issues in other areas of mathematics which such extensions will provide [Be1, Cn3, CM, RS1, Rs2]. In particular, one will study the index of elliptic operators on non-commutative differentiable manifolds. This is already explicitly indicated in Connes' original paper on non-commutative differential geometry [Cn1], and this is one of the ideas lying behind his work on index theorems for foliated manifolds [Cn2, CS].

Probably the most accessible interesting class of non-commutative differentiable manifolds are the non-commutative tori; they are surely the best understood, although many questions about them still remain open. In this report I will try to survey much of what is known, and indicate some of the open questions. It is my hope that readers of this report will come to feel that non-commutative tori are not exotic objects, but rather are attractive well-behaved objects closely associated with classical situations.

1. THE DEFINITION OF NON-COMMUTATIVE TORI. Non-commutative tori arise naturally in a number of different situations. For example, they play a certain universal role in the representation theory of Lie groups [Pg], and they

---

1980 Mathematics Subject Classification (1985 revision): 46L80, 22D25.
*Supported in part by National Science Foundation grant DMS 8601900.

provide a convenient framework within which to study Schrodinger operators with quasi-periodic potentials [Bel]. But here we will choose to approach them from the direction which most strongly suggests their relationship to ordinary manifolds, namely from the direction of deformation quantization [Rf8, Rf9].

Let T^n be an ordinary n-torus. We will often use real coordinates for T^n, that is, view T as R/Z. Let $C^\infty(T^n)$ be the algebra of infinitely differentiable complex-valued functions on T^n. Pick a real skew-symmetric n×n matrix θ. Then we can use θ to define a Poisson bracket on $C^\infty(T^n)$ by

$$\{f, g\} = \Sigma\, \theta_{jk}(\partial f/\partial x_j)(\partial g/\partial x_k)$$

for $f, g \in C^\infty(T^n)$. The idea of deformation quantization is to seek to deform the pointwise product of $C^\infty(T^n)$ to a one-parameter family of associative products, $*_\hbar$, in such a way that

$$\lim_{\hbar\to 0} (f*_\hbar g - g*_\hbar f)/i\hbar = \{f, g\} .$$

To this end, use the Fourier transform to carry $C^\infty(T^n)$ to $S(Z^n)$, the space of complex-valued Schwartz functions on Z^n. Then the Fourier transform carries the pointwise multiplication on $C^\infty(T^n)$ to convolution on $S(Z^n)$. A simple computation shows that it also carries the Poisson bracket to

$$\{\phi, \psi\}(p) = -4\pi^2 \sum_q \phi(q)\psi(p - q)\gamma(q, p - q)$$

for $\phi, \psi \in S(Z^n)$ and $p, q \in Z^n$, where

$$\gamma(p, q) = \Sigma\, \theta_{jk}p_j q_k ,$$

and where the factor $4\pi^2$ comes from our convention that the Fourier transform, $\hat{f}$, for $f \in C^\infty(T^n)$ is defined by

$$\hat{f}(p) = \int_{T^n} \exp(-2\pi i x\cdot p)f(x)dx .$$

For every $\hbar \in R$ define a bicharacter, $\sigma_\hbar$, on Z^n by

$$\sigma_\hbar(p, q) = \exp(-\pi i\hbar\gamma(p, q)) ,$$

and then set

$$(\phi*_\hbar\psi)(p) = \sum_q \phi(q)\psi(p - q)\sigma_\hbar(q, p - q) .$$

Define the involution on $S(Z^n)$, independent of $\hbar$, to be that coming from complex conjugation on $C^\infty(T^n)$, so that

$$\phi^*(p) = \bar{\phi}(-p) .$$

For each $\hbar$ define the norm $\|\ \|_\hbar$ on $S(Z^n)$ to be the operator norm for the action of $S(Z^n)$ on $\ell^2(Z^n)$ given by the same formula as used above to define the product $*_\hbar$. Let $C_\hbar$ be $C^\infty(T^n)$ but with product, involution, and norm obtained by pulling back through the inverse Fourier transform the product $*_\hbar$, involution, and norm $\|\ \|_\hbar$. Then one can show [Rf8, Rf9] that the completions of the $C_\hbar$'s form a continuous field of C^*-algebras, and that for $f, g \in C^\infty(T^n)$ one has

$$\|(f*_\hbar g - g*_\hbar f)/i\hbar - \{f, g\}\|_\hbar \longrightarrow 0$$

as $\hbar \longrightarrow 0$. Since C_0 is just $C^\infty(T^n)$ with its usual pointwise multiplication and supremum norm, this all means that the $C_\hbar$'s form a strict deformation quantization of $C^\infty(T^n)$ in the direction of the Poisson bracket coming from θ, as defined in [Rf8].

We denote the algebra for $\hbar = 1$ by A_θ. Since A_θ is obtained by deforming $C^\infty(T^n)$, it is natural to call A_θ a non-commutative n-torus — the "C^∞" version. Let $\bar{A}_\theta$ denote the norm completion of A_θ, so that $\bar{A}_\theta$ is a C^*-algebra. This will be a deformation of $C(T^n)$, and so is considered the "topological" version of the non-commutative torus determined by θ.

It will be important for us at various points that, as is easily seen, the action of T^n by translation on $C^\infty(T^n)$ is also an action by continuous $*$-algebra automorphisms for the product $*_\hbar$ (that is, the deformation quantization constructed above is T^n-invariant), and so gives an action of T^n on $\bar{A}_\theta$, called the dual action. This dual action is easily seen to be ergodic in the sense that the only invariant elements of $\bar{A}_\theta$ are the scalar multiples of the identity element. In fact, it is shown in [OPT] that the $\bar{A}_\theta$'s are exactly all unitial C^*-algebras admitting ergodic actions of T^n (with full spectrum if the action is to be faithful).

Now for each $p \in Z^n$ the function $t \longmapsto \exp(2\pi i t\cdot p)$ will correspond to a unitary operator, U_p, in $\bar{A}_\theta$, and the mapping $p \longmapsto U_p$ will be a projective unitary representation of Z^n. Thus $\bar{A}_\theta$ can be viewed as the C^*-algebra generated by this representation. Alternatively, if we let $U_1, \cdots, U_n$ denote the unitary operators corresponding to the standard basis for Z^n, then they already generate $\bar{A}_\theta$ and satisfy the relation

$$U_k U_j = \exp(2\pi i\theta_{jk}) U_j U_k ,$$

and it is not difficult to show that $\bar{A}_\theta$ is (isomorphic to) the universal C^*-algebra generated by n unitary operators which satisfy the above relations.

When $n = 2$, the skew-symmetric matrix θ is just determined by a real number, again denoted θ, and A_θ will be isomorphic to the crossed product C^*-algebra for the action of Z on the circle T coming from rotation by angle $2\pi\theta$. For this reason these algebras are often called "rotation algebras" [AP], or "irrational rotation algebras" when θ is irrational [Rf2].

Going back to the general case, let us define a functional, τ, on $S(Z^n)$ by

$$\tau(\phi) = \phi(0) \quad .$$

Note that τ is just the Fourier transform of Lebesgue measure on T^n. Let α denote the dual action of T^n on $\bar{A}_\theta$ defined earlier. Then it is easily seen that τ is invariant for α, and in fact that

$$\tau(\phi) = \int_{T^n} \alpha_t(\phi)dt \quad ,$$

where the left-hand side should initially be viewed as $\tau(\phi)I$, where $I = U_0$ is the identity element of $\bar{A}_\theta$. From this, one sees that τ extends to a trace τ on $\bar{A}_\theta$, which is faithful. And that the only α-invariant traces on $\bar{A}_\theta$ are the scalar multiples of τ [Sl, Gr, OPT]. The Hilbert space obtained by applying the GNS construction [KR] to τ is easily identified with $\ell^2(Z^n)$, with the U_p's as an orthonormal basis.

For every $a \in \bar{A}_\theta$ one can define its "Fourier coefficients" $\{a_p\}$ by

$$a_p = \tau(aU_p^*) \quad .$$

Because τ is faithful, an element of $\bar{A}_\theta$ will be determined by its "Fourier coefficients", though, just as with $C(T^n)$, it can be difficult to tell when a given function on Z^n is the set of "Fourier coefficients" for an element of $\bar{A}_\theta$. But at any rate, one can do for non-commutative tori many of the same maneuvers which one does for ordinary Fourier series in n variables. This is additional justification for using the term "non-commutative tori".

Because the non-commutative tori are so closely related to ordinary tori, it is natural to expect that many of their properties will be similar to those for ordinary tori. In the next sections we will first discuss this question at the "topological" level, and then at the "C^∞" level.

2. "TOPOLOGICAL" PROPERTIES OF NON-COMMUTATIVE TORI. If θ has sufficient irrationality so that whenever $\theta(p, Z^n) \subset Z$ for some $p \in Z^n$ it follows that $p = 0$, then $\bar{A}_\theta$ is simple (as is A_θ), i.e. has no proper two-sided ideal [Sl, Gr, OTP]. In this sense A_θ is then as far away as possible from being an algebra of functions on a topological space. One can show in this case that the only traces on $\bar{A}_\theta$ are the scalar multiples of τ [Sl, Gr, OTP].

If θ is rational, then $\bar{A}_\theta$ is a bundle of full matrix algebras over a T^n, the matrix size depending on the denominators of θ [Gr, HS, OTP, Rf4]. In particular, $\bar{A}_\theta$ is strongly Morita equivalent to $C(T^n)$ [Rf3]. But these bundles are almost never trivial (i.e. product) bundles [HK, Rf4, DB].

Even though $\overline{A}_\theta$ is far from being a function algebra when it is simple, we will see in section 4 that it is still important to extend to it the notion of the classical dimension of a compact space [Pr]. This can be done in the following way, described in more detail in [Rf4]. Let M be a compact space. Then a standard theorem from classical dimension theory (proposition 3.3.2 of [Pr]) says that the classical dimension of M is the least integer n such that every continuous function f from M into R^{n+1} can be approximated arbitrarily closely by functions which do not contain the origin in their ranges. Now such a function f is an (n+1)-tuple, $f_1, \cdots, f_{n+1}$, of real-valued functions, and the condition that f miss the origin is just the condition that the f_i's nowhere vanish simultaneously. Let $C_R(M)$ denote the Banach algebra of real-valued continuous functions on M. Then this last condition just says that the ideal of $C_R(M)$ generated by the f_i's all together is $C_R(M)$. To generalize this to non-commutative algebras, one must choose to use either left or right ideals, though for algebras with involution this choice will make no difference. For A an algebra with identity element, we let $Lg_n(A)$ denote the set of n-tuples of A which generate A as a left ideal. When A is a Banach algebra, the theorem from classical dimension theory stated above indicates that we are interested in when $Lg_n(A)$ is dense in A^n. But we are interested in algebras over the complex numbers, and a complex-valued function on a compact space M will correspond to two real-valued functions. Because of this, it will not be appropriate to use the term "dimension", and so instead we will use the term "topological stable rank".

2.1 Definition [Rf4]. Let A be a Banach algebra with identity element. By the ***left topological stable rank*** of A, denoted ltsr(A), we mean the least integer n such that $Lg_n(A)$ is dense in A^n (= ∞ if no such integer exists).

Then for a compact space M we will have $tsr(C(M)) = [\dim(M)/2] + 1$ where [ ] denotes "integer part of". (We use *tsr* instead of *ltsr* for algebras with involution, where the choice of left or right ideals gives the same result.)

An upper bound on the tsr of non-commutative tori can be obtained as follows. Any non-commutative torus can be constructed as a succession of crossed products by actions of the group Z, starting from a trivial action on the one-dimensional algebra $\mathbb{C}$ [El2, Rf6]. But a basis result [Rf4] about tsr is that if α is an action of Z on a C^*-algebra A, then

$$tsr(A \times_\alpha Z) \leq tsr(A) + 1 \quad .$$

Thus we see that if $\overline{A}_\theta$ is a non-commutative torus based on Z^n, then $tsr(\overline{A}_\theta) \leq n + 1$. Actually, when θ is not rational one can show that $tsr(\overline{A}_\theta)$ is no larger than 2, and must be equal to 2 if $\overline{A}_\theta$ is not simple.

But Riedel [Rdl, AP] has shown that, surprisingly, for some simple $\overline{A}_\theta$ one actually has $tsr(\overline{A}_\theta) = 1$. How often this happens is unclear. That is, a very open question is:

2.2 QUESTION: Is $tsr(\overline{A}_\theta) = 1$ whenever $\overline{A}_\theta$ is simple?

Let us mention that having $tsr(A) = 1$ is equivalent to the invertible elements being dense in A [Rf4]. While I was making final corrections to this manuscript, Ian Putnam told me by telephone that he believes he has a proof that the answer to Question 2.2 is always affirmative for irrational rotation C*-algebras.

Related to the above is the very recent result of Choi and Elliott [CE] that for a dense set of θ's, any self-adjoint element of the irrational rotation C*-algebra $\overline{A}_\theta$ can be approximated in norm by self-adjoint elements which have finite spectrum.

3. VECTOR BUNDLES AND K-THEORY. Let M be a compact space and let E be a complex vector-bundle over M. Then the space, $\Gamma(E)$, of continuous cross-sections of E is a finitely generated projective module over M, and all finitely generated projective C(M)-modules arise in this way, according to a theorem of Swan [Sw, Rf3]. Since we think of C*-algebras with identity element as being "non-commutative compact spaces", it is then natural to view projective modules over them as being the generalization of vector bundles. (For brevity we will say "projective" when we mean "finitely generated projective".)

Given any ring A with identity element, it is of great interest to classify the isomorphism classes of projective A-modules. With the operation of forming direct sums, these classes form an Abelian semigroup, which we will denote by S(A). So one would like to describe S(A). But many examples show that S(A) need not satisfy the cancellation law, including the case of $A = C(T^n)$ for $n \geq 5$. This tends to make it difficult to describe S(A). A potentially easier problem is first to describe the cancellative semigroup, C(A), formed as the quotient of S(A) by forcing cancellation, that is, by decreeing that two elements, r and s, of S(A) are equivalent if there is a $t \in S(A)$ such that $r + t = s + t$. But I believe that this problem also is unsolved for $C(T^n)$ for large n — it is certainly known that the situation becomes complicated [Rf6, Bl]. Finally, C(A) embeds in its enveloping Abelian group ("Grothendieck groups"), denoted $K_0(A)$. This group is part of a homology theory periodic of period 2. The other group, $K_1(A)$, is defined to be the inductive limit of the groups $GL_k(A)/GL_k^0(A)$, where $GL_k(A)$ is the group of invertible $k{\times}k$ matrices over A while $GL_k^0(A)$ is its connected component, and $GL_k(A)$ is embedded in $GL_{k+1}(A)$ by $T \longmapsto T\oplus 1$. See [B1]. When X is a compact space and $A = C(X)$, these groups are just the groups of topological K-theory.

The first major break-through in finding techniques for calculating the K-groups of non-commutative C^*-algebras was made by Pimsner and Voiculescu [PV1, B1]. They obtained a periodic exact sequence for the K-groups of a crossed product of form $A\times_\alpha Z$, in terms of the K-groups of A and of the effect of α on the K-groups of A. Because, as mentioned in the last section, a non-commutative torus $\bar{A}_\theta$ can be constructed as a succession of crossed products for actions of Z, it is easy to deduce from the Pimsner-Voiculescu exact sequence that

$$K_0(\bar{A}_\theta) \cong Z^{2^{n-1}} \cong K_1(\bar{A}_\theta) \quad ,$$

just as happens for the topological K-groups of an ordinary n-torus T^n. Note in particular that the K-groups do not distinguish between the algebras $\bar{A}_\theta$ for different θ.

Another notable result of Pimsner and Voiculescu [PV2] of the same vintage as their exact sequence, is that an irrational rotation algebra $\bar{A}_\theta$ can be embedded in a specific AF C^*-algebra, B_θ, in such a way that the corresponding map on the K_0-groups is an isomorphism. We recall that an AF C^*-algebra is just an inductive limit of finite dimensional C^*-algebras, and that the AF C^*-algebras are considered to be the non-commutative analogues of Cantor sets. (Note that $\bar{A}_\theta$ itself is not an AF C^*-algebra since the K_1 group of any AF C^*-algebra is trivial.) That this embedding is remarkable is seen by observing that the corresponding statement for spaces would say that one has a compact space which is the quotient of a Cantor set in such a way that the quotient map gives an isomorphism of their K_0 groups. This seldom happens, and in particular does not happen for ordinary tori.

To carry the story further, Kumjian has shown [Km1] that B_θ can be embedded (unitally) in $\bar{A}_\theta$. By iterating this embedding and that of Pimsner and Voiculescu, he concludes in [Km2] that the AF algebra B_θ is the increasing limit of a sequence of subalgebras each of which is isomorphic to $\bar{A}_\theta$, with the inclusion maps all giving isomorphisms of the K_0 groups.

4. NON-STABLE K-THEORY. The K-groups can be viewed as defined by suitable stabilizations. The study of what happens before stabilizing, that is, of the semigroup $S(A)$ itself, and of the groups $GL_k(A)/GL_k^0(A)$ before taking the limit, is often called non-stable K-theory. Although for ordinary n-tori T^n the semigroup $S(T^n)$ is badly behaved and not yet fully understood, for large n, it turns out that as soon as θ has any irrational entries, $S(\bar{A}_\theta)$ is well-

behaved, and can be fully described [Rf6]. To explain this, we first mention that the canonical trace τ on $\bar{A}_\theta$ (the "Lebesgue measure") determines a group homomorphism, denoted again by τ, of $K_0(\bar{A}_\theta)$ into R, obtained by extending τ to a trace on matrices over $\bar{A}_\theta$ in the evident way, and then by evaluating this extended trace on the projection matrices which define projective modules. The main result of [Rf6] says that as soon as θ has at least one irrational entry, then cancellation holds in $S(\bar{A}_\theta)$ (so $S(\bar{A}_\theta) = C(\bar{A}_\theta)$), and that the embedding of $S(\bar{A}_\theta)$ into $K_0(\bar{A}_\theta)$ identifies $S(\bar{A}_\theta)$ with exactly the set of elements of $K_0(\bar{A}_\theta)$ on which τ is positive. Furthermore, it is shown in this case how to construct all projective modules over $\bar{A}_\theta$, up to isomorphism (theorems 6.1 and 7.1 and corollary 7.2 of [Rf6]). The proof depends in a crucial way on the information about $\mathrm{tsr}(\bar{A}_\theta)$ described in the previous section, as well as the differential geometric techniques which we will discuss shortly.

The following is a prototypical example of a (non-free) projective module for a non-commutative two-torus [Cn1, Cn3, Rf2, Rf5]. In this case θ is specified by just one real number, which we again denote by θ. Let $\lambda = \exp(2\pi i\theta)$, and let U and V be two unitary generators for $\bar{A}_\theta$ satisfying the commutation relation $VU = \lambda UV$. Let $S(R)$ denote the space of Schwartz functions on R. We let U act on $S(R)$ by translation by θ, and V act by multiplication by $t \longmapsto \exp(2\pi it)$, and extend these actions to finite sums of products of powers of U and V. In the next section we will indicate how to define an inner-product on $S(R)$ with values in $\bar{A}_\theta$, and a corresponding norm. When $S(R)$ is completed for this norm, it becomes a projective $\bar{A}_\theta$-module which is not free [Cn1, Cn3, Rf2, Rf5]. This is certainly not an exotic object. The projective modules over non-commutative n-tori for higher n, for θ not rational, can all be decomposed as finite direct sums of projective modules which are suitable higher dimensional generalizations of the above module (corollary 7.2 of [Rf6]).

Already for 2-tori the contrast in non-stable K-theory between the rational and irrational cases is quite interesting. For T^2 the cancellation property does hold (by theorem 1.5 of chapter 8 of [Hs]), and $S(T^2)$ defines a positive cone in $K^0(T^2)$, as does $S(\bar{A}_\theta)$ in $K_0(\bar{A}_\theta)$. But for T^2, if one identifies $K^0(T^2)$ ($\cong Z^2$) with the integer lattice points in the plane, this positive cone can be identified with the integer lattice points in the upper half plane. In particular, $K^0(T^2)$ is not totally ordered by $S(T^2)$. But for θ irrational it can be shown [PV1, Rf2] that τ identifies $K_0(\bar{A}_\theta) \cong Z^2$ with the dense subgroup $Z + Z\theta$ of the real line, and that then $S(\bar{A}_\theta)$ is identified with the elements of $Z + Z\theta$ which are positive real numbers. In particular, $K_0(\bar{A}_\theta)$

is totally ordered by $S(\bar{A}_\theta)$.

One important consequence of this information about the positive cone is an answer to the question of when, given two different irrational numbers θ and θ', the corresponding algebras are isomorphic (a question which had remained open for some years). The answer is that they are isomorphic if and only if $\theta' = \pm(\theta + k)$ for some integer k [PV1, PV2, Rf2]. This answer was one of the first striking applications of K-theoretic methods to C^*-algebras.

For higher dimensional non-commutative tori these techniques are not powerful enough to settle the isomorphism question, and the situation remains tantalizingly unclear in spite of considerable effort [CEGJ, DEKR, Th], that is:

4.1 QUESTION: Given skew $n \times n$ matrices, θ and θ', when are $\bar{A}_\theta$ and $\bar{A}_{\theta'}$ isomorphic?

The strongest partial results I am aware of are given in [BCEN], which also contains partial results concerning when the corresponding smooth algebras A_θ are isomorphic. For information on the case when θ is rational see [DEKR, Br3, Db, Rf5, Ym]. See also [Rh1, Rh2].

While the range of the trace on $K_0(\bar{A}_\theta)$ does not give enough information to answer the isomorphism question for $n > 2$, it is still of much interest. Elliott [El2] has given the following elegant description of the range of the trace. View θ as a (nilpotent) element of the even exterior algebra $\Lambda^e R^n$, so that we can form the element $\exp(\theta)$ of $\Lambda^e R^n$. Let D denote the integral lattice in L^*, so that we can view $\Lambda^e D$ as the integral lattice in $\Lambda^e L^*$. Then the range of the trace is obtained by the pairing $\langle \exp(\theta), \Lambda^e D \rangle$. A proof of this within the context of Connes theory of n-traces is contained in Pimsner's paper [Pm].

The non-stable K-theory of non-commutative tori for θ not rational has further attractive properties. For example, the projective submodules of $\bar{A}_\theta$ as right module over itself already generate $K_0(\bar{A}_\theta)$ (corollary 7.10 of [Rf5]), and any two projections in a matrix algebra $M_k(\bar{A}_0)$ which determine isomorphic projective modules will be in the same path component of the set of projections in $M_k(\bar{A}_\theta)$ (theorem 8.13 of [Rf5]). For K_1 one finds that the natural map from $GL_k(\bar{A}_\theta)/GL_k^0(\bar{A}_\theta)$ to $K_1(\bar{A}_\theta)$ is an isomorphism for all $k \geq 1$ (theorem 8.3 of [Rf5]).

For any C^*-algebra A the group $K_1(A)$ is closely related to the homotopy groups of the groups $GL_k(A)$ [Bl]. By using the results on the non-stable K-theory of A_θ described above, one can show [Rf7] that for θ not rational one has

$$\pi_m(GL_k(\bar{A}_\theta)) \cong Z^{2^{n-1}}$$

for all integers $m \geq 0$ and $k \geq 1$.

5. HERMITIAN METRICS. Just as it is useful to equip a vector space with an inner-product, it is useful to equip a vector bundle, E, with inner-products on each fiber chosen in a continuous way, that is, with a Hermitian metric. We need a similar structure for projective modules over a C^*-algebra [Cnl]. To see what this should be, we note that it is natural to consider, for $\xi, \eta \in \Gamma(E)$, the function $\langle\xi, \eta\rangle_A$ on M defined by

$$\langle\xi, \eta\rangle_A(m) = \langle\xi(m), \eta(m)\rangle \quad .$$

It will be continuous, so in $A = C(M)$. Then $\langle\ ,\ \rangle_A$ can be considered to be an A-valued inner product on $\Xi = \Gamma(E)$. If the inner products on the fibers are chosen to be linear in the second variable, as will be convenient, then $\langle\ ,\ \rangle_A$ satisfies

1) $\langle\xi, \eta a\rangle_A = \langle\xi, \eta\rangle_A a$

2) $\langle\xi, \eta\rangle_A^* = \langle\eta, \xi\rangle_A$

3) $\langle\xi, \xi\rangle_A \geq 0$

for $\xi, \eta \in \Xi$ and $a \in A$. One will also have definiteness, that is, if $\langle\xi, \xi\rangle_A = 0$ then $\xi = 0$. But this would still be satisfied if the fiber inner-product at one non-isolated point of M were zero while all other fiber inner-products were definite. Thus we need to consider the stronger property of self-duality, which holds exactly if all the fiber inner-products are definite, namely:

4) For any linear map, ϕ, from Ξ to A such that $\phi(\xi a) = \phi(\xi)a$ for all $\xi \in \Xi$ and $a \in A$, there is an $\eta \in \Xi$ such that $\phi(\xi) = \langle\eta, \xi\rangle_A$ for all $\xi \in \Xi$.

5.1 **Definition.** Let A be a unital C^*-algebra, and let Ξ be a projective right A-module. By a Hermitian metric on Ξ we mean a bi-additive A-valued function $\langle\ ,\ \rangle_A$ on $\Xi\times\Xi$ which satisfies properties 1 to 4 above.

A projective module can always be equipped with a Hermitian metric (in many ways) by viewing it as a summand of a free module and restricting to it the standard Hermitian metric on the free module. Given a Hermitian metric on a projective right A-module Ξ, it is natural to define a norm on Ξ by

$$\|\xi\| = \|\langle\xi, \xi\rangle_A\|^{\frac{1}{2}} \quad .$$

As an example, consider the projective module defined in the previous section. With the notation used there, we want to define a Hermitian metric on S(R) (and then complete for the corresponding norm). For $\xi, \eta \in S(R)$ we can hope to write $\langle\xi, \eta\rangle_A$ as a finite sum

$$\langle \xi, \eta \rangle_A = \Sigma \langle \xi, \eta \rangle_A(m, n) U^m V^n$$

for suitable coefficients $\langle \xi, \eta \rangle_A(m, n)$. It turns out that the appropriate formula is

$$\langle \xi, \eta \rangle_A(m, n) = \int \overline{\xi}(r) \eta(r - m\theta) \exp(-2\pi inr) dr \quad .$$

Motivation for this formula can be found in [Rf1]. I hope that the reader will consider this formula to be nothing especially exotic, but rather quite similar to formulas found in traditional harmonic analysis.

6. SMOOTH STRUCTURE. As mentioned earlier, the action of T^n by translation on $C^\infty(T^n)$ gives an action, α, on $\overline{A}_\theta$, which leaves the canonical trace τ invariant. On the Fourier space $S(Z^n)$ this action is given by

$$(\alpha_t(\phi))(p) = \exp(2\pi it \cdot p)\phi(p)$$

for $t \in T^n$, $p \in Z^n$ and $\phi \in S(Z^n)$. From this it is not difficult to see that the space of C^∞-vectors for this action on $\overline{A}_\theta$ is just $C^\infty(T^n) \sim S(Z^n)$ itself, that is, our original A_θ. Let δ_k denote differentiation on A_θ in the kth direction of T^n, that is, the infinitesimal generator for the action of translation on A_θ in the kth direction of T^n. Then δ_k is given by

$$(\delta_k(\phi))(p) = 2\pi ip_k\phi(p) \quad .$$

Each δ_k is a $*$-derivation of A_θ, that is, satisfies

1) $(\delta_k(a))^* = \delta_k(a^*)$

2) $\delta_k(ab) = \delta_k(a)b + a\delta_k(b)$.

These directional derivatives can be used to form "partial differential operators" on A_θ. For example, the Laplace operator is

$$\Delta = \sum_k \delta_k^2 \quad .$$

Note that Δ coincides in its action on the linear space A_θ, with the usual Laplace operator on $C^\infty(T^n)$. Hence, as Connes has indicated [C1], $(1 - \Delta)^{-1}$ is a compact operator on $L^2(\overline{A}_\theta, \tau)$. One can also define linear partial differential operators with "non-constant coefficients". For example a first order operator would be of the form

$$\Sigma\, a_k \delta_k$$

for $a_k \in A_\theta$. Connes indicates [C1] that one can also develop an appropriate calculus of pseudodifferential operators, index theorems for elliptic operators, etc. In a different direction, Ji has studied [Ji] a generalization to non-commutative tori of Toeplitz operators.

A very important property of A_θ as a subalgebra of $\overline{A}_\theta$ is that it is closed under the holomorphic functional calculus of $\bar{A}_\theta$, in the sense that if $a \in A_\theta$ and if f is a function analytic in a neighborhood of the spectrum of a viewed as an element of $\bar{A}_\theta$, then $f(a) \in A_\theta$ (and similarly for the $k \times k$ matrix algebras). This is an immediate consequence of results in the appendix of [Cn2]. Much of its importance lies in the fact that it implies that the embedding of A_θ in $\bar{A}_\theta$ gives an isomorphism of their K-theory [Cn2].

Another attractive property of A_θ as a "smooth" algebra is [BEJ] that any derivation, δ, of A_θ into itself has a decomposition $\delta = \delta_0 + \tilde{\delta}$ where δ_0 is a linear combination of the generators $\delta_1, \cdots, \delta_n$ with coefficients in the center of A_θ, while $\tilde{\delta}$ is an approximately inner derivation in an appropriate sense. Even more, if θ satisfies a suitable diophantine approximation condition, then $\tilde{\delta}$ must, in fact, be inner [BEJ]. These results are used in [BEGJ] to classify the possible smooth actions of Lie groups on non-commutative two-tori. See also [Jr].

Just as diffeomorphisms of ordinary manifolds are of interest, so should be those of non-commutative tori. Now the (generalization of) homeomorphisms consist just of the automorphisms of $\bar{A}_\theta$, while the (generalization of) diffeomorphisms consist of the automorphisms which carry A_θ onto itself. Elliott [El3] has shown that for non-commutative two-tori for which θ satisfies a suitable diophantine approximation condition, any diffeomorphism of A_θ is the product of an inner automorphism coming from a unitary in A_θ, a diffeomorphism coming from the action of T^2 on A_θ, and the diffeomorphism coming from an element of $SL(2, Z)$ acting on the generators in the natural way [Br1, Br2]. But Kodaka [Kd] has shown that this very nice description can fail when θ does not satisfy Elliott's diophantine approximation condition. The "entropy" of the diffeomorphisms coming from elements of $SL(2,Z)$ has been discussed by Watatani in [Wt].

In section 1 we used the fact that ordinary tori carry natural Poisson structures. In [Xu], a definition is given of a Poisson structure on a non-commutative algebra, and then it is shown how non-commutative two-tori carry such Poisson structures, and their Poisson cohomology is calculated.

7. DeRHAM HOMOLOGY. Connes has shown [Cn4] that the way to generalize to non-commutative algebras the deRham homology (defined in terms of currents) of a differentiable manifold, is by means of his cyclic cohomology. A k-cochain for the cyclic cohomology of an algebra A is a $(k + 1)$-multilinear functional ϕ on A which satisfies the cyclic condition

$$\phi(a_0, \cdots, a_k) = (-1)^k \phi(a_1, \cdots, a_k, a_0) \quad .$$

Since Hochschild cohomology is defined in terms of general multi-linear functionals, one can apply the Hochschild coboundary operator to cyclic cochains. One finds then that the cyclic cochains form a subcomplex of the Hochschild complex. The cohomology groups of this subcomplex, denoted $HC^k(A)$, are by definition the cyclic cohomology groups of A. There are natural maps from $HC^k(A)$ to $HC^{k+2}(A)$ for each k, and the limit groups for these maps of the even groups and of the odd groups are the even and odd parts of the non-commutative deRham cohomology of A, and their direct sum is the non-commutative deRham cohomology of A, which we denote by $H_{dR}(A)$. Although the definition of cyclic cohomology makes sense for C^*-algebras, it is not for C^*-algebras that it has been of primary interest so far, but rather it is the cyclic cohomology of suitable dense subalgebras ("smooth structures") on which interest has focused.

Connes calculated the cyclic cohomology and deRham cohomology of (the smooth version of) non-commutative 2-tori in [Cn4], and more recently Nest [Ns] has calculated the cyclic cohomology of general non-commutative tori A_θ. One obtains

$$H_{dR}(A_\theta) \cong \wedge L \cong C^{2^{n-1}}$$

where L is the complexified Lie algebra of T^n, so $L \cong C^n$. Just as happens for an ordinary n-torus, every cohomology class can be represented by cocycles which are invariant under the action of T^n. To describe these invariant cocycles, define δ_X for any $X \in L$ by

$$\delta_X = \sum c_k \delta_k$$

where the c_k's are the coefficients of X in the standard basis for L. Then for any $\mu = X_1 \wedge \cdots \wedge X_m$ with the X_j's in L, define ϕ_μ on A_θ^{m+1} by

$$\phi_\mu(a_0, \cdots, a_m) = \sum \operatorname{sgn}(\sigma)\tau(a_0\delta_{X_1}(a_{\sigma(1)})\cdots\delta_{X_m}(a_{\sigma(m)}))$$

where σ runs over the permutations of m elements. Then the ϕ_μ's are invariant, and finite sums of them represent all cohomology classes.

Let us mention that Connes shows in [Cn4] that the Hochschild cohomology groups of a non-commutative two-torus depend on the diophantine approximation properties of θ. We mentioned earlier that the diophantine approximation properties are also involved in whether the invertible elements are dense, whether approximately inner derivations are inner, and in the structure of diffeomorphisms. It would be interesting to find closer relations between these four phenomena.

The cyclic homology of an algebra A (corresponding to the deRham cohomology of a manifold defined in terms of differential forms rather than currents), is the homology of a quotient of the complex for Hochschild homology, which in turn has as k-chains elements of the (k + 1)-fold tensor product of A with itself. Thus elements of cyclic homology will have such tensors as representatives. The deRham homology group $H^{dR}_{.}(A)$ is again defined as a limit of the cyclic homology groups.

As far as I know, no paper has yet appeared which gives an explicit calculation of the cyclic or deRham homology of non-commutative tori. But there is little doubt that the deRham homology group will again be $C^{2^{n-1}}$. In section 8 we will describe many elements of this group, in which it appears in the guise of $\wedge L^*$, where L^* is the vector space dual of L.

8. CONNECTIONS AND CURVATURE. Given a projective module, we would like to attach to it a Chern character, having values in deRham homology. If the module, say $\tilde{\Xi}$, is over $\bar{A}_\theta$, we need to find a smooth version of it, that is, an A_θ-module Ξ such that $\tilde{\Xi} = \Xi \otimes_{A_\theta} \bar{A}_\theta$. This can always be done [Cn1, Rf6] because of the fact that A_θ is closed under the holomorphic functional calculus of $\bar{A}_\theta$.

Connes showed [Cn1, Cn4] that the Chern character can be constructed by the Chern-Weil approach in terms of connections and curvature. (See also [Ka]). If Ξ is a projective A_θ-module, then a connection for Ξ is a linear map, ∇, of Ξ into $L^* \otimes \Xi$ satisfying the Leibnitz rule

$$\nabla_X(\xi a) = (\nabla_X \xi)a + \xi \delta_X(a)$$

for all $X \in L$, $\xi \in \Xi$ and $a \in A_\theta$, where δ_X is as defined in the previous section. Connections always exist, for one can use δ component-wise on free modules, and then compress this to direct summands of free modules. If ∇ and ∇' are connections on Ξ, then a simple calculation shows that $\nabla_X - \nabla'_X$ is in $\mathrm{End}_{A_\theta}(\Xi)$ for all $X \in L$, so that the connections form an affine space over the linear maps from L to $\mathrm{End}_{A_\theta}(\Xi)$.

The curvature of a connection measures, in the context of [Cn1], the extent to which the connection fails to be a Lie algebra homomorphism. Since here L is Abelian, this means that the curvature, R^∇, of a connection ∇ is defined simply by

$$R^\nabla(X, Y) = [\nabla_X, \nabla_Y]$$

for $X, Y \in L$. It is easily seen that the values of R^∇ are in $\mathrm{End}_{A_\theta}(\Xi)$, so that it is a skew bilinear map on L with values there.

For the projective module S(R) over a non-commutative two-torus described in section 4, a connection can be defined [Cn1] by defining its values, ∇_1 and ∇_2, on the standard basis for L, to be

$$(\nabla_1 \xi)(t) = (d\xi/dt)(t)$$

$$(\nabla_2 \xi)(t) = (2\pi i t/\theta)\xi(t) \quad .$$

Its curvature is then determined by its value on the wedge of the standard basis vectors, that is, by

$$[\nabla_1, \nabla_2] = (2\pi i/\theta) I \quad ,$$

where I is the identity operator on $S(R)$. Analogous connections on many projective modules over higher-dimensional non-commutative tori are explicitly constructed in [Rf6], and their curvatures calculated.

9. CHERN CHARACTER. Let Ξ be a projective A_θ-module, and let ∇ be a connection on Ξ with curvature R^∇. We wish to associate with this data a (non-homogeneous) even cycle on A_θ. This cycle will pair with the deRham cohomology, which we saw was ΛL, and so determines an element, ch^∇, in $\Lambda^e L^*$. We will specify ch^∇ by describing how its components, ch_k^∇, pair with the components, $\Lambda^{2k} L$, of the even deRham cohomology of A_θ. To this end we note that when Ξ is equipped with a Hermitian metric, then the canonical trace, τ, on A_θ, determines [Rf2] a (non-normalized) trace, τ', on $E = End_{A_\theta}(\Xi)$ such that

$$\tau'(\langle \xi, \eta \rangle_E) = \tau(\langle \eta, \xi \rangle_A) \quad ,$$

where $\langle \xi, \eta \rangle_E$ is the element of E defined by $\langle \xi, \eta \rangle_E \zeta = \xi \langle \eta, \zeta \rangle_A$. Next, let $(R^\nabla)^{\wedge k}$ be the exterior (wedge) k^{th} power of R^∇, so that it is an alternating $2k$-form with values in E. Then ch_k^∇ is defined by

$$ch_k^\nabla = \tau'((R^\nabla/2\pi i)^{\wedge k})/k! \quad .$$

Thus ch_k^∇ is in $\Lambda^{2k} L^*$. Connes shows [Cn1] that ch_k^∇ is independent of ∇, so depends only on Ξ, and can thus be denoted by $ch_k(\Xi)$. Then he defines the total Chern character, $ch(\Xi)$, of Ξ by

$$ch(\Xi) = \oplus ch_k(\Xi) \quad ,$$

an element of $\Lambda^e L^*$.

For the projective module $\Xi = S(R)$ over a non-commutative two-torus described earlier, straightforward computation shows [Cn1] that

$$ch_1(\Xi) = \pm \bar{Z}_1 \wedge \bar{Z}_2$$

where $\{\bar{Z}_j\}$ is the dual basis to the standard basis of L, and where the sign depends on the orientation chosen for the basis of L. As Connes showed [Cn1, Cn4] the Chern character of a projective module depends only on the class of the module in $K_0(A)$, and taking Chern characters gives a homomorphism from all of $K_0(A)$ into the even cyclic homology group. Elliott [El2] showed that for non-commutative tori this homomorphism is injective, and he gave an elegant description of its range inside $\Lambda^e L^*$, as follows. As earlier, view θ as a (nilpotent) element of the even exterior algebra $\Lambda^e L$, so that we can form $\exp(\theta)$ in this algebra. Then contraction of elements of $\Lambda^e L^*$ by $\exp(\theta)$ defines an automorphism, $\exp(\theta)\lrcorner$, of the exterior algebra $\Lambda^e L^*$. With D the integral lattice in L^*, Elliott shows that the range of the Chern character on $K_0(A_\theta)$ is exactly

$$\exp(\theta)\lrcorner\Lambda^e D$$

inside $\Lambda^e L^*$. This fact is crucial for the proof of the results described in section 4 concerning the non-stable K-theory of non-commutative tori.

10. YANG-MILLS. The Yang-Mills problem can be posed and studied in the context of non-commutative tori [CR, Rf10, Sp]. Let Ξ be a projective A_θ-module, equipped with a Hermitian metric. A connection ∇ on Ξ is said to be compatible with the Hermitian metric if it satisfies the Leibnitz rule

$$\delta_X(\langle\xi,\ \eta\rangle_A) = \langle\nabla_X\xi,\ \eta\rangle_A + \langle\xi,\ \nabla_X\eta\rangle_A \quad .$$

We will let $CC(\Xi)$ denote the space of compatible connections on Ξ. It is easily seen to be an affine space over the linear maps from L into E_s, where E_s denotes the elements of $\mathrm{End}_{A_\theta}(\Xi)$ which are skew-adjoint for the Hermitian metric.

We wish to define a functional on $CC(\Xi)$ which measures the "strength" of a connection. For this purpose we need a "Riemannian metric" on A_θ. Since L is playing the role of the tangent space of A_θ, this means that we must choose an inner-product on L. This inner-product then determines an E-valued inner-product, $\{\ ,\ \}$, on the space of alternating E-valued 2-forms on L. We can then define a non-negative real-valued non-linear functional, YM, on $CC(\Xi)$ by

$$YM(\nabla) = -\tau'(\{R^\nabla,\ R^\nabla\}) \ .$$

The Yang-Mills problem is then to determine the minima and critical points for YM. The Yang-Mills equations are the Euler-Lagrange equations for the critical points.

Let UE denote the group of elements of E which are unitary (for the Hermitian metric). This is the gauge group for our context. It acts on $CC(\Xi$ by conjugation, and simple calculations show that YM is invariant under this action. The set $MC(\Xi)$ of minima for YM is thus invariant under the action of UE. By definition the orbit space $MC(\Xi)/UE$ is the moduli space for the minima of YM. One has other moduli spaces for various families of critical points of YM.

For non-commutative two-tori all the above can be calculated. In this case every projective module is of the form Ξ^d where Ξ is not a multiple of any other projective module. Then it is shown in [CR] that $MC(V)/UE$ is homeomorphic to $(T^2)^d/S_d$ where S_d is the permutation group on d elements. (A somewhat different approach has recently been given in [Sp].) The moduli spaces for critical points have a similar but slightly more complicated description [Rf10].

For higher dimensional non-commutative tori there are certain projective modules to which the results for two-tori readily extend (the modules of form $V \otimes \Xi$ in theorem 5.6 of [Rf6] where V is an "elementary" module as in definition 4.2 of [Rf6]). However, for more general modules the situation is unclear at present, though it should be very interesting to investigate. I have obtained some very partial results, which indicate the usefulness of considering Einstein-Hermitian vector bundles, generalizing those defined, for example, in [Kb].

11. RELATIONS WITH MATHEMATICAL PHYSICS. The simplest discrete Schrodinger operators with almost-periodic potential, the almost Mathieu operators, are closely related to non-commutative two-tori [Be1, Be2, Be3, BLT], and so the latter have provided a convenient setting for their study. As earlier, let U and V be generators for a non-commutative two-torus satisfying $VU = \exp(2\pi i\theta)UV$. Let β be a real coupling constant. Then the corresponding almost Mathieu operator is

$$H = U + U^* + \beta(V + V^*) \quad .$$

Clearly H is contained in A_θ. The main questions about H have to do with its spectrum, and in particular with how often its spectrum is a Cantor set. The K-theory of A_θ is relevant to this question because the spectral projections of H for intervals whose endpoints are in gaps in the spectrum of H will be elements of A_θ, and so will contribute to the K-theory of A_θ, and can be labeled by their Chern characters.

The strongest results to date which use A_θ were obtained very recently by Choi, Elliott and Yui [CEY], and state that for $\beta = 1$ and for θ a Liouville number the spectrum of H is indeed a Cantor set, and provide information on the labeling of the gaps in the spectrum of H. Other work in the framework of A_θ has been done by Riedel [Rd2, Rd3, Rd4], who has developed techniques which perhaps will eventually be able to be used to obtain examples of almost Mathieu operators whose spectrum is not a Cantor set.

Let α be the automorphism of A_θ which carries U to U^* and V to V^*. It is clear that H is invariant under α, and so is contained in the fixed-point algebra, A_θ^α, of α. It is thus very desirable to understand A_θ^α, but this has so far proved to be surprisingly elusive. In particular, it is not known at present how to calculate the K-theory of A_θ^α. Some very partial results about A_θ^α are contained in [BEEK].

Another situation in which non-commutative tori have been related to mathematical physics is in quantum diffusions. See [Ap1, Ap2, Ap3, HR1, HR2] and the references therein.

BIBLIOGRAPHY

[AP] Anderson, J. and Paschke, W., "The rotation algebra," *Houston Math. J.*, to appear.

[Ap1] Applebaum, D., "Quantum stochastic parallel transport on non-commutative vector bundles," in *Quantum Probability and Applications, III* Lecture Notes in Math **1303** (1988), 20-37.

[Ap2] ———, "Stochastic evolution of Yang-Mills connections on the non-commutative two-torus," *Lett. Math. Phys.* **16** (1988), 93-99.

[Ap3] ———, "Quantum diffusions on involutive algebras," preprint.

[Be1] Bellissard, J., "K-theory of C*-algebras in solid state physics," *Statistical Mechanics and Field Theory, Mathematical Aspects*, Lecture Notes in Physics **257** (1986), 99-156.

[Be2] ———, "Almost periodicity in solid state physics and C*-algebras," preprint.

[Be3] ———, "C*-algebras in solid state physics, 2D electrons in a uniform magnetic field," preprint.

[BLT] Bellissard, J., Lima, R., and Testard, D., "Almost Schrodinger operators," pp.1-64, *Mathematics and Physics, Lectures on Recent Results,* World Scientific, Singapore, 1985.

[Bl] Blackadar, B., *K-Theory for Operator Algebras*, MSRI Pub. 5, Springer-Verlag, New York, 1986.

[BEEK] Bratteli, O., Elliott, G.A., Evans, D.E., and Kishimoto, A., "Non-commutative spheres, I," preprint.

[BEGJ] Bratteli, O., Elliott, G.A., Goodman, F.M., and Jorgensen, P.E.T., "Smooth Lie group actions on noncommutative tori," *Nonlinearity,* to appear.

[BEJ] Bratteli, O., Elliott, G.A., and Jorgensen, P.E.T., "Decomposition of unbounded derivations into invariant and approximately inner parts," *J. reine angew. Math.* **346** (1984), 166-193.

[Br1] Brenken, B.A., "Representations and automorphisms of the irrational rotation algebra," *Pacific J. Math.* **111** (1984), 257-282.

[Br2] ———, "Approximately inner automorphisms of the irrational rotation algebra," *C. R. Math. Rep. Acad. Sci. Canada* **7** (1985), 363-368.

[Br3] ———, "A classification of some noncommutative tori," preprint.

[BCEN] Brenken, B.A., Cuntz, J., Elliott, G.A., and Nest, R., "On the classification of non-commutative tori, III," pp.503-526 *Operator Algebras and Mathematical Physics*, Contemporary Math. **60**, American Mathematical Society, Providence, 1986.

[CE] Choi, M.-D. and Elliott, G.A., "Density of the self-adjoint elements with finite spectrum in an irrational rotation C*-algebra," preprint.

[CEY] Choi, M.-D., Elliott, G.A., and Yui, N., "Gauss polynomials and the rotation algebra," *Invent. Math.*, to appear.

[Cn1] Connes, A., "C*-algèbres et géométrie differentielle," *C. R. Acad. Sci. Paris* **290** (1980), 599-604.

[Cn2] ———— , "An analogue of the Thom isomorphism for crossed products of a C*-algebra by an action of R," *Adv. Math.* **39** (1981), 31-55.

[Cn3] ———— , "A survey of foliations and operator algebras," in *Operator Algebras and Applications*, pp.521-628 (ed. R. V. Kadison), Proc. Symp. Pure Math. **38**, American Mathematical Society, Providence, 1982.

[Cn4] ———— , "Non-commutative differential geometry," *Publ. Math. IHES* **62** (1986), 94-144.

[CM] Connes, A. and Moscovici, H., "Cyclic cohomology and the Novikov conjecture," IHES preprint.

[CR] Connes, A. and Rieffel, M.A., "Yang-Mills for non-commutative two-tori," *Contemporary Math.* **62** (1987), 237-266.

[CS] Connes, A. and Skandalis, G., "The longitudinal index theorem for foliations," *Publ. RIMS Kyoto Univ.* **20** (1984), 1139-1183.

[CEGJ] Cuntz, J., Elliott, G.A., Goodman, F.M. and Jorgensen, P.E.T., "On the classification of non-commutative tori, II," *C.R. Math Rep. Acad. Sci. Canada* **7** (1985), 189-194.

[DB] DeBrabanter, M. "The classification of rational rotation C*-algebras," *Arch. Math.* **43** (1984), 79-83.

[DEKR] Disney, S., Elliott, G.A., Kumjian, A., and Raeburn, I., "On the classification of non-commutative tori," *C.R. Math. Rep. Acad. Sci. Canada* **7** (1985), 137-141.

[El1] Elliott, G.A., "Gaps in the spectrum of an almost periodic Schrodinger operator," *C.R. Math. Rep. Acad. Sci. Canada* **4** (1982), 255-259.

[El2] ———— , "On the K-theory of the C*-algebra generated by a projective representation of a torsion-free discrete abelian group," pp.159-164, *Operator Algebras and Group Representations*, vol. 1, Pitman, London, 1984.

[El3] ———— , "The diffeomorphism groups of the irrational rotation C*-algebra," *C. R. Math. Rep. Acad. Sci. Canada* **8** (1986), 329-334.

[El4] ———— , "Gaps in the spectrum of an almost periodic Schrodinger operator, II," pp.181-191, *Geometric Methods in Operator Algebras* (ed. H. Araki and E.G. Effros), Pitman, London, 1986.

[Gr] Green, P., "The structure of twisted covariance algebras," *Acta Math.* **140** (1978), 191-250.

[HS] Hoegh-Krohn, R. and Skjelbred, T., "Classification of C*-algebras admitting ergodic actions of the two-dimensional torus," *J. reine angew. Math.* **328** (1981), 1-8.

[HR1] Hudson, R.L. and Robinson, P., "Quantum diffusions and the noncommutative torus," *Letters Math. Phys.* **15** (1988), 47-53.

[HR2] ———— , "Quantum diffusions on the non-commutative torus and solid state physics," Proceedings XVII International Conference on Differential Geometric Methods in Theoretical Physics, ed. A. Solomon, (World Scientific, Singapore, 1989).

[Hs] Husemoller, *Fibre Bundles*, Springer-Verlag, New York, Heidelberg, Berlin, 1966.

[HK] Husemoller, D. and Kassel, C., "Notes on cyclic homology," preprint.

[Ji] Ji, R., "Toeplitz operators on non-commutative tori and their real valued index," preprint.

[Jr] Jorgensen, P.E.T., "Approximately inner derivations, decompositions, and vector fields of simple C*-algebras," preprint.

[KR] Kadison, R.V. and Ringrose, J.R., *Fundamentals of the Theory of Operator Algebras, I*, Academic Press, New York, 1983.

[Ka] Karoubi, M., *Homologie Cyclic et K-Théorie*, *Astérisque* **149**, Soc. Math. France, 1987.

[Kb] Kobayashi, S., *Differential Geometry of Complex Vector Bundles*, Princeton Univ. Press, New Jersey, 1987.

[Kd] Kodaka, K., "A diffeomorphism of an irrational rotation C*-algebra by a non generic rotation," preprint.

[Km1] Kumjian, A., "On localizations and simple C*-algebras," *Pacific J. Math.* **112** (1984), 141-192.

[Km2] ———, "A sequence of irrational rotation algebras," *C.R. Math. Rep. Acad. Sci. Canada* 3 (1981), 187-189.

[Ns] Nest, R., "Cyclic cohomology of non-commutative tori," preprint.

[OPT] Olesen, D., Pedersen, G.K. and Takesaki, M., "Ergodic actions of compact Abelian groups," *J. Operator Theory* **3** (1980), 237-269.

[Oc] O'uchi, M., "On C*-algebras containing irrational rotation algebras," preprint.

[Pr] Pears, A.R., *Dimension Theory of General Spaces*, Cambridge Univ. Press, 1975.

[Pm] Pimsner, M.V., "Range of traces on K_0 of reduced crossed products by free groups," *Operator Algebras and Their Connections with Topology and Ergodic Theory*, pp.374-408, Lecture Notes Math. **1132**, Springer-Verlag, Heidelberg, 1985.

[PV1] Pimsner, M.V. and Voiculescu, D., "Exact sequences for K-groups and Ext-groups of certain crossed-product C*-algebras," *J. Operator Theory* **4** (1980), 93-118.

[PV2] ———, "Imbedding the irrational rotation algebras into an AF algebra," *J. Operator Theory* **4** (1980), 201-210.

[Po] Poguntke, D., "Simple quotients of group C*-algebras for two step nilpotent groups and connected Lie groups," *Ann. Sci. Ec. Norm. Sup.* **16** (1983), 151-172.

[Rd1] Riedel, N., "On the topological stable rank of irrational rotation algebras," *J. Operator Theory* **13** (1985), 143-150.

[Rd2] ———, "Point spectrum for the almost Mathieu equation," *C.R. Math. Rep. Acad. Sci. Canada* **8** (1986), 399-403.

[Rd3] ———, "Almost Mathieu operators and rotation C*-algebras," *Proc. London Math. Soc.* **56** (1988), 281-302.

[Rd4] ———, "On spectral properties of almost Mathieu operators and connections with irrational rotation C*-algebras," preprint.

[Rf1] Rieffel, M.A., "Strong Morita equivalence of certain transformation group C*-algebras," *Math. Ann.* **222** (1976), 7-22.

[Rf2] ———, "C*-algebras associated with irrational rotations," *Pacific J. Math.* **93** (1981), 415-429.

[Rf3] ———, "Morita equivalence for operator algebras," pp.299-301, *Operator Algebras and Applications* (ed. R.V. Kadison), Proc. Symp. Pure Math. **38**, American Mathematical Society, Providence, 1982.

[Rf4] ———, "Dimension and stable rank in the K-theory of C*-algebras," *Proc. London Math. Soc.* **46** (1983), 301-333.

[Rf5] ———, "The cancellation theorem for projective modules over irrational rotation C*-algebras," *Proc. London Math. Soc.* **47** (1983), 285-302.

[Rf6] ———, "Projective modules over higher-dimensional non-commutative tori," *Can. J. Math.* **40** (1988), 257-338.

[Rf7] ———, "The homotopy groups of the unitary groups of non-commutative tori," *J. Operator Theory* **17** (1987), 237-254.

[Rf8] ———, "Deformation quantization of Heisenberg manifolds," *Comm. Math. Phys.*, to appear.

[Rf9] ———, "Deformation quantization and operator algebras," preprint.

[Rf10] ———, "Critical points of Yang-Mills for non-commutative two-tori," *J. Diff. Geom.*, to appear.

[Rs1] Rosenberg, J., "C*-algebras, positive scalar curvature, and the Novikov conjecture, III," *Topology* **25** (1986), 319-336.

[Rs2] ———, "K-theory of group C*-algebras, foliation C*-algebras, and crossed products," *Index Theory of Elliptic Operators, Foliations, and Operator Algebras*, pp. , Contemporary Math. **70**, American Mathematical Society, Providence, 1988.

[Rh1] Rouhani, H.A., "Classification of certain non-commutative tori," Ph.D. thesis, Dalhousie University, 1988.

[Rh2] ———, "Quasi-rotation C*-algebras," preprint.

[Sl] Slawny, J., "On factor representations and the C*-algebra of canonical commutation relations," *Comm. Math. Phys.* **24** (1972), 151-170.

[Sp] Spera, M., "Yang-Mills equations and holomorphic structures on C* dynamical systems," preprint.

[Sw] Swan, R., "Vector bundles and projective modules," *Trans. Amer. Math. Soc.* **105** (1962), 264-277.

[Th] Thomsen, K., "A partial classification result for noncommutative tori," *Math. Scand.* **61** (1987), 134-148.

[Wt] Watatani, Y., "Toral automorphisms on irrational rotation algebras," *Math. Jap.* **26** (1981), 479-484.

[Xu] Xu, P., "Non-commutative Poisson algebras," preliminary version.

[Ym] Yim, H.S., "A simple proof of the classification of rational rotation C*-algebras," *Proc. Amer. Math. Soc.* **98** (1986), 469-470.

DEPARTMENT OF MATHEMATICS
UNIVERSITY OF CALIFORNIA
BERKELEY, CALIFORNIA 94720

rieffel@math.berkeley.edu

Contemporary Mathematics
Volume **105**, 1990

ANALYTIC AND COMBINATORIAL TORSION[1]

Mel Rothenberg[2]

INTRODUCTION

The invariants which we will refer to as combinatorial torsion were discovered and studied a half century ago by Reidemeister, Franz, and de Rham, in their study of finite group actions on simplicial complexes. Whitehead refined and generalized these invariants in defining the torsion of a homotopy equivalence in 1950, and his work after lying fallow for a number of years, was to play a crucial role in the development of geometric topology and algebraic K-theory in the 60's. The survey article of Milnor [**M**] remains an invaluable introduction to this material. Such invariants, and their refinements, now encoded in the surgery theoretic framework, remain central to any serious study of discrete group actions from the point of view of geometric topology.

In 1971, Ray and Singer [**RS**] introduced an analytic torsion associated with the de Rham complex of forms with coefficients in a flat bundle over a compact Riemannian manifold, and conjectured it was the same as the combinatorial torsion associated with the action of the fundamental group on the covering space,and the representation associated with the flat bundle. The Ray-Singer conjecture was established independently by Cheeger [C] and Müller [**M**] a few years later.

The insight of Ray-Singer was to analyze the combinatorial torsion in

[1]1980 Mathematics Subject Classification (1985 Revision). 58 G 05, 57 S 17
[2]Partially supported by NSF Grant DMS 8903168)

terms of the product of determinants of combinatorial Laplacians, and to show how to define a determinant for the smooth Laplacian with analogous properties. Subsequently Ray and Singer also defined a torsion in the complex analytic case. Their method of defining the determinant essentially used heat equation methods to study the ζ function of the Laplacian. In principle this was a generalization of the famous trace formula of Selberg, and this has recently been exploited to study closed geodesics on Hyperbolic manifolds [**Fi**]. This method of defining determinants was subsequently exploited by Quillen [**Q**] to define a holomorphic metric on the determinant line bundle over the space of holomorphic structures on a complex bundle over a Riemann surface. The geometry of such determinant line bundles have long been of interest to Algebraic geometers, and more recently has been related to theoretical physics, particularly by Witten (see [**F**] for a discussion) and Quillen's work had a dramatic impact. It was subsequently generalized by Bismut-Freed, who used it to prove a conjecture of Witten [**BF**]. The point to to be made here is that both the combinatorial and analytic torsion have become imbedded in sophisticated and complex mathematical machinery. While this is the inevitable fate of any really fundamental notions, it has the effect of making them less accessible to the non-specialist.

The purpose of this paper is to return to foundations and give a more or less self-contained exposition of the basic ideas behind combinatorial and analytic torsion. In essence these ideas are both extremely simple, yet have a certain elegance. They are, in principle, the working-out of the invariance properties of the ordinary determinant. It turns out that the combinatorial and analytic torison are not always the same in the situation we study. In fact, they are the same in essentially half the cases, and one of the concrete goals of this paper is to publicize the actual relationship.

The machinery we invokle in this presentation is undergraduate linear algebra and some elementary functional analysis. We utilize the asymptotic expansion of the heat kernel for the Laplacian on compact manifolds, which is not quite elementary, but which at this point has several nice expositions ([**JR**] is our favorite). We refer to one deep theorem, the Cheegar-Müller theorem, but we only indicate vaguely a proof.

In order to tempt the curious but uncommitted we keep our presentation brief and focus on foundations, we do not make any calculations, and we do not discuss further applications of these notions, except casually in passing. It should be kept in mind that the formal properties of torsion we develop make it very simple to manipulate and elementary to calculate in the combinatorial case. Our point of view is that the combinatorial torsion is but a slight generalization of the Euler characteristic, still the most important topological invariant. The analytic torsion reflects more deeply the geometry, and the relationship between between the two is a way of putting topological restraints on geometry. This is the focus of our approach.

It should be emphasized that this paper is essentially expository. Although the form of presentation has some novelty, the basic ideas are either classical, or are already contained in the author's joint work with John Lott [**LR**] which we have drawn on frequently, and to which we refer for proofs. The author wishes to express appreciation to W. Lück for many clarifying discussions, and to recommend his interesting and sophisticated treatment of torsion [**L**] to the reader. We also thank John Lott for reviving our interest in these questions and Dan Freed for teaching us about Dirac operators.

I. BASIC DEFINITIONS

V is a finite dimensional vector space over a field F. $\underline{\Lambda}(V)$ is the exterior algebra of V. $\underline{\Lambda}(V) = \sum_k \Lambda^k(V)$. We define $\Lambda(V) = \Lambda^n(V)$, where n is the dimension of V.

Of course, $\Lambda(V) \cong F$, but not canonically. Such an isomorphism is the same as choosing a non-zero element of $\Lambda(V)$, which we will call a *polarization* of V.

We will consider the family of 1-dimensional vector spaces over F with the natural identifications as a category $L = L(F)$, with the maps being linear isomorphisms. L is an Abelian group under tensor product with F being the identity and the dual space operator, denoted as usual by $*$, yielding the inverse. Note (exercise) categorical composition of maps in the usual composition if the latter is defined. If we let γ denote the category of finite-dimensional vector spaces over F with maps being linear isomorphisms, then Λ is a functor form $\gamma \to L$ with $\Lambda(a) = \det(a)$ for a map a.

The functor Λ may seem too simple to be interesting, but the whole subject of torsion, analytic and combinatorial, consists in generalizing its properties. The following properties are the basic ones:

1.1 Properties of Λ

I. $V = (0)$ and $V = F$ have natural polarizations, and thus $\Lambda(0) = \Lambda(F) = F$.

II. Whitney Sum. $\Lambda(V + W) = \Lambda(V) \cdot \Lambda(W)$

III. $\Lambda(V + V^*) = \Lambda(V)\Lambda(V)^* = \Lambda(V) \cdot \Lambda(V)^* = F$

IV. Given an exact sequence $0 \to V^1 \xrightarrow{i} V^2 \xrightarrow{j} V^2 \to 0$ there is defined an isomorphism $\Lambda(V^2) \cong \Lambda(V^1) \cdot \Lambda(V^3)$. The point here is that this isomorphism depends on i and h, but *not* on splitting j.

Property IV admits an immediate and crucial generalization as follows: Suppose we are given a complex (V,d),

$$0 \longrightarrow V^j \underset{d_j}{\longrightarrow} V^{j+1} \dots V^{j+2} \dots V^{j+k} \longrightarrow 0.$$

That is, we assume $d_r d_{r-1} = 0$. We write as usual:

$$Z^r = \ker d_r, \quad B^r = \operatorname{image} d_{r-1}, \quad H^r = Z^r/B^r,$$
$$V^e = \sum_r V^{2r}, \text{ and } V^{\mathrm{od}} = \sum_r V^{2r+1}.$$

Then

$$\Lambda(V^e) = \prod_r \Lambda(V^{2r}) = \prod_r \Lambda(Z^{2r}) \prod_r \Lambda(B^{2r+1})$$
$$= \prod_r \Lambda(H^{2r}) \prod_r \Lambda(B^{2r})\Lambda(B^{2r+1})$$

and

$$\Lambda^{\mathrm{od}} = \prod_r \Lambda(Z^{2r+1}) \prod_r \Lambda(B^{2r}) = \prod_r \Lambda(H^{2r+1}) \prod_r \Lambda(B^{2r+1}) \prod_r \Lambda(B^{2r}).$$

Hence we have an isomophism

$$\Lambda(V^e)\Lambda(H^{\mathrm{od}}) \cong \Lambda(V^{\mathrm{od}})\Lambda(H^e)$$

which depends on the complex (V,d).

1.2 DEFINITION. *This isomorphism will be denoted by*

$$\det(V) : \Lambda(V^e)\Lambda(H^{\mathrm{od}}) \overset{\cong}{\longrightarrow} \Lambda(V^{\mathrm{od}})\Lambda(H^e).$$

It is convenient to identify maps in L as follows:The maps form a monoid under tensor product. If we identify all maps $W \overset{\mathrm{id}}{\longrightarrow} W$ to the identity map $F \xrightarrow{\text{id}} F$, this monoid becomes an Abelian group, which we denote by $\mathrm{Hom}(L)$. This identifies all maps with same range and domain to elements of F^*. There is then a natural functor of L to the Abelian group $\mathrm{Hom}(L)$ itself considered as a category in the usual way, i.e., one object, the morphisms being the elements of $\mathrm{Hom}(L)$. We can and will consider det then as taking values in $\mathrm{Hom}(L)$.

Considered as such $\det(V) : \Lambda(V^e)\Lambda(H^{\mathrm{od}}) \xrightarrow{\sim} \Lambda(V^{\mathrm{od}})\Lambda(H^e)$ can be multiplied on the left by $\mathrm{Id}(\Lambda(V^{e*}))$ and on the right by $\mathrm{Id}(\Lambda(H^{\mathrm{od}\,*}))$ and we get

$$\begin{aligned}\Lambda(H^{\mathrm{od}})\Lambda^*(H^e) &= \Lambda^*(V^e)\Lambda(V^e)\Lambda(H^{\mathrm{od}})\Lambda^*(H^e)\\ &\to \mathrm{Id}\,\Lambda^*(V^e)\det(V)\,\mathrm{Id}\,\Lambda^*(H^e)\\ &\to \Lambda^*(V^e)\Lambda(V^{\mathrm{od}})\Lambda(H^e)\Lambda^*(H^e)\\ &= \Lambda^*(V^e)\Lambda(V^{\mathrm{od}}).\end{aligned}$$

In $\mathrm{Hom}(L)$ we have not changed anything, and so we can and will consider $\det(V) : \Lambda(H^{\mathrm{od}})\Lambda^*(H^e) \to \Lambda(V^{\mathrm{od}})\Lambda^*(V^e)$.

The following terminology is useful and will become more and more so as we proceed.

1.3 DEFINITION. *For any mod 2 graded vector space* $V = V^{\mathrm{od}} + V^e$ *we define* $\operatorname{detline} V = \Lambda^*(V^e)\Lambda(V^{\mathrm{od}}) = \Lambda(V^{*e} + V^{\mathrm{od}}) = \Lambda(\overline{V})$. *Here we write* $\overline{V} = V^{*e} + V^{\mathrm{od}}$.

Thus when the grading of V comes from a complex V, $\det(V)$: $\operatorname{detline}(H) \to \operatorname{detline}(V)$. When either $\overline{H}$ or $\overline{V}$ is polarized (e.g. $H = 0$), $\det(V)$ polarizes the other. When H is polarized, $\det(V)(1) \in \operatorname{detline}(V) = \mathrm{Hom}(\Lambda(V^e), \Lambda(V^{\mathrm{od}}))$ When $\overline{V}$ is polarized $\det(V)(1) \in \operatorname{detline}(H) =$

$\mathrm{Hom}(\Lambda(H^e)\Lambda(H^{od}))$. If both are polarized, it determines an element in $F^* = F - (0)$, the usual determinant. When $d = 0$, i.e., $H = V$ then $\det(V) = 1$.

In particular, we consider the case when F is a subfield of the real numbers $\mathbf{R}$ and V or H carries an inner product. Then $\overline{V}$ or $\overline{H}$ are polarized up to a factor of ∓ 1, i.e., we then have an isomorphism detline( )$/\mp 1 = F^*/\mp 1 = F_{>0}$. If both, carry an inner product $|\det V| \in F$ is well defined. We will call an isomorphism of detline$(V)/\pm 1$ with $F^*/\mp 1$ an almost polarization of V. *An important observations is that this is equivalent to putting a norm or inner product on* detline V.

The following generaliztion of property IV of 1.1 is probably the basic device for calculating these generalized determinants. Consider an exact sequence of complexes: $0 \to \underline{V_1} \to \underline{V_2} \to \underline{V_3} \to 0$. then detline $V_2 =$ detline V_1 detline V_3. We also have the acyclic complex H,

$$H^k(V_1) \to H^k(V_2) \to H^k(V_3) \to H^{k+1}(V_1)$$

with

$$H^e = H(V_1)^e + H(V_2)^{od} + H(V_3)^e \text{ and}$$
$$H^{od} = H(V_1)^{od} + H(V_2)^e + H(V_3)^{od}$$

and

$$\text{detline } H = \text{detline } H(V_1) \text{ detline}(H(V_3)) \, \textit{detline}^* H(V_2)$$

An easy calculation yields the following basic formula.

$$\det(V_2)\det(H)^{-1} = \det(V_3)\det(V_1) \tag{1.4}$$

THE COMBINATORIAL TORSION

We now are in a position to define combinatorial torsion. Let G be a discrete group. A Euclidean G complex of finite type is a complex of not necessarily finite dimensional vector spaces over F, a subfield of $\mathbf{R}$, $0 \longrightarrow V_j \xrightarrow{d_j} V_{j-1} \xrightarrow{d_j-1} \cdots \longrightarrow V_{j-k} \longrightarrow 0$ such that each V_j is an $F(G)$-module along with a G invariant inner product, and each d_j is a G morphism. We further assume that for each i, $F \otimes_{F(G)} V_i$ is finite dimensional.

Now suppose we are given an orthogonal action of G on a finite dimensional F inner product space W, i.e., a representation $\chi : G \longrightarrow O(W)$. The inner product on V_i induces maps $V_i \otimes W \longrightarrow V_i^* \otimes W \longrightarrow \mathrm{Hom}_F(V_i\ W)$ which are morphisms of $F(G)$-modules. Passing to the G invariants we have maps $(V_i \otimes W)^G \longrightarrow (V_i^* \otimes W)^G \longrightarrow \mathrm{Hom}_{F(G)}(V_i, W)$. The first of these spaces has an inner product coming from that of V_i, W, and because of the finite dimensional hypothesis on $F \otimes_{F(G)} V_i$ these maps are all bijections. Thus $\mathrm{Hom}_{F(G)}(V_i, W)$ is finite-dimensional and almost polarized and the complex (D, δ) with $D_i = \mathrm{Hom}_{F(G)}(V_i, W)$ and $\delta = d^*$ is an slmost polarized complex. We define

$$\mathcal{T}(V, \chi) = |\det(D)^{-1}(1)| \in \text{detline } H_G^*(V; W)/\mp 1$$

The example of interest to topologists is when G acts cellularly on a CW-complex X with X/G compact. The cellular chains of X, $C(X)$, over any field have a natural G invariant inner product with the cells as an orthonormal basis and we write $\mathcal{T}(X, \chi)$ for $\mathcal{T}(C(X), \chi)$. This is combinatorial torsion. It lies in detline $H_G^*(X, W)/\mp 1$. This is an interesting invariant because [**LR**].

THEOREM 2.2. *$\mathcal{T}(X, \chi)$ is invariant under subdivision of X.*

Given a smooth manifold and smooth action, there is a canonical G triangulation, unique up to isomorphism and subdivision, which makes it a G-CW complex, hence this defines an invariant of smooth actions, which is a generalization of the Reidemeister torsion.

The reader should do the simple calculation of $\mathcal{T}$ for $\mathbf{Z}/n$ acting by rotations and reflections on S^1 and compare it to the character. This calculation along with the exactness formula, (see section 7) give an algorithm for calculating $\mathcal{T}$ for any linear action of any finite group on a sphere. The order of difficulty of calculating $\mathcal{T}$ for linear actions on spheres is the same as for calculating characters, which are in fact determined by $\mathcal{T}(R)$.

CALCULATIONS WITH THE LAPLACIAN

We seek to connect up our determinant with geometry, and in particular the de Rham complex of a manifold. The beauty of the de Rham complex is that it admits natural operators which encode most of the geometry. The difficulty with it from the torsion point of view is that it is infinite-dimensional. It was the great insight of Ray and Singer **[R-S]** to first observe that one could use sharp control over the Laplace operator to define a sort of determinant for this unbounded operator on a Hilbert space, and thus link up the combinatorial torsion with analytic invariants. In this section we begin the transition to the analytic situation by developing the relationship between the determinant and the Laplacian in the finite dimensional situation.

Thus suppose we are given a complex (V, d) over $F \subset \mathbf{R}$ as follows:

$$0 \longrightarrow V^j \xrightarrow{d_j} V^{j+1} \dots V^{j+i} \longrightarrow 0, \quad j \geq 0$$

and we assume that each V^r is provided with an inner product $\langle \quad \rangle_r$.

We can then form the Laplacian $\Delta_r = d_r^* d_r + d_{r-1} d_{r-1}^*$ and we want to calculate $\det(V)$ in terms of the Δ_r.

We assume for the moment that V is acyclic. We consider $Z^i = \ker d_i$ and the orthogonal complement $Z_i^\perp$. Because V is acyclic $d_i X_i^\perp \cong Z^{i+1}$. We define the complex $\overline{Z}$, where $\overline{Z}^{2r} = Z_{2r}^\perp$ and $\overline{Z}^{r+1} = Z^{2r+1}$, and the differential $\overline{d}$ on $\overline{Z}$ by $\overline{d}_{2r} = d_{2r}, \overline{d}_{2r+1} = 0$. We have an exact sequence $0 \to \overline{Z} \to V \to \overline{\overline{Z}} \to 0$ where $\overline{\overline{Z}}$ is the complex,

$$\overline{\overline{Z}}^{2r+1} = X_{2r+1}^\perp, \quad \overline{\overline{Z}}^{2r} = Z^{2r}, \quad \overline{\overline{d}}_{2r+1} = d_{2r+1}, \quad \overline{\overline{d}}_{2r} = 0$$

1.4 and the fact that all the complexes involved are acyclic yields:

$$\det(V) = \det(\overline{Z}) \det(\overline{\overline{Z}}) \tag{3.1}$$

The complexes $\overline{Z}$ and $\overline{\overline{Z}}$ are very simple and we can calculate the determinants explicitly as follows. Let $d_e = \sum_r d_{2r} : \sum_r Z_{2r}^\perp \xrightarrow{\cong} \sum_r Z^{2r+1}$. Then $\det d_e = \prod_r \det d_{2r} = \det(\overline{Z})$ and similarly $\det d_{\mathrm{od}}^{-1} = \prod_r (\det d_{2r+1})^{-1} = \det(\overline{\overline{Z}})$ where for these acyclic complexers we are considering $\det : \Lambda(\ _e) \to \Lambda(\ _{\mathrm{od}})$. Thus we have

$$\det(V) = \prod_r \det\ d_{2r} \prod_r (\det\ d_{2r+1})^{-1} \tag{3.2}$$

Now for any isomorphism of vector spaces a, $\det(a) = \det(a^*)$ and this suggests we can replace d_r with d_r^*, but we must exercise care in identifying spaces with their duals. Let $\lambda_i : V^i \to V^{*i}$ be the isomorphism induced by $\langle\ \rangle_i$ and $\lambda_i^0 = \lambda_i | Z^i : Z^i \to Z^{*i}$, $\lambda_i^1 = \lambda_i | Z_i^+ : Z_i^+ \to Z_i^{\perp *}$. Then $\lambda_i = \lambda_i^0 + \lambda_i^1$. We then have $d_{i+1}^{**} : Z^{i+1} \to Z_i^\perp$ given by $d_{i+1}^{**} = (\lambda_i^1)^{-1} d_i^* \lambda_{i+1}^0$ and thus

$\det(d^{**}_{i+1}) = \det \lambda^{i-1}_i \det\ d_i \det\ \lambda^0_{i+1}$. By definition $\Delta_i = d^{**}_{i+1}d_i + d_{i+1}d^{**}_1$ and thus

$$\begin{aligned}\det \Delta_1 &= \det d^{**}_{i+1} \det d_i \det d_{i-1} \det d^{**}_i \\ &= \det{}^2 d_i \det{}^2 d_{i-1} \det(\lambda^1_i)^{-1} \det \lambda^0_{i+1} \det(\lambda^1_{i-1})^{-1} \det \lambda^0_i .\end{aligned}$$

Now a routine calculation yields:

$$\prod_r \det(\Delta_r)^{(-1)^r r} = \prod_r \det{}^2 d_{2r+1} \prod_r \det{}^{-2} d_{2r} \prod_r \overset{-1}{\det}(\lambda_{2r+1}) \prod_r \det(\lambda_{2r})$$

Thus we conclude

$$\det{}^2(V) = \prod_r \det \Delta_r^{(-1)^{r+1} r} \prod_r \det \lambda_{2r} \prod_r \det{}^{-1}(\lambda_{2r+1})$$

We can express this more elegantly as follows: Let $\lambda = \lambda_e + \lambda_{\text{od}} V^e + V^{\text{od}} \to W^e + W^{\text{od}}$ be a graded isomorphism. Let $\overline{\lambda} = \lambda_e^{-1} + \lambda_{\text{od}} : W^e + V^{\text{od}} \to V^e + W^{\text{od}}$. Then 3.3 translates into

$$\det{}^2(V) \det(\overline{\lambda}) = \prod_r \det(\Delta_r)^{(-1)^{r+1} r} \tag{3.4}$$

One should not be surprised that one gets the square of the determinant in terms of the Laplacian and metric rather the determinant itself. This is all one can expect due to the fact that the metric gives only an almost polarization of the complex.

Now let us drop the assumption that V is acyclic. Let H be the subcomplex of V consisting of the kernel of the Laplacian (the harmonics) which we identify with the cohomology and W the quotient complex. Note that $W_r = H_r^{\perp}$ and we have the exact sequence $0 \longrightarrow H \longrightarrow H \longrightarrow V \longrightarrow\longrightarrow 0$, where W is acyclic and H has trivial differential. 1.4 then yields $\det(V) = \det(W)$. If we put the metric on W as a subspace of V we can apply the above to calculate $\det(W)$ in terms of $\Delta(W)$ and $\lambda(W)$. These in turn are determined by $\Delta(V)$ and $\lambda(V)$ as follows:

Let $\Delta' = \Delta'(V) = \Delta(V)+$projection onto H. Then $\det \Delta'_r = \det \Delta_r(W)$. We will refer to both Δ and Δ' as the Laplacian using the $'$ to distinguish them when there is a possibility of confusion. Let $\lambda_i(H) : H^i \longrightarrow H^{*i}$ come from the inner product on H induced by the inner product on V. Then $\det \lambda_i(V) = \det \lambda_i(H) \cdot \det \lambda_i(W)$. Thus 3.3 and 3.4 yield:

(3.5)
$$\det{}^2(V) = \prod_r \det(\Delta'_r)^{(-1)^{r+1}r} \prod_r \det \lambda_{2r}(V) \prod_r \det \lambda_{2r+1}(H)$$
$$\prod_r \det{}^{-1} \lambda_{2r}(H) \prod_r \det{}^{-1}_{\lambda_{2r+1}}(V)$$

or more elegantly

$$\det{}^2(V) \det \overline{\lambda}(V) = \prod_r \det(\Delta'_r)^{(-1)^{r+1}r} \det \overline{\lambda}(H) \tag{3.6}$$

Observe that $\det \overline{\lambda}(H)$ is in fact exactly a metric on detline H. Taking λ as almost polarizing V and thus identifying $\det \overline{\lambda}(V)$ to 1 we can interpret , via 3.6, $\det^2(V)$ as another metric on detline H, namely the one gotten by distorting $\det \overline{\lambda}(H)$ by the products of powers of the determinants of the various Laplacians.

The other interesting aspect of 3.6 is that it exhibits the dependency of the determinant product of the Laplacians on the underlying metrics. Specifically, suppose we work over a family of metrics such that $\det \overline{\lambda}(V)$

does not vary. Then since $\det^2(V)$ is metric independent, the product of the determinants of the Laplacian depends only on the induced metrics on the homology.

Here is an interesting example of a family of metrics whose determiants do not vary. Let W be a finite dimensional vector space of dimension n, and consider any complex W of the form:

$$0 \longrightarrow \Lambda^0(W) \xrightarrow{d_0} \Lambda^1(W) \cdots \longrightarrow \Lambda^n(W) \longrightarrow 0. \tag{3.7}$$

Given an inner product on W, this induces an inner product on $\Lambda^k(W)$ and thus a corresponding isomorphism $\lambda V \to V^*$. If $\lambda_i : V \to V^*$ is the isomorphism induced by another inner product on W we want to compare $\det \overline{\lambda}$ with $\det \overline{\lambda}_i$. This is done by calculating $\det \overline{\lambda_i^{-1}\lambda}$ which is actually a number since $\lambda_1^{-1}\lambda$ is an automorphism of W. We let $\alpha_k = \lambda_i^{-1}\lambda :$ $\lambda^k(W) \to \Lambda^k(W)$. Then $\alpha_k = \Lambda^k(\alpha_i)$. Now $\det \Lambda^k(\alpha_1) = \det(\alpha_1)^{\binom{k-1}{n-1}}$. Thus for $n > 1$,

$$\det(\overline{\lambda_1^{-1}\lambda}) = \det(\overline{\alpha}_1) = \det(\alpha_1)^{\sum_{k=1}^{n}\binom{k-1}{n-1}} = (\det \alpha_1)^0 = 1.$$

Thus

3.8 THEOREM. *For V of the form 3.7 with metric induced from that on W, $\det \overline{\lambda}(V)$ is independent of the metric when dimension of $W > 1$ and , in particular, the left-hand side of 3.6 is independent of the initial metric on W.*

NOTE: Curiously it is simple to construct counter examples for dimension $W = 1$.

In the cases in which we are interested, i.e., the de Rham complex, the metric on V is related to a Hodge Star **operator**, which we will denote by $\#$, as follows: There exists isomorphisms $k\# : V^k \longrightarrow V^{n-k}$ and $k^\wedge : V^k \longrightarrow V^{n-k}{}_*$ with $(n-k)^\wedge = k^{\wedge *}$ such that $\lambda_k = (n-k)^\wedge k^*$, where, as above λ_k is the isomorphism associated with the metric on V^k. Making this substitution in 3.5 we note that when n is odd so that k and $n-k$ have different parity mod 2 the terms involving $k^\wedge$ drop out and we can replace λ_k by k^* in 3.5. If we let ${}^*e$ be the sum of the k^* over all even k, and similarly for ${}^*\,\mathrm{od}$, we have ${}^*e + {}^*\,\mathrm{od} : V^e + V^{\mathrm{od}} \longrightarrow V^{\mathrm{od}} + V^e$, and we can replace λ by ${}^*e + {}^*\,\mathrm{od}$ in 3.6. In the de Rham complex for odd dimensional manifolds we also have the relation $(n-k)^* k^* = \mathrm{Id}$. This means $\det {}^*e + {}^*\,\mathrm{od} = \det^2 {}^*e$. Hence summing up, under the hypothesis of this paragraph, with n odd.

$$\det{}^2(V)\det{}^2({}^*e) = \prod_r \det(\Delta_r')^{(-1)^{r+1}r} \det{}^{2*}e(H) \tag{3.9}$$

where ${}^e(H) : H^e \longrightarrow H^{\mathrm{od}}$ is induced by * on V.

In the case n is even , and with the hypothesis of this paragraph, the terms involving k^* drop out and we get formulas involving $k^\wedge$. In this case "Poncaré duality" forces $\prod_r \det(\Delta_r')^{(-1)^{r+1}r} = 1$ and $\det^2 *_e(H)$ is metric invariant, determined by the Poincaré duality map.

3.10 THEOREM. *Let X be a G CW-complex as in section 2, and $C_G^*(X; W)$ the complex of invariant cochains in an orthogonal representation, W, with the metric defined as in section 2. Then by definition, the left-hand side of 3.6 is the square of combinatorial torison. Thus so is the right-hand side, which is thus a combinatorial invariant by 2.2.*

WARNING: It is not true that $\det \overline{\lambda}(H) = \overline{\lambda}'(H)$, where λ' corresponds to the metric induced from a subdivision of X, X'. The product of the determinants of the combinatorial Laplacians is not invariant under subdivision. It is only the composite expression on the right-hand side of 3.6 that is invariant under subdivision. However, in the case of even dimensional manifolds and orientation preserving action $\det \overline{\lambda}(H)$ is determined by the Poincaré duality and both expressions are independent of metric (see [**LR**], Prop. 17). This is an important point which we do not develop in this article. See, however, section 7.

GLOBALIZING PREVIOUS RESULTS

We begin by globalizing the previous construction in the most elementary manner.

Instead of vector spaces we consider finite dimensional vector bndles V over a fixed space X. $\underline{\Lambda}(V)$ is then a vector bundle, and $\Lambda(V)$ a line bundle over X. We consider complexes of vector bundles, i.e., sequences of bundles and bundle morphisms, with $d_{r+1}d_r = 0$

$$0 \longrightarrow V^0 \xrightarrow{d_0} V^1 \xrightarrow{d_1} V^2 \longrightarrow \cdots \longrightarrow V^k \longrightarrow 0$$

There is only one technical obstacle to pushing through all previous constructions and results in this more general category. The problem is that the kernel and cokernel of the morphisms d_r may not be vector bundles because of jumps in dimension from fiber to fiber, i.e., dimension ker $d_r(x)$; $d_r(x) : V^r(x) \longrightarrow V^{r+1}(x)$ may vary with x. We adjust for this by constructing an *enlargement*.

First write V^k as quotient of a trivial vector bundle $\gamma_k : R^{n_k} \rightarrow V^k$. We now have a complex:

$$0 \rightarrow V^0 \xrightarrow{d_0} V^1 \rightarrow \cdots \rightarrow V^{k+1} + R^{n_k} \xrightarrow{d+\gamma_k} V^k \rightarrow 0$$

Since $d + \gamma_k$ is onto, its kernel is a vector bundle B^{k-1}. Again we write this bundle as a quotient of a trivial bundle $\gamma_{k-1} : R^{n_{k-1}} \rightarrow B^{k-1}$, add $R^{n_{k-1}}$ to V^{k-2} and form a new complex. Continuing in this way yields a complex:

$$0 \rightarrow V^0 + R^{n_1} \xrightarrow{d+\gamma_1} V^1 + R^{n_2} \xrightarrow{d_2+\gamma_2} V^2 + R^{n_3} \rightarrow \cdots \rightarrow 0$$

where each kernel and thus each cokernel is a vector bundle.

We can now make all necessary constructions, and since all the trivial bundles R^{n_k} are polarized, and since we can enlarge any two enlargements of a given complex to a common third enlargement, the various determinants and determinant lines constructed are independent of the enlargement chosen.

Specifically, given a complex V as above we can take an enlargement and get a cohomology complex $H(V)$. This depends on the enlargement but the associated determinant line bundle, detline $H(V)$ does not, nor does the morphism $\det(V) : \text{detline}\, H(V) \rightarrow \text{detline}\, V$. When each V^r has an inner product as a vector bundle the Laplacians are defined and their determinants are now functions from $X \rightarrow \mathbf{R}$. All the previous calculations go through routinely, the various formulae of section 3, and the metric dependence and independence results are valid under the analogous hypothesis.

This works neatly but is not the globalization we are really interested in. What we want is to replace the V^i by $C^\infty(V^i)$ and consider complex

$$0 \longrightarrow C^\infty(V^0) \xrightarrow{d_0} C^\infty(V^1) \xrightarrow{d_1} \cdots \longrightarrow 0$$

where the $d_r : C^\infty(V^r) \to C^\infty(V^{r+1})$ are differential operators. We might hope to generalize our construction to get a determinant: $C^\infty(\text{detline}\, H) \to C^\infty(\text{detline}\, V)$. The case above is exactly the case when d_r is a 0-th order operator, i.e., comes from a vector bundle map $V^r \to V^{r+1}$. In this case we have the map on C^∞ induced form the bundle map $\det V$: detline $H \to$ detline V

While the above generalization is too simple it is not totally without interesting examples. Consider, the case of 4.1 where the complex is elliptic over a compact manifold. We then have the principle symbol complex $\sigma(V)$, which is exact over the cotangent sphere bundle $ST^*(X)$,

$$0 \longrightarrow \pi^*(V^0) \xrightarrow{\sigma(d_0)} \pi^*(V^1) \xrightarrow{\sigma(d_1)} \cdots$$

We can then form $\det(\sigma(V))$ which since $\sigma(V)$ is exact can be considered an element of $C^\infty(\text{detline}\, \sigma(V))$.

The reader might work through the case of the de Rham complex, observing that by the results of section 3, this is a situation where for dimension $X > 1$, the product of the Laplacian determinants is metric indpendent and in fact depends only on the dimension of the manifold.

DETERMINANTS FOR HIGHER ORDER OPERATORS

The theory of analytic torsion begins with the discovery of Ray and Singer that one could develop a useful notion of determinant for certain differential operators, even though they are unbounded operators on infinite dimensional spaces. In our setup one could proceed as follows: Suppose we consider a linear map $d_1 : V^1 \to V^2$ where we are now allowing V^1, V^2 to be infinite dimensional, say Frechet, spaces. If we assume d Fredholm, so that $\ker d$ and $\operatorname{coker} d$ are finite dimensional, we can form the detline $H = \Lambda^*(\ker d)\Lambda(\operatorname{coker} d)$. When V^1 and V^2 are finite dimensional inner product spaces, and we use the induced almost polarization of V, 3.6 determines $\det^2(V)$ in terms of the metric on H and the numerical determinants of the Laplacians. Thus if we could define reasonable numerical determinants for the Laplacians in the infinite dimensional case, we could use 3.6 as a definition of $\det^2(V)$. Ray and Singer first realized that it was possible to use heat equation methods to define such determinants. Here is the procedure:

Let A be a positive operator on a Hilbert space, for example, the Laplacian acting on L^2 forms on a compact manifold. Suppose $A = e^B$, where B is of trace class. Proceeding formally, we could write $\det A = \det e^B = e^{\operatorname{tr} B}$. Thus $\log\det A = \operatorname{tr}(B)$. We can write for complex s, $A^s = e^{sB}$. Then $dA^s/ds|(0)$ is B, and thus $\log\det \operatorname{tr}(dA^s/ds)(0) = d \ \operatorname{tr} \ A^s/ds(0)$. Replace s by $-s$, and we get $-\log\det A = d \ \operatorname{tr} A^{-s}/ds(0)$. Now $\operatorname{tr} A^{-s} = \sum_i (\lambda_i)^{-s}$, where λ_i runs over the eigenvalues of A counted with multiplicity is usually written $\zeta_A(s)$, and if we replace d/ds by $'$ we get the expression

$$-\log\det A = \zeta_A'(0) \tag{5.1}$$

Unfortunately the operators A we are interested in are not of trace class and we cannot define ζ_A near 0 as an eigenvalue sum. What Ray and Singer, following Seeley, did was find a more sophisticated way of defining it.

The first observation is that while A is not of trace class, for the operators we are intersted in e^{-tA}, $t > 0$, is of trace class. This suggests we might be able to use the Mellin formula:

$$\lambda^{-s} = 1/\Gamma(s) \int_0^\infty e^{-t\lambda} t^{s-1}\, dt,$$

valid for $\mathrm{Re}(s) > 0$, $\mathrm{Re}(\lambda) > 0$. It is then natural to define

$$A^{-s} = 1/\Gamma(s) \int_0^\infty e^{-tA} t^{s-1}\, dt$$

and thus

$$\mathrm{tr}(A^{-s}) = 1/\Gamma(s) \int_0^\infty \mathrm{tr}(e^{-tA}) t^{s-1}\, dt. \tag{5.2}$$

Differentiating this at $s = 0$ would then yield a definition of $-\log\det A$. Unfortunately, the integral will almost certainly not exist near $s = 0$. For general operators A there is not much we can do. We are interested in Laplace operators reflecting the geometry of a manifold. The class of operators of Dirac type seem to be a fundamental family, which includes most of the familiar universal operators. We will not define them here. There is a very nice treatment in John Roe's book [JR]. We note that the Laplacians we use are of the form $(d + d^*)^2$ for d the exterior derivative, and $d + d^*$ is an operator of Dirac type.

To fix notation, we let $D : C^\infty(S) \to C^\infty(S)$ be an operator of Dirac type, where S is a Clifford bundle over the compact Riemannian manifold M of dimension n. S then comes with a metric and D is an elliptic, first order, self-adjoint, differential operator. We consider D as an unbounded, densely defined operator on the Hilbert space of L^2 sections.s The basic analytic fact we utilize is:

5.3 THEOREM. [**JR**], 5.17 *Let K_t be the Schwartz Kernel of e^{-tD^2}, $t > 0$. Then $K_t \in C^\infty(M \times M, S \otimes S)$ and restricted to the diagonal*

$$K_t(m,m) \sim 1/((4\pi t)^{n/2})(\theta_0(m) + t\theta_1(m) + t^2\theta_2(m) + \cdots)$$

where $\theta_j \in C^\infty(M, \mathrm{End}(S))$ and $\sim$ means the following: Let A_k be the part of the expression on the right involving t^j, for $j \leq k$, then $|K_t - A_k| = o(t^k)$

This theorem is actually true for more general elliptic differential operators, (S). The advantage of restricting to Dirac type operators is that the argument is more elementary and uses geometry, where the more general case involves more sophisticated analysis. One can also show,

5.4 THEOREM. ([**JR**], p. 94.5) For $\mathbf{Re}(s) \gg 0$ (in fact, by Seeley, for $\mathbf{Re}(s) > n/2$) the integral in 5.2 converges to a function ζ_{D^2} holomorphic in s.

In any case e^{-tD^2} is of trace class for $t > 0$, and tr $e^{-tD^2} = \int \mathrm{tr}\, K_t(m,m) d\,\mathrm{Vol}\, M$

D^2 is Fredholm. Let $H = \ker D^2$. Let $m = \dim H$. As in section 3, set $\Delta' = D^2$ restricted to $H^\perp$. Define $e^{-t\Delta'} = e^{-tD^2} - P(H)$, where $P(H)$ is projection on H. Then $\mathrm{tr}(e^{-t\Delta'}) = \mathrm{tr}(e^{-tD^2}) - m$. We now extend $\mathrm{tr}(\Delta'^{-s})$ to a meromorphic function of s on the whole complex plane as follows:

$$\mathrm{tr}(\Delta'^{-s}) = 1/\Gamma(s)\int_1^\infty \mathrm{tr}(e^{-t\Delta'})t^{s-1}dt + 1/\Gamma(s)\int_0^1 \mathrm{tr}(e^{-t\Delta'})t^{s-1}\,dt.$$

The first integral is regular for all s, as comes from the elementary estimates, $t \geq t_0 > 0$, then $|\,\mathrm{tr}(e^{(-t\Delta')}| < ce^{-t\lambda_0}$, where $\lambda_0 > 0$ is the smallest nonzero eigenvalue of D^2.

As for the second term, fix s and choose k such that $k + \mathbf{Re}(s) > 0$. Write $\mathrm{tr}(e^{-t\Delta'}) = (\mathrm{tr}(e^{-tD^2}) - \mathrm{tr}(A_k)) + \mathrm{tr}(A_k) - m$. Make the substitution in the second integral above. We get the sum of three terms. By 5.3 the first term

$$\frac{1}{\Gamma(s)} \int_0^1 \Big(\mathrm{tr}(e^{-tD^2}) - \mathrm{tr}(A_k(t))\Big) t^{s-1} dt$$

is regular for s such that $k + \mathbf{Re}(s) > 0$. The second term involving A_k integrates out to

$$\frac{1}{\Gamma(s)}\Bigg|_0^1 \theta_0 t^{s-n/2}/s - n/2 + \theta_1 t^{s-n/2+1}/s - n/2 + 1 + \cdots + t^{s+k}/s + k.$$

Throw away the evaluation at $t = 0$, i.e., set $t = 1$. This does not change anything when $s > n/2$.

As a function of s, taking account of the fact that the poles of Γ are exactly the nonpositive integers, this is meromorphic with simple poles, which occur exactly at numbers of the form $n/2 -$ positive integer which are not nonpositive integers. In particular this function is regular at 0. The third term integrates out to $m/s\Gamma(s)$ for $s > 0$ and we use this expression to extend meromorphically to the plane. It is again regular at 0. This does not change anything when $s > n/2$.

We have now given a meromorphic continuation of the function $\mathrm{tr}((\Delta')^{-s})$, which is regular for $s > n/2$, to the whole complex plane, such that 0 is a regular polint. Taking its derivative at 0 then defines $-\log\det(\Delta')$.

6. THE ANALYTIC TORSION

We now have the machinery available to define the analytic torsion. It involves the determinant of certain Laplacians on the De Rham complex.

Its crucial property is that it is independent of metrics in a certain sense, meaning that it has topological significance.

We will consider only compact manifolds, where the definitions are most transparent. This is surely too restrictive, since one wants at least to do covering spaces of compact manifolds, and more general complete manifolds. The original Ray-Singer approach does handle covering spaces of compact manifold by considering the de Rham complex of sections of flat bundles on the quotient, but this is not convenient for not necessarily free group actions. Presumably one can extend our formalism at least to covers of compact manifolds by using the approach of Atiyah's Γ index theorem A, but the real goal should be even greater generality, and this remains work to be done.

The data involves a closed, n-dimensional, oriented, Riemannian manifold M. We consider the de Rham complex of forms with values in a finite dimensional inner product space W over R, $\Omega(M;W)$, and their L^2 completions. We write Δ_k for the Laplacian on k forms, and $\Delta = \sum_k \Delta_K$. Let F be the operator which multiplies each k form by k, and $(-1)^F$ which multiplies each k form by $(-1)^k$, Δ'_k, and $e^{-t\Delta'}$ will mean what they did in section 5. Recall that in section 5 we defined:

$$(6.0) \qquad -\log\det(\Delta'_k) = d/ds|_{s=0} \frac{1}{\Gamma(s)} \int_0^\infty t^{s-1} \operatorname{tr}\ e^{-t\Delta'_k}\, dt$$

where the right-hand side converges for $s > n/2$ and is extended meromorphically to the complex plane, regular at 0. Looking at the formula's involving the Laplacian in sectin 3 the exponential of the quantity T occurs, where T is defined by the formula:

(6.1)

$$T = -\Big(\sum_{k=0}^{n} k \log \det(\Delta'_k) = d/ds|_{s=0} \frac{1}{\Gamma(s)} \int_0^\infty t^{s-1} \operatorname{tr}((-)^F F e^{-t\Delta'} dt$$

We will consider a generalization where we permit a twisting by $\Pi = \sum_k \pi_k$, where π_k is a bounded operator on L^2_k commuting with d and the Hodge map $*$. Thus we define:

$$(6.2) \qquad T(\Pi) = d/(s)|_{s=0} \frac{1}{\Gamma(s)} \int_0^\infty t^{s-1} \operatorname{tr}(-)^F F \Pi e^{-t\Delta'}\, dt.$$

If n is even, an easy calculation shows that $T(\Pi)$ is 0. For n odd this is not true. What is true is that this is metric independent in the following sense. Suppose we are given a smooth one-parameter family of metrics on M, $g(\varepsilon)$. Then $*$ and Δ become a one-parameter family of operators.

Assume II commutes with all of them. Then $T(II)$ becomes a smooth function of ε. We have the following formula:

$$(6.3) \qquad d/d\varepsilon\, T(\Pi) = -\operatorname{tr} | \ker \Delta\ (-1)^F\ \Pi(d/d\varepsilon^*)^{*-1}.$$

This is proved in **[LR]** based on an argument of **[R-S]**.

To analyze the right hand side of 6.3, observe: let $f(\varepsilon) : V_1 \to V_2$ be any one-parameter family of nonsingular maps. Choose a basis of V_1 and V_2 and suppose with respect to these bases $f(\varepsilon) = a(\varepsilon)$, a one-parameter family of matrices of positive determinant. Then $d/d\varepsilon \log\det a = \operatorname{tr}((da/d\varepsilon)a^{-1})$ $= \operatorname{tr}(df/d\varepsilon f^{-1})$ is independent of basis chosen.

Consider first, the case where Π is the identity. Let $H(*)$ be $*$ restricted to the kernel of Δ, $H, H(*) = H(*e) + H(*\,\mathrm{od})$, where the summands are the restriction of H(*) to the even and odd forms. Recall that for n odd, $H(*\,\mathrm{od})^{-1} = H(*e)$. Now taking into account the sign operator $(-)^F$ and the last sentence of the previous paragraph we see that the right-hand side of 6.3 is exactly $d/d\varepsilon \log \det^2 H(*e)$. Now recall from 3.8 that $\det^2 H(*e)$ is exactly the determinant of the metric restricted to H, and 6.3 says that $e^{-T} \det^2 H(*e)$ is independent of the metric. As in section 3 we can consider this as a metric on the detline(H).

More generally, suppose we have a finite group, G, acting on M via orientation preserving isometries, and W is the space of an orthogonal G representation. Then we have actions of G on $\Omega(M, W)$,$L^2(M, W)$, and we can consider the invariant subspaces $\Omega^G(M, W)$, $L^2_G(M, W)$, and we take Π to be the projection operator. The left and the right hand side of 6.3 both involve traces of operators, commuting with the projection operator, composed with the projection operator. These are the same as traces of the operators restricted to the range of the projection operator, in this case L^2_G, and $H^G = H^{\bar{*}}_G(X; W)$. Thus reasoning as above we see that 6.3 yields a metric on detline $H^*_G(X, W)$.

6.4 DEFINITION. *We define* $T(M, G, W) = e^{-T(\Pi)} \det^2 H(^*e)$, *considered as a metric on* detline $H^*_G(X; W)$. *This is the analytic torsion.*

The notation has been set up to highlight the following theorem.

6.5 THEOREM. *For M closed, oriented, odd dimensional, and G acting through orientation preserving isometries* $T(M, G, W) = \mathcal{T}(M, G, W)$. *In other words, the analytaic torsion equals the combinatorial torsion.*

A form of this theorem was first conjectured by Ray and Singer. That form of it was proved independently by Cheeger [C] and Müller [M]. Theorem 6.5 is proved in **[L-R]** by generalizing the techniques of Cheeger and Müller.

Let us first make some comments on the hypothesis, or rather on what happens when you try to relax them. When M is even dimensional and the action of G orientation preserving, the analytic invariants defined become trivial, while the comnbinatorial torsion is nontrivial. As we show in **[LR]** the combinatorial torsion in this case is very simple , can be defined independent of the metric, and is computable. It has properties of a Euler characteristic, and should in fact have a nice analytic expression, perhaps as a trace of a naturally associated operator. Finding such an expression is an interesting problem.

When M is even dimensional and G acts through orientation reversing isometries then 6.5 is true. This is a bit awkward in our present context since the only group that can act this way is $\mathbf{Z}/2$. This becomes more interesting when we produce a formula, as we will, which allows one to calculate in terms of the action of the individual group elements. This is important when one comes to consider manifolds with boundary. By passing to the double, and introducing the flip involution, which is orientation reversing one can reduce the analytic theory for manifolds with boundary to that of closed manifolds. This was Cheeger's idea. Formulae relating the combinatorial torsion for manifolds with boundary to analytic invariants using this technique will be given in Section 7.

The only known proofs of 6.5 at present all utilize the following strategy: Show that the difference between the analytic and combinatorial torsion is constant under surgery, and using surgery reduce it to the case of linear group actions on the sphere, where you calculate explicitly the difference to be 0. This is not a satisfactory argument. It is too calculational, and in the case we consider in 6.5, where it is necessary to do equivariant surgery, rather complicated. More important it is not the type of argument that generalizes naturally to the non compact situation. Recently, F. Tangermann [T] has proposed a new technique, using a method of Witten, where one perturbs the Laplacian by a Morse function, which allows one to localize the differences of the two torsions near the critical points. While this does not yet dispense with the surgery, it is a promising development.

THE GENERALIZED CHARACTER

The results of this section are essentially contained in [**LR**], where complete proofs can be found.

Our basic setup for torsion, involves two components. The action of a group on a manifold, and an orthogonal representation of the group. It is obvious that the reprentation enters in a relatively simple way. We can make this explicit as follows: Let c denote the character of the representation of G on W, i.e., $c(g) = \operatorname{tr}\chi(g)$, $\chi(g) \in O(W)$. Then we can write

$$T(\Pi) = 1/|G| \sum_g c(g)t(g)$$

where

$$t(g) = d/d(s)|_{s=0}\, 1/\Gamma(s) \int_0^\infty t^{s-1} \operatorname{tr} g^*(-)^F F e^{-t\Delta'} dt.$$

Here g^* denotes the action of the isometry induced by g on the forms.

The function on the group t should be regarded as a "character" for the action of the group G on M. As with the ordinary character $t(g)$ depends only on the action of g and not on the full group action. When M is a linear G sphere, t is closely related to the ordinary character, and the calculation of these t by Ray [**Ra**], for key examples, is a crucial element in all proofs of the Ray-Singer conjecture. Further the fact that these t determine the usual character (R) in the basis of all proofs of de Rham's theorem that smooth equivalence implies linear equivalence for linear actions of groups on spheres. The great advantage of these t over the usual character is that they are defined much more generally than for representations.

We observe that the definition of t and $T(\Pi)$ makes perfectly good sense if instead of using the de Rham complex, one uses the complex of cellular cochains with values in W, and the metric given in sectin 2. 3.10 implies that the product of e^{-T} with the metric induced ont he determinant line of the invariant cohomology is the combinatorial torsion. The problem is that in odd dimensions or if G reverses orientation, neither of these factors by themselves is a combinatorial invariant, and therefore the functions t so defined are not topological invariants. The way to adjust for this is to define a combinatorial T^C, coming from a combinatorial t^C as above,where t^C is considered as a function of two variables, the first being an element of the group, and the second λ, a metric on the detline $H^*_G(X; W)$. t^C is uniquely determined by the condition that $t^C(g, \ )$ be defined by 7.0 for the cellular metric, and $e^{-T(\lambda)}\lambda$ is independent of the metric λ. An explicit formula for t^C is given in **[LR]** (6.24). For a Riemannian manifold we always use the metric induced by the Riemannian structure, and we suppress its designation. We can now refine 6.5.

7.5 THEOREM. *Let MN be an oriented closed Riemannian manifold, g an isometry of finite order. If M is odd dimensional and g preserves orientation, or if M is even dimensional and g reverses orientation then $t(g) = t^C(g)$. For M even and g orientation preserving (and for n odd and g orientation reversing) $t(g) = 0$. $t^C(g) \neq 0$ in general, and in these cases* *is independent of metric.*

We remarked before that by use of the double construction, orientation reversing actions are the key to analysing torsion on manifolds with boundary. On a Riemannian manifold with boundary one has two associated analytic torsions depending on whether one uses forms with absolute (Neuman) or relative (Dirichlet) boundary conditions [**RS**]. The analytic theory goes through in both cases, and we can define t and t_∂ exactly as in section 6.

The relations between the torsions in the simplest case, when the invariant cohomology of M, ∂M is 0, is given by [[**LR**] (see [**L**] for more general form)]:

7.2. *Let G act by orientation preserving isometries. IfM is even dimensional*

$$T(M) = -T_\partial(M) = T(\partial M)/2.$$

If M is odd dimensional

$$T(M) = T^C(M) - T^C(\partial M)/2 = T_\partial(M) = T^C(M, \partial M) + T^C(\partial M)/2.$$

We now give some product formulae for t and T. Plugged into 6.4 or 3.10 they yield product formulae for the analytic and combinatorial torsion. Note they hold for combinatorial torsion in the even dimensional situation where there is no analytic torsion. It is interesting that in the combinatorial case, direct proofs of these formulae, using purely combinatorial techniques, are not evident.

In all the following assertions all manifolds will be closed and oriented, but we permit even and odd dimensions.

7.3. *Let g_i be an isometry of M_i, $i = 1, 2$. Suppose that the product isometry $g_1 \times g_2$ of $M_1 \times M_2$ is orientation preserving. Then*

$$t(g_1, g_2) = t(g_1)L(g_2) + L(g_1)t(g_2),$$

where $L(g_i)$ denotes the Lefschetz number.

7.4 THEOREM. *Let W_i be an orthogonal representation of G_i which acts on M_i, $i = 1, 2$. We suppose the product action on $M_1 \times M_2$ is orientation preserving. Then*

$$\begin{aligned} T(M_1 \times M_2 W_1 \otimes W_2) = & \, T(M_1, W_1)\chi(H^*_G(M_2; W_2)) \\ & + T(M_2, W_2)\chi(H^*_G(M_1; W_1)) \end{aligned}$$

where χ is the Euler characteristic.

7.5. *If G acts on M_1 and M_2 then for the diagonal action on $M_1 \times M_2$*

$$T(M_1 \times M_2, W) = T(M_1, W \otimes H^*(M_2; W)) + T(M_2, W \otimes H^*(M_1; W))$$

CONCLUSION

As a conclusion we would like to indulge in some speculation on one direction where the further study of torsion would be worthwhile.

Torsion on compact manifolds, at least from a general, foundational point of view seems to be now well understood, although there still remain some outstanding problems even here. Perhaps the most obvious one is to find a nice analytic expression for the combinatorial torsion on even dimensioal manifolds. As is shown in **[LR]** this is a very simple Euler characteristic type invariant, and should have a direct characteristic class expression.

More significant perhaps is the quesion of extending the theory to the non compact case. A guiding motivation here is the result of Milnor [M] whcih shows that one can interpret the Alexander polynomial of a knot as a torsion of a Z covering of its complement. Defining an analytic torsion here would lead to an analytic expression for the Alexander polynomial.

There is some interesting work in this direction of Carey and Mathai, [C] using Banach algebra methods to define a combinatorial torsion for certain non compact manifolds. This is strikingly similar to, and was inspired by, Atiyah's [**A**] use of such methods to extend the index theorem to coveringing spaces of compact manifolds. As remarked earlier it is not unreasonable to hope to extend these methods to define analytic torsion for nice open manifolds. The delicate point here is to discover the right conditions at infinity which blends nicely the topology and geometry.

One variant of this problem is as follows: We are given a "smooth" object X, e.g., a Riemannian manifold, and a group of symmetries G. We are given a compactification of X, X^C, and an extension of the action of G to X^C. X^C is no longer smooth but has a combinatorial structure, e.g., a G cell structure. X^C then has a combinatorial torison. The problem is to define an analytic torsion on X, depending on the geometry of X, which yields the combinatorial torison on X^C. This problem should be solvable in quite general situations.

REFERENCES

[M] J. Milnor, "Whitehead torsion," BAMS 173 (1966), p.358.

[RS] D. B. Ray and I. M. Singer, "R-torsiona dn the Laplacian on Riemannian manifolds," Adv. in Math., 7 (1971), p. 145.

[C] J. Cheeger, "Annalytic torsions and the heat equation," Ann. of Math., 109, (1979), p. 259.

[M] W. Müller, "Analytic torsion and R-torsion of Riemannian manifolds," Adv. in Math., 28, (1978), p. 233.

[R] M. Rothenberg, "Torsion invariants and finite transformations groups," Proc. of Symp. in Pure Math., 32, AMS (1978).

[LR] J. Lott and M. Rothenberg, "Analytic torison for group actions,", in preparation.

[L] J. Lott, "Analytic torsion for group actions," Cont. Math. Proc. of AMS Special Session in Differential Geometry, Los Angeles, 1987, to appear.

[Ra] D. B. Ray, "Reidemeister torsion and the Laplacian on lens spaces," Adv. in Math., 4 (1970), p. 109.

[Fi] D. Fried, "Analytic torsion and closed geodesics on hyperbolic manifolds," Inv. Math., 84 (1986), p. 523.

[DP] S. della Pietra and V. della Pietra, "Analytic torsion and finite group actions," IAS preprint, 1988.

[JR] J. Roe, *Elliptic Operators, Topology, and Asymptotic Methods*, Research Notes in Mathematics, Longman House, UK, 1988.

[BF] J. M. Bismut and D. S. Freed, "The analysis of elliptic families $\cdots$, I,II. Comm. Math. Phys., 106 (1986), 139–176.

[F] D.S.Freed, "Determinants, torsions and stings," Commun. Math. Phys. 107 (1986), 483–513.

[M] J. Milnor, "A duality theorem for Reidemeister torsion," Ann. of Math., 76 (1962), p. 137.

[CM] A Carey and V.Mathai, "L-acyclicity and L-torsion invariants," preprint.

[L] W. Lück, "Analytic and topological torsion for manifolds with boundary and symmetries," preprint, 1989.

[T] F.Tangerman, "Reidemeister torsion and analytic torsion," preprint.

[Q] D. Quillen, "Determinants of Cauchy-Riemann operators over a Riemann surface," Funct. Anal. Appl., 19 (1985) 31–34.

[A] M. Atiyah, "Elliptic operators, discrete groups, and von Neumann algebrals," Asterisque 33 (1976), 43–72.

[S] R.S. Seeley, "Complex powers of an elliptic operator," Proc. of Symp. on Pure Math., AMS, 10 (1967).

Contemporary Mathematics
Volume 105, 1990

PSEUDODIFFERENTIAL OPERATORS AND K-HOMOLOGY, II

Michael E. Taylor [1]

1. Introduction.
2. Approach to identities in $K^1(\Psi^0)$ via the (co)boundary map.
3. Identities involving the Szegö projector.
4. The intersection product $K_1(\Psi^0)\times K^1(\Psi^0)\to \mathbb{Z}$.

1. Introduction.

This work is a continuation of [15], particularly of §3-4 of that paper. The first two sections of [15] recalled work of the author with Baum and Douglas [5] on relative K-homology classes in $K_j(\Omega,\partial\Omega)$ defined by elliptic operators on a manifold with boundary. Central to that work were identities in relative K-homology, which, via the boundary map $\partial : K_j(\Omega,\partial\Omega)\longrightarrow K_{j-1}(\partial\Omega)$, led to identities in $K_{j-1}(\partial\Omega)$. Of particular interest is the following result, proved in [5]. Let Ω be a strongly pseudoconvex complex manifold. Then there is the Szegö projector S acting on $L^2(\partial\Omega)$, and also $\partial\Omega$ has a natural spin^c-structure, with associated Dirac operator D. The identity is

$$[D] = [S] \quad \text{in } K_1(\partial\Omega). \tag{1.1}$$

As explained in [5], this provides a refinement of Boutet de Monvel's index theorem (and also an extension, as this result generalizes to a large class of weakly pseudoconvex manifolds). The identity (1.1) has also further

1980 *Mathematics Subject Classification* (1985 *Revision*). Primary 47G05; Secondary 19K33.

[1] Research supported by NSF grant DMS8801457

implications for index theory.

In particular (1.1) applies to $\Omega = B^*M$, the ball bundle of a smooth compact manifold M. With the identity (1.1) in hand, it is possible to prove

$$[D] = \wp_M \text{ in } K_1(S^*M), \tag{1.2}$$

where $\wp_M \in K_1(S^*M)$ is the "pseudodifferential operator extension," which associates to any $p \in C(S^*M)$ the pseudodifferential operator with principal symbol p, which is well defined modulo the ideal $\mathcal{K}$ of compact operators on $L^2(M)$, and hence gives a $*$-homomorphism $C(S^*M) \to \mathcal{Q}(L^2(M))$, the Calkin algebra. One way in which the identity (1.2) is useful is the following. There is the commutative diagram, analyzed in [3]

$$\begin{array}{ccc} K^0(S^*M) & \xrightarrow{\ \rho\ } & \\ {\scriptstyle \cap \wp_M}\downarrow \approx & \searrow & \\ K_1(S^*M) & \xrightarrow[\pi_*]{} & K_1(M) \end{array} \tag{1.3}$$

with the following property. If $A: C^\infty(M,E_0) \to C^\infty(M,E_1)$ is an elliptic self adjoint pseudodifferential operator defining a class $[A] \in K_1(M)$, and if $E_+ \to S^*M$ is the vector subbundle of $\pi^* E_0 \to S^*M$ which is the direct sum of the positive eigenspaces of the symbol of A, then

$$\pi_*(E_+ \cap \wp_M) = [A] \quad \text{and} \quad \rho([E_+]) = [A]. \tag{1.4}$$

By (1.2) we can replace $\wp_M$ by [D] in (1.3). Using the Bott map, it is then established in [5] that there is a commutative diagram of the form:

$$\begin{array}{ccc} K^0(\hat{M}) & \xrightarrow{\ i_a\ } & \\ {\scriptstyle \cap [D]}\downarrow \approx & \searrow & \\ K_0(\hat{M}) & \xrightarrow[\pi_*]{} & K_0(M) \end{array} \tag{1.5}$$

$\hat{M}$ is the double of the ball bundle in T^*M. A vector bundle E over $\hat{M}$ is obtained from a pair of bundles $E_0, E_1 \to M$ via the clutching construction, yielding a symbol over S^*M of an elliptic pseudodifferential operator $P: L^2(M,E_0) \to L^2(M,E_1)$, and

$$i_a([E]) = [P] \text{ in } K_0(M). \tag{1.6}$$

Every elliptic pseudodifferential operator P is obtained, and amongst other things (1.5) implies that the index of any such pseudodifferential operator is equal to the index of an explicitly described twisted Dirac operator.

We note here another implication of (1.5); namely this can be used to prove that every element of $K_0(M)$ is represented by an elliptic pseudodifferential operator. Indeed, since the double of the ball bundle $\hat{M} \to M$ has a section, the map $\pi_*: K_0(\hat{M}) \to K_0(M)$ is surjective. Since $\cap[D]$ is an isomorphism, i_a is also surjective, so the result is established.

It is interesting to compare this with the following commutative diagram, a variant of (1.3):

(1.7)
$$\begin{array}{ccc} K^1(S^*M) & & \\ {\scriptstyle \cap \wp_M} \downarrow \approx & \searrow^{\rho} & \\ K_0(S^*M) & \xrightarrow[\pi_*]{} & K_0(M) \end{array}$$

Here, if $\Phi: S^*M \to GL(k,\mathbb{C})$ is a smooth map, defining a class $[\Phi] \in K^1(S^*M)$, then $\rho([\Phi])$ is the class of a pseudodifferential operator on M whose principal symbol is given by Φ. Thus the image of

(1.8) $$\pi_*: K_0(S^*M) \to K_0(M)$$

consists precisely of classes in $K_0(M)$ which can be represented by elliptic operators acting on <u>trivial</u> bundles, $P: L^2(M,\mathbb{C}^k) \to L^2(M,\mathbb{C}^k)$. When M has a nonvanishing vector field, the map (1.8) is surjective, but it is not always surjective. For example, with $M = S^2$, one has

(1.9) $$K_0(S^2) = \mathbb{Z} \oplus \mathbb{Z}, \quad K_0(S^*S^2) = K^1(RP^3) = \mathbb{Z},$$

so π_* is certainly not surjective in this case.

Sections 3 and 4 if [15], upon which we will build in this paper, dealt with the K-cohomology of the C^*-algebra $\Psi^0(M)$, the L^2-operator norm closure of the algebra $OPS^0(M)$ of classical scalar pseudodifferential operators of order 0 on a compact manifold M. The K-cohomology groups $K^j(\Psi^0(M))$ are closely related to the K-homology groups of the sphere bundle S^*M, $K_j(S^*M)$. As noted in §3 of [15], the general K-cohomology exact sequence easily yields the following, with exact rows and commuting triangles:

$$(1.10)\qquad \begin{array}{ccccccc} 0 \to & K_0(S^*M) & & \xrightarrow{\sigma^*} & & K^0(\Psi^0(M)) & \to 0 \\ & & \pi_* \searrow & & \swarrow \mu^* & & \\ & & & K_0(M) & & & \end{array}$$

and

$$(1.11)\qquad \begin{array}{ccccccc} 0 \to \mathbb{Z} \xrightarrow{\cdot \rho_M} & K_1(S^*M) & & \xrightarrow{\sigma^*} & & K^1(\Psi^0(M)) & \to 0 \\ & & \pi_* \searrow & & \swarrow \mu^* & & \\ & & & K_1(M) & & & \end{array}$$

In concert with Poincare duality, we hence have isomorphisms

$$(1.12)\qquad K^0(\Psi^0(M)) \approx K^1(S^*M), \quad K^1(\Psi^0(M)) \approx \tilde{K}^0(S^*M).$$

One of our motivations for looking at $K^j(\Psi^0(M))$ is that elliptic pseudodifferential operators, which define cycles for classes in $K_j(M)$, may also define classes in $K^j(\Psi^0(M))$ which have more structure. For example, if $A: C^\infty(M,E) \to C^\infty(M,E)$ is a self adjoint elliptic operator in $OPS^m(M)$, let Q denote the orthogonal projection onto the positive spectral space of A, $Q \in OPS^0(M)$, acting on $L^2(M,E)$. Let $\mathcal{R}$ denote its range. Then

$$(1.13)\qquad f \mapsto QM_f\big|_{\mathcal{R}} \quad (\text{mod } \mathcal{K})$$

defines a $*$-homomorphism $C(M) \to \mathcal{Q}(\mathcal{R})$, hence an element $[A] \in K_1(M)$. (Generally, $\mathcal{K}$ will denote the space of compact operators on a Hilbert space, and $\mathcal{Q}$ the Calkin algebra.) Now also

$$(1.14)\qquad P \mapsto Q\tilde{P}\big|_{\mathcal{R}} \quad (\text{mod } \mathcal{K}),$$

where $P \in OPS^0(M)$ is scalar and $\tilde{P}$ is any operator on $L^2(M,E)$ with the same (scalar) principal symbol as P, defines a $*$-homomorphism $\Psi^0(M) \to \mathcal{Q}(\mathcal{R})$, hence an element $\{A\} \in K^1(\Psi^0(M))$, satisfying

$$(1.15)\qquad \mu^*\{A\} = [A].$$

The extra structure in $\{A\}$ arises because its construction uses the fact that pseudodifferential operators of order 0 (such as Q) have compact commutator with general 0 order pseudodifferential operators with scalar principal symbol, not just with multiplication operators, and it is reflected in the fact that the map $\mu^*: K^1(\Psi^0(M)) \to K_1(M)$ is not generally injective, though it is always surjective.

In case $M = \partial\Omega$ is the boundary of a strongly pseudoconvex manifold, the Szegö projector S belongs to $OPS^0_{\frac{1}{2},\frac{1}{2}}(M)$, so it has compact commutator with elements of $OPS^0(M)$. It follows that the map

$$P \mapsto SP\big|_{\text{Range } S} \pmod{\mathcal{K}} \tag{1.16}$$

gives a $*$-homomorphism $\Psi^0(M) \to \mathfrak{Q}(\text{Range } S)$, hence an element $\{S\} \in K^1(\Psi^0(\partial\Omega))$, as noted in [15]. Since the Dirac operator D on $\partial\Omega$ also gives an element $\{D\} \in K^1(\Psi^0(M))$, it is natural to ask if we can specify a relation between these two elements, refining the result (1.1). We will show in §3 of this paper that in some cases the identity (1.1) can be strengthened to $\{D\} = \{S\}$ in $K^1(\Psi^0(\partial\Omega))$.

An elliptic pseudodifferential operator $P: C^\infty(M,E_0) \to C^\infty(M,E_1)$ gives rise to an element $[P]$ of $K_0(M)$, but not necessarily an element of $K^0(\Psi^0(M))$. To see this, we recall how $[P]$ arises. If P has order 0, it intertwines mod $\mathcal{K}$ the $*$-representations of $C(M)$ on $L^2(M,E_0)$ and $L^2(M,E_1)$, given by scalar multiplication. Now there are not such $*$-representations of $\Psi^0(M)$ on $L^2(M,E_j)$ in general. If E_j is trivial there is, but as noted in §3 of [15], an obstruction to this for nontrivial E_j is given by a map

$$\tau : K^0(M) \to K^1(\Psi^0(M)), \tag{1.17}$$

characterized as follows. If $E \to M$ is a vector bundle, $P \in OPS^0(M)$ is scalar, then $\tilde{P}: L^2(M,E) \to L^2(M,E)$ with scalar principal symbol equal to the symbol of P is well defined mod $OPS^{-1}(M)$, and this gives rise to a $*$-homomorphism $\Psi^0(M) \to \mathfrak{Q}(L^2(M,E))$, defining $\tau(E) \in K^1(\Psi^0(M))$. $\tau(E) = 0$ when E is trivial, so (1.17) factors through a map

$$\tilde{\tau} : \tilde{K}^0(M) \to K^1(\Psi^0(M)). \tag{1.18}$$

As shown in §3 of [15], ker τ consists of $[E] \in K^0(M)$ such that $\pi^*[E]$ is stably trivial over S^*M, where $\pi^* : K^0(M) \to K^0(S^*M)$ is the natural map. Whenever π^* is injective, for example when M has a nonvanishing vector field, the map $\tilde{\tau}$ in (1.18) is injective.

The map τ also helps measure the failure of μ^* in (1.11) to be injective, as one clearly has $\mu^*\tau = 0$. In §3 of [15] it was shown that the sequence

$$\tilde{K}^0(M) \xrightarrow{\tilde{\tau}} K^1(\tilde{\Psi}^0(M)) \xrightarrow{\mu^*} K_1(M) \longrightarrow 0 \tag{1.19}$$

is exact whenever M has trivial tangent bundle. There are indications that this exactness holds more generally, though I have not been able to establish the exactness for all M. We recall from (3.33) of [15] that this is equivalent to exactness of

$$K^0(M) \xrightarrow{\pi^*} K^0(S^*M) \xrightarrow{\rho} K_1(M) \longrightarrow 0 \tag{1.20}$$

with ρ as in (1.3).

Returning to $K^0(\Psi^0(M))$, we conclude that an elliptic pseudodifferential operator $P: C^\infty(M, E_0) \longrightarrow C^\infty(M, E_1)$ determines an element $\{P\} \in K^0(\Psi^0(M))$ provided E_j are trivial bundles. There is an essential converse to this result, which is derived as follows. The commutative diagram (1.7) can be enlarged to the diagram:

$$\begin{array}{ccc} K^1(S^*M) & \xrightarrow{\rho} & K_0(M) \\ \cap\mathcal{P}_M \downarrow \approx & & \uparrow \mu^* \\ K_0(S^*M) & \xrightarrow[\sigma^*]{\approx} & K^0(\Psi^0(M)) \end{array} \tag{1.21}$$

satisfying the property that, if $P \in OPS^0(M, \mathbb{C}^k)$ is elliptic, with symbol $\Phi: S^*M \to GL(k, \mathbb{C})$, defining $[\Phi] \in K^1(S^*M)$, then

$$\sigma^*([\Phi] \cap \mathcal{P}_M) = \{P\} \quad \text{in } K^0(\Psi^0(M)). \tag{1.22}$$

Thus, complementing the conclusion drawn after (1.7), we see that every element of $K^0(\Psi^0(M))$ is of the form $\{P\}$ where P is an elliptic operator on a trivial bundle over M.

We remark that, if P is an elliptic operator between sections of trivial bundles E_0 and E_1, the class of P depends on a choice of trivializations of these bundles. Different trivializations might yield different K-cohomology classes.

In addition to (1.21), we also have the following enlargement of the diagram (1.3):

(1.23)
$$\begin{array}{ccccccc} & & & K^0(S^*M) & \xrightarrow{\rho} & K_1(M) & \\ & & \nearrow & \approx \big\downarrow \cap \mathcal{P}_M & & \big\uparrow \mu^* & \\ 0 \longrightarrow & \mathbb{Z} & \longrightarrow & K_1(S^*M) & \xrightarrow[\sigma^*]{} & K^1(\Psi^0(M)) & \longrightarrow 0 \end{array}$$

the bottom row being exact. In particular, the identity (1.4) has the refinement

(1.24) $$\sigma^*(E_+ \cap \mathcal{P}_M) = \{A\} \text{ in } K^1(\Psi^0(M)).$$

Indeed, when the bundle $E_0 \to M$ on whose sections A acts is trivial, the proof of (1.4) given in Lemma 4.13 of [3] actually implies (1.24). To establish (1.24) is general it remains to note that when $A = I$ on $L^2(M,E)$, so $E_+ = \pi^* E$, then

(1.25) $$\sigma^*(\pi^* E \cap \mathcal{P}_M) = \tau(E),$$

which is easily verified; compare Proposition 3.1 of [15].

As we stated above, one of the main goals of this paper is to derive identities in $K^j(\Psi^0(M))$ refining previously obtained identities in $K_j(M)$. The identities (1.22) and (1.24) are elementary examples of this. In the next section we set up some machinery to derive more subtle identities. As in [5], these are obtained by applying the (co)boundary map to certain relative cohomology classes. In §2 we set up a C^*-algebra Ξ of operators on a compact manifold Ω with boundary $\partial\Omega = M$, and a $*$-ideal Ξ_0 such that $\Xi/\Xi_0 \approx \Psi^0(M)$. The pair Ξ, Ξ_0 also has other technical properties, required to carry out the necessary program. In §3 we derive extensions of (1.1), for certain types of pseudoconvex domains. We see this as the first look into a phenomenon that might be quite subtle.

Identities in K-(co)homology lead to index formulas, via the intersection product. In §4 we make note of some properties of the intersection product $K_j(\Psi^0(M)) \times K^j(\Psi^0(M)) \to \mathbb{Z}$, and how this relates to identities obtained in §3.

As results of this paper focus on K-cohomology and K-homology of algebras of pseudodifferential operators, we note that papers of Brylinski-Getzler [9] and Wodzicki [17] deal also with related subjects. These papers focus on rather different phenomena from those considered here, and there seems to be almost no overlap with the present paper, though future developments might incorporate these various perspectives.

2. Approach to identities in $K^1(\mathfrak{X}^0)$ via the (co)boundary map.

Here we develop further the technique of using the coboundary map

$$\delta : K^0(\mathfrak{A}, \mathfrak{J}) \longrightarrow K^1(\mathfrak{A}/\mathfrak{J}) \tag{2.1}$$

to produce identities in $K^1(\mathfrak{A}/\mathfrak{J})$, which figured into the analysis in [5]. First we recall the general set-up. If $\mathfrak{A}$ is a C^*-algebra with unit and $\mathfrak{J}$ a $*$-ideal, a cycle defining a class in $K^0(\mathfrak{A}, \mathfrak{J})$ consists of a pair (σ, T). Here $\sigma = \sigma_0 \oplus \sigma_1$ is a representation of $\mathfrak{A}$ on a sum $H_0 \oplus H_1$ of Hilbert spaces, and $T: H_0 \rightarrow H_1$ is a bounded map, assumed to have closed range and be a partial isometry mod $\mathcal{K}$, the space of compact operators. We require in addition the following two conditions,

$$\sigma_1(f)T = T\sigma_0(f) \quad \text{mod } \mathcal{K}, \text{ for all } f \in \mathfrak{A}, \tag{2.2}$$

and

$$\sigma_j(f)P_j \in \mathcal{K} \quad \text{for all } f \in \mathfrak{J}, \tag{2.3}$$

where P_j is the orthogonal projection onto Ker T or Ker T^*, for $j = 0$ or 1. Given such a cycle, the image $\delta[(\sigma, T)] \in K^1(\mathfrak{A}/\mathfrak{J})$ is defined as follows. We have $*$-homomorphisms

$$\tau_j : \mathfrak{A}/\mathfrak{J} \rightarrow \mathcal{Q}(\text{Range } P_j) \tag{2.4}$$

given by

$$\tau_j(f) = P_j\, \sigma_j(\tilde{f})P_j \quad (\text{mod } \mathcal{K}), \tag{2.5}$$

$\tilde{f} \in \mathfrak{A}$ being any preimage of $f \in \mathfrak{A}/\mathfrak{J}$. The hypothesis (2.3) guarantees that (2.5) is well defined mod $\mathcal{K}$. Thus we have classes $[\tau_j] \in K^1(\mathfrak{A}/\mathfrak{J})$, and

$$\delta[(\sigma, T)] = [\tau_0] - [\tau_1] \quad \text{in } K^1(\mathfrak{A}/\mathfrak{J}). \tag{2.6}$$

A pair (σ, T) as above satisfies more than enough conditions to define also a class in the Kasparov group $KK(\mathfrak{J}, \mathbb{C})$. In [4] it was proved that there is an isomorphism $KK(\mathfrak{J}, \mathbb{C}) \approx K^0(\mathfrak{A}, \mathfrak{J})$, which makes (2.1) consistent with Kasparov's coboundary map $\delta : KK(\mathfrak{J}, \mathbb{C}) \longrightarrow K^1(\mathfrak{A}/\mathfrak{J})$. One consequence of this isomorphism is that the rich set of equivalence relations established amongst cycles in Kasparov K-theory can to a large degree be transferred to $K^0(\mathfrak{A}, \mathfrak{J})$. In particular one has the following result. Let (σ, T') define another cycle for $K^0(\mathfrak{A}, \mathfrak{J})$, with $\sigma = \sigma_0 \oplus \sigma_1$ as before. Then, provided

$$(2.7)\qquad \left.\begin{aligned} T\sigma_0(f) &= T'\sigma_0(f) \\ \sigma_1(f)T &= \sigma_1(f)T' \end{aligned}\right\} \bmod \mathcal{K}, \text{ for all } f \in \mathfrak{J},$$

we have

$$(2.8)\qquad [(\sigma, T)] = [(\sigma, T')] \text{ in } K^0(\mathfrak{A}, \mathfrak{J}).$$

Applications of the formula (2.6) for δ to identities of the form (2.8) yield identities in $K^1(\mathfrak{A}/\mathfrak{J})$ which are nontrivial. As mentioned in the introduction, the identity (1.1) is an example of this, established in [5].

We produce identities (2.8) by taking different closed extensions of an unbounded operator D from H_0 to H_1, with densely defined domain in H_0. Suppose D_e is one such closed extension. We set

$$(2.9)\qquad T = D_e(D_e^*D_e + 1)^{-\frac{1}{2}},$$

a bounded operator from H_0 to H_1. In §1 of [5] we established a sufficient condition that (σ, T) define a cycle for $K^0(\mathfrak{A}, \mathfrak{J})$, a variant of a result of Baaj-Julg. We suppose there is a dense *-subalgebra $\mathfrak{A}_0$ of $\mathfrak{A}$ such that

$$(2.10)\qquad \sigma_0(f) \text{ preserves } \mathcal{D}(D_e) \text{ for all } f \in \mathfrak{A}_0$$

and

$$(2.11)\qquad \sigma_1(f)D_e - D_e\sigma_0(f) \text{ extends from } \mathcal{D}(D_e) \text{ to a bounded operator from } H_0 \text{ to } H_1.$$

Suppose furthermore that

(2.12) <u>Either</u> $(D_e^*D_e+1)^{-1}$ is compact on H_0,

<u>or</u> $(D_eD_e^*+1)^{-1}$ is compact on H_1.

Our result is the following.

<u>Lemma 2.1</u>. Under hypotheses (2.10)-(2.11), T is a partial isometry mod $\mathcal{K}$ with closed range, satisfying the condition (2.2). Thus (σ, T) defines a cycle for $K^0(\mathfrak{A}, \mathfrak{J})$ provided (2.3) holds, equivalently provided

(2.13) $\sigma_0(f)(D_e^* D_e+1)^{-1}$ and $\sigma_1(f)(D_e D_e^*+1)^{-1}$ are compact, for all $f \in \mathcal{J}$.

This result was applied in [5] in the following situation. With $\bar{\Omega}$ a compact manifold with boundary, we took $\mathcal{O}\!\mathcal{l} = C(\bar{\Omega})$, $\mathcal{J} = C_o(\Omega)$, the space of functions $f \in C(\bar{\Omega})$ vanishing on the boundary $\partial\Omega$. Thus $\mathcal{O}\!\mathcal{l}/\mathcal{J} = C(M)$, $M = \partial\Omega$. D was a first order elliptic differential operator between sections of vector bundles, and we took $\mathcal{O}\!\mathcal{l}_0 = C^\infty(\bar{\Omega})$. The maps σ_j gave the usual scalar action on sections of vector bundles. In that case, we saw that (2.10)-(2.13) hold for certain closed extensions D_e of D; for example they hold for the maximal extension D_{max}. For a pair of closed extensions of D satisfying the hypotheses of Lemma 2.1, identities of the form (2.8) were proved in [5] via a finite propagation speed argument, of a sort which will be extended in the proof of Theorem 2.4 below.

We now construct a new algebra of operators on $L^2(\bar{\Omega})$, when $\bar{\Omega}$ is such a compact manifold with boundary, $\partial\Omega = M$. Impose a collaring $\mathcal{C} = [0,1) \times M$ on a neighborhood of $\partial\Omega$ in $\bar{\Omega}$; $\{0\} \times M = \partial\Omega$. Give $\bar{\Omega}$ a Riemannian metric, of product type on $\mathcal{C}$. We say an operator P belongs to $\mathcal{P}$ if it is of the following form:

$$Pu(x) = p(x)u(x), \qquad x \in \bar{\Omega} \setminus \mathcal{C}, \tag{2.14}$$

with $p \in C^\infty(\bar{\Omega} \setminus \mathcal{C})$, and, on $\mathcal{C}$, with $(s,y) \in [0,1) \times M$,

$$Pu(s,y) = P(s)u(s,y), \tag{2.15}$$

with $P(s) = p(s,y,D_y)$ a smooth function of $s \in [0,1]$ with values in $OPS^0(M)$, having the property that, for $s \in (\frac{1}{2},1)$, P(s) is a multiplication operator, matching up smoothly with p(x) in (2.14) at s = 1. Note that the L^2-operator norm of such P is given by

$$\|P\| = \sup \{ |p(x)|,\ x \in \bar{\Omega} \setminus \mathcal{C},\ \|P(s)\|,\ s \in [0,1] \}, \tag{2.16}$$

where P(s) denotes the operator norm of P(s) on $L^2(M)$. Let $\mathcal{P}_0$ denote the subset of $\mathcal{P}$ consisting of operators as just described such that P(0) = 0, and let $\mathcal{P}_{00}$ denote the subset consisting of those operators such that P(s) = 0 for s sufficiently close to 0.

Let Ξ be the L^2-operator norm closure of $\mathcal{P}$, and let Ξ_0 be the closure of $\mathcal{P}_0$. From (2.16) it easily follows that

(2.17) $$\Xi / \Xi_0 \approx \Psi^0(M)$$

and that

(2.18) $$\mathcal{P}_{00} \text{ is dense in } \Xi_0.$$

From (2.17), we have $\delta : K^0(\Xi, \Xi_0) \longrightarrow K^1(\Psi^0(M))$.

Now let $D: C^\infty(\bar{\Omega}, E_0) \longrightarrow C^\infty(\bar{\Omega}, E_1)$ be an elliptic first order differential operator between sections of vector bundles $E_j \rightarrow \bar{\Omega}$. Let D_e be a closed extension, such that $\mathcal{D}(D_{min}) \subset \mathcal{D}(D_e) \subset \mathcal{D}(D_{max})$. We make the assumption

(2.19) $$E_0 \text{ and } E_1 \text{ are \underline{trivial} bundles over } \mathcal{C},$$

which holds provided E_0 and E_1 are trivial over $\partial\Omega$. In fact, it suffices to assume E_0 is trivial over $\partial\Omega$, since the symbol of D evaluated at the conormal vector gives an isomorphism of E_0 and E_1 over $\partial\Omega$. Given trivializations of E_j over $\mathcal{C}$, the algebra Ξ acts on $L^2(\Omega, E_j)$ in a natural way; denote these actions by σ_j. Let us furthermore assume that

(2.20) $$\sigma_0(P) \text{ preserves } \mathcal{D}(D_e), \text{ for all } P \in \mathcal{P}.$$

It follows that

(2.21) $$\sigma_1(P)D_e - D_e\sigma_0(P) \text{ is bounded from } L^2(\Omega, E_0) \text{ to } L^2(\Omega, E_1)$$

for each $P \in \mathcal{P}$, since this commutator is multiplication by a smooth $End(E_0, E_1)$ valued section on $\bar{\Omega} \setminus \mathcal{C}$, and on $\bar{\mathcal{C}} = [0,1] \times M$ we have

(2.22) $$[p(s,y,D_y), D_{y_j}] = q_j(s,y,D_y),$$

a smooth function of s with values in $OPS^0(M)$, and

(2.23) $$[p(s,y,D_y), D_s] = (\partial p/\partial s)(s,y,D_y).$$

We can hence apply Lemma 2.1 to deduce the following.

<u>Proposition 2.2</u>. If hypotheses (2.19)-(2.20) hold and if either $D_e^* D_e$ or $D_e D_e^*$ has compact resolvent, then T, given by (2.9), is a partial isometry, mod $\mathcal{K}$, with closed range, and the pair (σ, T) defines a cycle for $K^0(\Xi, \Xi_0)$

<u>Proof</u>. It remains only to verify (2.13) in this case, with $f = P \in \mathcal{P}_{00}$, in view of (2.18). Since $(D_e^* D_e + 1)^{-1}$ and $(D_e D_e^* + 1)^{-1}$ both map $L^2(\Omega, E_j)$ to $H^2_{loc}(\Omega, E_j)$, this follows from Rellich's theorem.

We denote the relative cohomology class by

$$\{D_e\} \in K^0(\Xi, \Xi_0). \tag{2.24}$$

We note that, as long as (2.19) holds, (2.20) is always valid for $D_e = D_{max}$, so we have $\{D_{max}\} \in K^0(\Xi, \Xi_0)$ in such a case. We consider more generally the following class of extensions. Let F be a vector bundle over $\partial\Omega$, let $B \in C^\infty(\partial\Omega, \mathrm{Hom}(E_0, F))$ have constant rank, and let D_B be the closure of the restriction of D to

$$\{u \in C^\infty(\bar{\Omega}, E_0): Bu = 0 \text{ on } \partial\Omega\}. \tag{2.25}$$

(Note that $D_B = D_{max}$ if $B = 0$.) Then an action σ_0 of $\mathcal{P}$ on $C^\infty(\bar{\Omega}, E_0)$ can be defined so that $\mathcal{D}(D_B)$ is preserved provided the following condition is satisfied:

(2.26) Ker B is a trivial subbundle of $E_0|_{\partial\Omega}$, with a trivial complementary bundle.

Thus Proposition 2.2 yields the following.

Corollary 2.3. Let $D: C^\infty(\bar{\Omega}, E_0) \longrightarrow C^\infty(\bar{\Omega}, E_1)$ be a first order elliptic differential operator between sections of bundles which are trivial over $\partial\Omega$, and let D_B be the closure of D acting on the space (2.25). Assume that (2.26) holds. Then, provided either $D_B^* D_B$ or $D_B D_B^*$ has compact resolvent, the conclusion of Proposition 2.2 holds, and we have a class

$$\{D_B\} \in K^0(\Xi, \Xi_0). \tag{2.27}$$

We are now ready to state the main result of this section, giving rise to identities in $K^1(\Psi^0(M))$.

Theorem 2.4. Let D_B, D_C be two closed extensions of an elliptic operator D, satisfying all the hypotheses of Corollary 2.3. Then

$$\{D_B\} = \{D_C\} \text{ in } K^0(\Xi, \Xi_0). \tag{2.28}$$

Hence

$$\delta\{D_B\} = \delta\{D_C\} \text{ in } K^1(\Psi^0(M)). \tag{2.29}$$

Proof. The identity (2.28) is established by an argument similar to the proof of the identity (2.8) in [5]; compare the discussion following Proposition 1.1 in [15]. We set

$$B = \begin{pmatrix} 0 & D_B^* \\ D_B & 0 \end{pmatrix} \tag{2.30}$$

and define B' similarly, with D_B replaced by D_C; B and B' are self adjoint on $L^2(\Omega, E_0 \oplus E_1)$. It suffices to show that

$$\varphi(B)\sigma(P) = \varphi(B')\sigma(P) \mod \mathcal{K}, \text{ for all } P \in \mathcal{P}_{00}, \tag{2.31}$$

where

$$\varphi(\lambda) = \lambda/(\lambda^2 + 1)^{\frac{1}{2}}. \tag{2.32}$$

The operator $\varphi(B)$, defined by the Spectral Theorem, can be analyzed as

$$\varphi(B) = (2\pi)^{-\frac{1}{2}} \int \hat{\varphi}(t)\, e^{itB}\, dt. \tag{2.33}$$

Here, $\hat{\varphi} \in \mathcal{S}'(R)$ is singular only at $t = 0$; on $R \setminus 0$ it is smooth, and rapidly decreasing as $|t| \to \infty$. For any $\varepsilon > 0$, we can write $\varphi = \varphi_1 + \varphi_2$ with $\operatorname{supp} \varphi_1 \subset (-\varepsilon, \varepsilon)$ and $\varphi_2 \in \mathcal{S}(R)$. Thus, for any given $P \in \mathcal{P}_{00}$, we can write the difference of the two sides of (2.31) as

$$[\varphi_1(B) - \varphi_1(B')]\sigma(P) + [\varphi_2(B) - \varphi_2(B')]\sigma(P). \tag{2.34}$$

We can analyze the first of these two terms by exploiting finite propagation speed for e^{itB} and $e^{itB'}$, solution operators to symmetric hyperbolic systems. Thus, for any given $P \in \mathcal{P}_{00}$, if ε is picked small enough, the first term in (2.34) vanishes. The second term is the product of the compactly supported factor $\sigma(P)$ by a smoothing operator, and hence is compact. This proves the theorem.

We can calculate $\delta\{D_{max}\}$ by the same argument used to prove Proposition 4.4 of [5], in case D is an operator of "Dirac type," i.e., the symbol $\sigma_D(x,\xi): E_{0x} \to E_{1x}$ is proportional to an isometry, $\|\sigma_D(x,\xi)v\| = \|\xi\|\cdot\|v\|$. In this case, there is a self adjoint first order differential operator $D^{\#}$ on $\partial\Omega = M$, whose principal symbol is of the form

$$\tau(x)^{-1}\sigma_D(x,\xi): E_{0x} \to E_{0x}, \quad x \in \partial\Omega,$$
$$\tau(x) = (1/i)\,\sigma_D(x,\nu), \tag{2.35}$$

ν denoting the conormal to $\partial\Omega$, and $D^{\#}$ is of Dirac type. Precisely as in Proposition 4.4 of [5], we have the following result.

Proposition 2.5. Let $D:C^\infty(\bar{\Omega},E_0) \longrightarrow C^\infty(\bar{\Omega},E_1)$ be a first order elliptic differential operator between sections of bundles which are trivial over $\partial\Omega = M$. Then, with $D^\#$ specified above,

$$\delta\{D_{max}\} = \{D^\#\} \quad \text{in } K^1(\Psi^0(M)). \tag{2.36}$$

Consequently, if D_B is any other closed extension of D satisfying the hypotheses of Corollary 2.3,

$$\{D^\#\} = \delta\{D_B\} \quad \text{in } K^1(\Psi^0(M)). \tag{2.37}$$

At this point we recall the discussion following Proposition 4.4 of [5]. If $\bar{\Omega}$ is an even dimensional spin^c-manifold with boundary, $E \to \bar{\Omega}$ a smooth Hermitian vector bundle, and D_E an associated twisted Dirac operator, then, provided $E \otimes S_+$ and $E \otimes S_-$ are trivial over $\partial\Omega = M$ ($S_\pm$ being the bundles of spinors), then we have $\{D_{E,max}\} \in K^0(\Xi,\Xi_0)$, and the operator $(D_E)^\#$ described above is D_F, the twisted Dirac operator on $\partial\Omega$, with its induced spin^c-structure, with $F = E|_{\partial\Omega}$. In this case, (2.36) becomes

$$\delta\{D_{E,max}\} = \{D_F\} \quad \text{in } K^1(\Psi^0(M)). \tag{2.38}$$

3. Identities involving the Szegö projector.

Let Ω be a strongly pseudoconvex manifold. There is the standard $\bar\partial$-operator, $\bar\partial : \Lambda^{0,\text{even}}(\bar{\Omega}) \to \Lambda^{0,\text{odd}}(\bar{\Omega})$, and the operator

$$D = \bar\partial + \bar\partial^* : \Lambda^{0,\text{even}}(\bar{\Omega}) \to \Lambda^{0,\text{odd}}(\bar{\Omega}) \tag{3.1}$$

is elliptic. Furthermore it is the Dirac operator on Ω, with spin^c-structure induced from its complex structure. Throughout this section we will make the following hypothesis:

$$\Lambda^{0,\text{even}} \text{ is trivial over } \partial\Omega. \tag{3.2}$$

Note that this holds provided

$$\Lambda^{0,1} \text{ is trivial over } \partial\Omega; \tag{3.3}$$

of course, this holds whenever Ω is a domain in $\mathbb{C}^n$. Granted (3.2), it follows that D_{max} defines a class in $K^0(\Xi,\Xi_0)$ and, by Proposition 2.5 and the discussion following it,

$$\delta\{D_{max}\} = \{D_{\partial\Omega}\} \quad \text{in } K^1(\Psi^0(\partial\Omega)), \tag{3.4}$$

where $D_{\partial\Omega}$ is the Dirac operator on $\partial\Omega$, with its induced spin^c-structure.

We now recall the closed extension used in [5] to obtain the identity (1.1), defined by the zero-order part of the $\bar\partial$-Neumann boundary condition. D_N is the closure of its restriction to

(3.5) $$\{u \in \Lambda^{0,\mathrm{even}}(\bar\Omega) : \sigma_{\bar\partial}(x,dr)^* u = 0 \text{ on } \partial\Omega\}$$

where $\{r = 0\}$ defines $\partial\Omega$. In any case where the condition (2.26) holds, since $D_N D_N^*$ has compact resolvent by the subelliptic estimates for the $\bar\partial$-Neumann problem, Corollary 2.3 applies. In such a case, the same calculations used to establish Proposition 4.6 in [5] yield the identity

(3.6) $$\delta\{D_N\} = \{S\} \quad \text{in } K^1(\Psi^0(\partial\Omega))$$

where S is the Szegö projector, the orthogonal projection of $L^2(\partial\Omega)$ onto the space of L^2-boundary values of functions holomorphic in Ω. Whenever this holds, then, comparing (3.4) with (3.6) and appealing to Theorem 2.4, we have

(3.7) $$\{D_{\partial\Omega}\} = \{S\} \quad \text{in } K^1(\Psi^0(\partial\Omega)).$$

These calculations apply when Ω is a strongly pseudoconvex manifold which has the property

(3.8) $\quad$ Ker $\sigma_{\bar\partial}(x,dr)^*$ is a trivial subbundle of $\Lambda^{0,\mathrm{even}}\big|_{\partial\Omega}$, with trivial complementary bundle.

We now look at some cases when this holds, where Ω has complex dimension two. In this case we have

(3.9) $$\sigma_{\bar\partial}(x,dr)^* : \Lambda^{0,0}\big|_{\partial\Omega} \oplus \Lambda^{0,2}\big|_{\partial\Omega} \longrightarrow \Lambda^{0,1}\big|_{\partial\Omega},$$

and the kernel consists precisely of the bundle $\Lambda^{0,0}\big|_{\partial\Omega}$, which of course is trivial. We therefore have the following.

<u>Proposition 3.1</u>. If Ω is a strongly pseudoconvex manifold of dimension 2, with the property that the line bundle

(3.10) $$\Lambda^{0,2}\big|_{\partial\Omega} \text{ is trivial,}$$

then D_N defines a class in $K^0(\Xi, \Xi_0)$, and the identity (3.7) relating the Szegö projector and the Dirac operator holds.

Clearly (3.10) holds whenever (3.3) holds, in particular whenever Ω is a strongly pseudoconvex domain in $\mathbb{C}^2$. In fact, since $\Lambda^{0,0}|_{\partial\Omega} \oplus \Lambda^{0,2}|_{\partial\Omega}$ and $\Lambda^{0,1}|_{\partial\Omega}$ are equivalent in this case, (3.10) is actually equivalent to (3.3). The following gives another class of manifolds for which the identity (3.7) can be deduced.

Proposition 3.2. Let M be a compact orientable surface, $\Omega = B^*M$. Then (3.3) holds. Hence Proposition 3.1 applies to $\Omega = B^*M$, and thus (3.7) holds with $\partial\Omega = S^*M$.

Proof. For $p \in S^*M$, let $T_p\Omega$ denote the tangent space, as a real vector space of dimension 4, and $\mathfrak{I}_p\Omega$ the tangent space as a complex vector space, of complex dimension 2. Endow M with a Riemannian metric. Then $H_p\Omega$, the horizontal space, with respect to the Levi-Civita connection on M, is a totally real subspace of $\mathfrak{I}_p\Omega$ (of real dimension 2). There is a tautological section s_1 of $H\Omega$ over S^*M; an orientation of M then gives an orthogonal section s_2 of $H\Omega$ over S^*M. This gives a splitting $H\Omega|_{S^*M} = \Sigma_1 \oplus \Sigma_2$, a sum of two trivial line bundles over S^*M. Complexifying Σ_j gives a splitting

(3.11) $$\mathfrak{I}\Omega|_{S^*M} = \Sigma_1^{\#} \oplus \Sigma_2^{\#},$$

a sum of two trivial complex line bundles over S^*M. From this observation it is easy to construct a trivialization of $\Lambda^{0,1}\Omega$ over S^*M.

Remark. For $\Omega = B^*M$ as above, the tangent bundle of $\partial\Omega = S^*M$ is of the form $(s_1) + \mathcal{X}$, where (s_1) is the real linear span of s_1 and $\mathcal{X}$ is a complex line bundle, defining the CR-structure of $\partial\Omega$. Using s_2, we can construct a nonvanishing section, hence a trivialization, of $\mathcal{X}$. Thus the results of §4 of [15], given there for the case where Ω is a strongly pseudoconvex domain in $\mathbb{C}^2$, also apply to $\Omega = B^*M$.

We next consider the extent to which Propositions 3.1 and 3.2 refine the identity (1.1), i.e.,

(3.12) $$[D_{\partial\Omega}] = [S] \quad \text{in } K_1(\partial\Omega).$$

Of course, the identity (3.7) refines (3.12) precisely when the natural map $\mu^*: K^1(\mathfrak{F}^0(\partial\Omega)) \longrightarrow K_1(\partial\Omega)$ is not injective. Since $\partial\Omega$ has a nonvanishing vector field, we see from the discussion following (1.18) that μ^* is not injective whenever $\tilde{K}^0(\partial\Omega) \neq 0$.

When $\bar{\Omega}$ is diffeomorphic to the ball in $\mathbb{C}^2$, $\partial\Omega$ is diffeomorphic to S^3. Now $\tilde{K}^0(S^3) = 0$; moreover, as shown in (3.20) of [15], we have

(3.13) $$K^1(\Psi^0(S^3)) = K_1(S^3) = \mathbb{Z} ,$$

so (3.7) gives no improvement over (3.12) in this case.

We claim that, for $\Omega = B^*M$, $\partial\Omega = S^*M$, as in Proposition 3.2, (3.7) is always stronger than (3.12). For example, in the case $M = \mathbb{T}^2$, $\partial\Omega = \mathbb{T}^3$, and as we have seen in (3.11) of [15],

(3.14) $$K_1(\mathbb{T}^3) = \mathbb{Z}^4, \quad K^1(\Psi^0(\mathbb{T}^3)) = \mathbb{Z}^7.$$

In this case, $\tilde{K}^0(\mathbb{T}^3) = \mathbb{Z}^3$.

In the case $M = S^2$, $S^*M = RP^3$, and it is a well known calculation ([1], compare the use in (3.38) of [15]) that $\tilde{K}^0(RP^3) = \mathbb{Z}_2$. To compare $K_1(S^*S^2)$ and $K^1(\Psi^0(S^*S^2))$, we first note that, by Poincare duality,

(3.15) $$K_1(S^*S^2) = K^0(S^*S^2) = \mathbb{Z} \oplus \mathbb{Z}_2.$$

Since S^*S^2 has trivial tangent bundle, Proposition 3.2 of [15] implies that the sequence

(3.16) $$\begin{array}{ccccccccc} 0 \longrightarrow & \tilde{K}^0(S^*S^2) & \xrightarrow{\tilde{\tau}} & K^1(\Psi^0(S^*S^2)) & \xrightarrow{\mu^*} & K_1(S^*S^2) & \longrightarrow 0 \\ & \| & & & & \| & \\ & \mathbb{Z}_2 & & & & \mathbb{Z} \oplus \mathbb{Z}_2 & \end{array}$$

is exact.

Now consider a surface M of genus $g \geq 2$. Rather than calculate $\tilde{K}^0(S^*M)$ directly to show it is nonzero (we suspect this calculation is well known but haven't seen it), we bring in some connections with K-homology of M. Recall from (1.12) that $\tilde{K}^0(S^*M)$ is isomorphic to $K^1(\Psi^0(M))$, which surjects by μ^* onto $K_1(M)$, so it suffices to show that $K_1(M) \neq 0$ in this case. Since M has a spin^c-structure, we have $K_1(M) \approx K^1(M)$. But the Chern character produces an isomorphism

(3.17) $$\mathrm{Ch}: K^1(M) \otimes \mathbb{Q} \xrightarrow{\approx} H^1(M, \mathbb{Q}) = \mathbb{Q}^{2g},$$

so this point is established.

We discuss briefly the condition (3.8) for Ω of higher dimension. First, using exactness of the symbol sequence for $\bar{\partial}^*$, we have (3.8) if and only if the analogous condition holds, with $\Lambda^{0,\mathrm{even}}|_{\partial\Omega}$ replaced by $\Lambda^{0,\mathrm{odd}}|_{\partial\Omega}$. Then by duality and another use of exactness, we deduce that (3.8) is equivalent to the following condition on

(3.18) $$\sigma_{\bar{\partial}}(x,dr): \Lambda^{0,\mathrm{odd}}|_{\partial\Omega} \longrightarrow \Lambda^{0,\mathrm{even}}|_{\partial\Omega},$$

namely

(3.19) Ker $\sigma_{\bar{\partial}}(x,dr)$ is a trivial subbundle of $\Lambda^{0,\mathrm{odd}}|_{\partial\Omega}$, with trivial complementary bundle.

This is also equivalent to the same sort of condition with $\Lambda^{0,\mathrm{odd}}|_{\partial\Omega}$ replaced by $\Lambda^{0,\mathrm{even}}|_{\partial\Omega}$. Furthermore, condition (3.19) holds provided

(3.20) $(\bar{\partial}\, r)$ has a trivial complementary bundle in $\Lambda^{0,1}|_{\partial\Omega}$,

where $(\bar{\partial}\, r)$ denotes the complex line bundle spanned by $\bar{\partial} r$, over $\partial\Omega$. We formalize this:

Proposition 3.3. The identity (3.7) holds whenever Ω is a strongly pseudoconvex manifold satisfying the condition (3.20).

This condition is a strong topological restriction in higher dimensions A family of examples is provided by the following straightforward extension of Proposition 3.2.

Proposition 3.4. Let M be a compact manifold, $\mathcal{H} \to S^*M$ the pull-back of T^*M. Suppose $\mathcal{H}$ is trivial and furthermore that the tautological line subbundle of $\mathcal{H}$ has a trivial complementary bundle. Then $\Omega = B^*M$ satisfies the condition (3.20), and hence the identity (3.7) holds.

An example of a class of manifolds M to which Proposition 3.4 applies is the following. Let M be a compact 4-dimensional manifold with a quarternionic action on each cotangent space T^*_xM, so that $S^*M \to M$ gets the structu of a principal SU(2)-bundle. Then applying the quarternions i,j,k to any $v \in S^*_xM$ provides a trivialization of the complementary bundle to the tautological bundle in $\mathcal{H}$, so (3.7) applies to $\Omega = B^*M$, $\partial\Omega = S^*M$, in such a case.

We end this section by noting a special property possessed by the class $\{S\}$ of the Szegö projector in $K^1(\Psi^0(\partial\Omega))$. This is is connected to the special section of $S^*(\partial\Omega)$, corresponding to the natural contact structure on $\partial\Omega$, and the fact that the wave front relation of S is contained in this contact ray bundle.

More generally, suppose M is a compact manifold such that S^*M has a section γ. This gives rise to a map

$$\gamma_*: K_1(M) \longrightarrow K_1(S^*M), \tag{3.21}$$

which then fits into the diagram (1.11). If $S \in OPS^0_{\frac{1}{2},\frac{1}{2}}(M)$ is a projection, defining

$$[S] \in K_1(M), \quad \{S\} \in K^1(\Psi^0(M)), \tag{3.22}$$

and if the wave front relation of S is contained in the cone generated by γ, then

$$\{S\} = \sigma^* \gamma_*[S] \quad \text{in } K^1(\Psi^0(M)). \tag{3.23}$$

We can restate this as $\{S\} = \sigma^* \gamma_* \mu^* \{S\}$. Now we have $\mu^*(\sigma^* \gamma_*) = \pi_* \gamma_* = \text{id.}$ on $K_1(M)$, so

$$\Pi_\gamma = \sigma^* \gamma_* \mu^* \quad \text{is a projection on } K^1(\Psi^0(M)); \tag{3.24}$$

note that Id. $- \Pi_\gamma$ is a projection onto the kernel of $\mu^*: K^1(\Psi^0(M)) \longrightarrow K_1(M)$. Then (3.23) is equivalent to the relation

$$\{S\} = \Pi_\gamma \{S\} \text{ in } K^1(\Psi^0(M)). \tag{3.25}$$

Now, for $M = \partial\Omega$, the boundary of a strongly pseudoconvex manifold, since (3.12) holds generally, we see that the refinement (3.7) holds if and only if

$$\{D_{\partial\Omega}\} = \sigma^* \gamma_*[D_{\partial\Omega}], \tag{3.26}$$

equivalently if and only if

$$\{D_{\partial\Omega}\} = \Pi_\gamma \{D_{\partial\Omega}\}, \tag{3.27}$$

γ being a section of $S^*(\partial\Omega)$ defining the contact structure on $\partial\Omega$.

4. The intersection product $K_1(\Psi^0) \times K^1(\Psi^0) \to \mathbb{Z}$.

One way to understand how identities in $K^\ell(\Psi^0(M))$ refine identities in $K_\ell(M)$ is to study the intersection product

$$K_j(\Psi^0(M)) \times K^\ell(\Psi^0(M)) \longrightarrow K^{j+\ell}(\Psi^0(M)), \tag{4.1}$$

followed by the index map $K^0(\Psi^0(M)) \to \mathbb{Z}$ if $j+\ell = 0 \bmod 2$. We begin this study by deriving a few elementary facts about the K-homology groups $K_j(\Psi^0(M))$, parallel to (1.10)-(1.12).

We start with a more general situation. Let $\mathfrak{A}$ be any C^*-algebra, acting on a Hilbert space H, containing the set $\mathcal{K}$ of compact operators. From the exact sequence $0 \to \mathcal{K} \to \mathfrak{A} \to \mathfrak{A}/\mathcal{K} \to 0$ there arises the K-homology exact sequence. Since $K_0(\mathcal{K}) = \mathbb{Z}$ and $K_1(\mathcal{K}) = 0$, this takes the form

$$0 \to K_1(\mathfrak{A}) \to K_1(\mathfrak{A}/\mathcal{K}) \xrightarrow{\delta} \mathbb{Z} \to K_0(\mathfrak{A}) \to K_0(\mathfrak{A}/\mathcal{K}) \to 0. \tag{4.2}$$

Here $\delta : K_1(\mathfrak{A}/\mathcal{K}) \to \mathbb{Z}$ is given by Kasparov product with $[\mathfrak{A}/\mathcal{K}] \in K^1(\mathfrak{A}/\mathcal{K})$. Provided this map is surjective, we can break this exact sequence into two pieces:

$$\begin{array}{c} 0 \to K_1(\mathfrak{A}) \to K_1(\mathfrak{A}/\mathcal{K}) \xrightarrow{\delta} \mathbb{Z} \to 0 \\ \\ 0 \to K_0(\mathfrak{A}) \to K_0(\mathfrak{A}/\mathcal{K}) \to 0. \end{array} \tag{4.3}$$

Now, in the case $\mathfrak{A} = \Psi^0(M)$, $\mathfrak{A}/\mathcal{K} = C(S^*M)$, we have the commutative diagram (which can be compared with (1.21))

$$\begin{array}{ccc} K^1(S^*M) & \xrightarrow{\ \ \delta\ \ } & \mathbb{Z} \\ {\scriptstyle \cap P_M}\Big\downarrow \approx & \searrow^{\rho} & \Big\uparrow {\scriptstyle \text{ind.}} \\ K_0(S^*M) & \xrightarrow[\pi_*]{} & K_0(M) \end{array} \tag{4.4}$$

which makes it clear that $\delta : K^1(S^*M) \to \mathbb{Z}$ is surjective. Thus we have the following diagrams, with exact rows and commuting triangles:

$$\begin{array}{ccc} 0 \to K_0(\Psi^0(M)) & \xrightarrow{\ \sigma_*\ } & K^0(S^*M) \to 0 \\ {\scriptstyle \mu_*}\nwarrow & & \nearrow{\scriptstyle \pi^*} \\ & K^0(M) & \end{array} \tag{4.5}$$

and

$$(4.6)\qquad 0 \longrightarrow K_1(\Psi^0(M)) \xrightarrow{\sigma_*} K^1(S^*M) \xrightarrow{\delta} \mathbb{Z} \longrightarrow 0$$
$$\mu_* \nwarrow \; K^1(M) \; \nearrow \pi^*$$

In particular, we have the isomorphisms

$$(4.7)\qquad K_0(\Psi^0(M)) \approx K^0(S^*M), \quad K_1(\Psi^0(M)) \approx \text{Ker } \delta \subset K^1(S^*M).$$

Compare these with the isomorphisms implied by (1.10)-(1.11):

$$(4.8)\qquad K^0(\Psi^0(M)) \approx K_0(S^*M), \quad K^1(\Psi^0(M)) \approx K_1(S^*M)/(\wp_M).$$

It is useful to record the following fact about δ, whose proof is routine.

<u>Proposition 4.1</u>. Let $\Phi : S^*M \to GL(k,\mathbb{C})$ be a smooth map, defining a class $[\Phi] \in K^1(S^*M)$. Let $P \in OPS^0(M,\mathbb{C}^k)$ have principal symbol on S^*M given by Φ. Then

$$(4.9)\qquad \rho([\Phi]) = [P] \text{ in } K_0(M).$$

Thus $\delta([\Phi])$ is equal to the index of P.

In light of (4.7)-(4.8), it is reasonable to analyze the product $K_1(\Psi^0(M)) \times K^1(\Psi^0(M)) \to \mathbb{Z}$ in terms of a product

$$(4.10)\qquad [\text{Ker } \delta \subset K^1(S^*M)] \times K^1(\Psi^0(M)) \longrightarrow \mathbb{Z}.$$

This product has the following description.

<u>Proposition 4.2</u>. Let $\Phi : S^*M \to GL(k,\mathbb{C})$ be a symbol of a pseudodifferential operator of index 0, defining a class $[\Phi] \in \text{Ker } \delta \subset K^1(S^*M)$. Consider $\{A\} \in K^1(\Psi^0(M))$, where $A \in OPS^m(M,E)$ is elliptic and self adjoint. Let Q denote the orthogonal projection of $L^2(M,E)$ onto the positive spectral space of A. Let $\Phi_E^{op} \in OPS^0(M,E\otimes\mathbb{C}^k)$ be any pseudodifferential operator with principal symbol $I \otimes Q$, and let $Q_{(k)} \in OPS^0(M,E\otimes\mathbb{C}^k)$ be the direct sum of k copies of Q. Then the product (4.10) satisfies

$$(4.11)\qquad [\Phi]\cdot\{A\} = \text{Index } Q_{(k)} \Phi_E^{op}\big|_{\mathcal{R}}$$

where $\mathcal{R}$ is the range of $Q_{(k)}$.

Note that a special case of this involves $A = I_E$, the identity operator on $C^\infty(M,E)$, where $E \to M$ is a nontrivial vector bundle. In that case we have $\{I_E\} = \tau(E)$, and, by (4.11),

$$[\Phi]\cdot\tau(E) = \text{Index } \Phi_E^{op}. \tag{4.12}$$

This last identity also follows from the commutative diagram

$$\begin{array}{ccc} K^0(M) & \xrightarrow{\ \tau\ } & K^1(\Psi^0(M)) \\ \pi^*\big\downarrow & & \big\uparrow \sigma^* \\ K^0(S^*M) & \underset{\approx}{\xrightarrow{\ \cap \wp_M\ }} & K_1(S^*M) \end{array} \tag{4.13}$$

discussed in §3 of [15], which implies

$$[\Phi]\cdot\tau(E) = [\Phi] \cap (\pi^*(E) \cap \wp_M). \tag{4.14}$$

We now discuss intersection products involving the class $\{S\}$ of the Szegö projector on $M = \partial\Omega$, when Ω is a strongly pseudoconvex manifold; $\{S\} \in K^1(\Psi^0(M))$. In analogy with (4.11), we have

$$[\Phi]\cdot\{S\} = \text{Index } S_{(k)}\, \Phi^{op}\big|_{\mathcal{H}^{(k)}} \tag{4.15}$$

where $S_{(k)} = S \otimes I$ on $L^2(M,\mathbb{C}^k)$, Φ^{op} is as above in the case of the trivial line bundle E, and $\mathcal{H}^{(k)}$ denotes the orthogonal sum of k copies of the range of S.

The fact that the wave front set of S is contained in the image of the section γ of S^*M defining the contact structure of $M = \partial\Omega$ has the following implication for the index calculation (4.15). Given any $\Phi : S^*M \to GL(k,\mathbb{C})$, define Φ_γ^b to be $\Phi\circ\gamma$, so $\Phi_\gamma^b : M \to GL(k,\mathbb{C})$. We also regard this map as $\Phi_\gamma : S^*M \to GL(k,\mathbb{C})$, constant on the fibers. We denote by Φ_γ^{op} the operator on $L^2(M,\mathbb{C}^k)$ obtained by matrix multiplication by Φ_γ^b. Then

$$S_{(k)}(\Phi^{op} - \Phi_\gamma^{op}) \in OPS^{-\frac12}_{\frac12,\frac12}(M),$$

hence is compact on $L^2(M)$. We conclude that, if Φ^{op} has index zero,

$$[\Phi]\cdot\{S\} = [\Phi_\gamma]\cdot\{S\}. \tag{4.16}$$

This identity is also related to properties of $\{S\}$ discussed at the end of §3.

In fact, since $[\Phi_\gamma] = \pi^*\gamma^*[\Phi]$ in $K^1(S^*M)$, functoriality of the intersection product gives

$$[\Phi_\gamma]\cdot\alpha = [\Phi]\cdot\Pi_\gamma\alpha \tag{4.17}$$

for all $\alpha \in K^1(\Psi^0(M))$, Π_γ being the projection given by (3.24). In view of this (4.16) then follows from the identity (3.25).

We furthermore note that, whenever the sequence (1.19) is exact, for any $\alpha \in K^1(\Psi^0(M))$ we can write

$$\alpha = \Pi_\chi \alpha + \tau(E) \text{ for some } E \in K^0(M). \tag{4.18}$$

When this holds, then (4.17) implies

$$\begin{aligned} ([\Phi] - [\Phi_\chi])\cdot\alpha &= [\Phi]\cdot(\alpha - \Pi_\chi\alpha) \\ &= [\Phi]\cdot\tau(E) \\ &= \text{Index } \Phi_E^{op}, \end{aligned} \tag{4.19}$$

the last identity following from (4.12).

Finally, let us relate these computations to the quesiton of whether $\{S\} = \{D_{\partial\Omega}\}$ in $K^1(\Psi^0(\partial\Omega))$, when $\partial\Omega$ is the boundary of a strongly pseudo-convex manifold. In view of (4.16), this would yield the following identity, also implied by (3.27):

$$[\Phi]\cdot\{D_{\partial\Omega}\} = [\Phi_\chi]\cdot\{D_{\partial\Omega}\} \tag{4.20}$$

for each $\Phi : S^*(\partial\Omega) \to GL(k,\mathbb{C})$ such that Φ^{op} has index zero. Now each side of (4.20) can in principle be calculated via the Atiyah-Singer index theorem, in light of the formula (4.11).

We close on the following speculative note. It seems rather likely that in general $\{S\}$ and $\{D_{\partial\Omega}\}$ are not identical in $K^1(\Psi^0(\partial\Omega))$, i.e., $(\text{Id.} - \Pi_\chi)\{D_{\partial\Omega}\} \neq 0$. It is tempting to guess that a further general refinement of Boutet de Monvel's index theorem would involve specifying this difference, in terms of the map τ.

Acknowledgment. Part of the work on this paper was done while I was visiting MSRI, in November 1988. During this time, I had several very stimulating conversations with R. Melrose on this material.

References.

1. M. Atiyah, K-Theory, Benjamin, N.Y., 1967.

2. P. Baum and R. Douglas, K-homology and index theory, Proc. Symp. Pure Math. 38, Part 1 (1982), 117-173.

3. ________________, Toeplitz operators and Poincare duality, Proc. Toeplitz Memorial Conf., Tel Aviv, Birkhauser, Basel, 1982, 137-166.

4. ________________, Relative K-homology and C^*-algebras, Preprint.

5. P. Baum, R. Douglas, and M. Taylor, Cycles and relative cycles in analytic K-homology, J. Diff. Geom., to appear.

6. B. Blackadar, K-Theory for Operator Algebras, Springer, N.Y., 1986.

7. L. Boutet de Monvel, On the index of Toeplitz operators of several complex variables, Invent. Math. 50(1979), 249-272.

8. L. Brown, R. Douglas, and P. Fillmore, Extensions of C^*-algebras and K-homology, Ann. Math. 105(1977), 265-324.

9. J.-L. Brylinski and E. Getzler, The homology of algebras of pseudodifferential symbols and the noncommutative residue, K-Theory 1(1987),385-403.

10. G. Folland and J.J. Kohn, The Neumann Problem for the Cauchy-Riemann Complex, Princeton Univ. Press, 1972.

11. L. Hörmander, The Analysis of Linear Partial Differential Operators, Springer, N.Y., Vols. 3 & 4, 1985.

12. G. Kasparov, Topological invariants of elliptic operators I: K-homology, Math. USSR Izv. 9(1975), 751-792.

13. __________, The operator K-functor and extensions of C^*-algebras, Math. USSR Izv. 16(1981), 513-572.

14. __________, Operator K-theory and its applications: elliptic operators, group representations, higher signatures, C^*-extensions. Chernogolovka, 1982.

15. M. Taylor, Pseudodifferential operators and K-homology, Proc. Symp. Pure Math., to appear.

16. _________, Pseudodifferential Operators, Princeton Univ. Press, 1981.

17. M. Wodzicki, Cyclic homology of differential operators, Duke Math. J. 54(1987), 641-647.

Department of Mathematics
University of North Carolina
Chapel Hill, NC 27514

Contemporary Mathematics
Volume **105**, 1990

THE HEAT FLOW ALONG THE LEAVES OF A RIEMANNIAN FOLIATION

by Philippe Tondeur and Jesús A. Alvarez López [1]

ABSTRACT. We prove a Hodge decomposition theorem for the leafwise Laplacian of a Riemannian foliation on a closed Riemannian manifold.

1. INTRODUCTION. Let $\mathcal{F}$ be a smooth foliation on a closed oriented manifold M. Let g_M be a Riemannian metric on M, which decomposes TM as an orthogonal direct sum $T\mathcal{F} \oplus Q$, where $Q = T\mathcal{F}^{\perp}$. This gives rise to the bigrading of the algebra of smooth differential forms $\Omega = \Omega_M$

$$(1.1) \qquad \Omega^{u,v} = \Gamma(\Lambda^v T\mathcal{F}^* \otimes \Lambda^u Q^*).$$

The exterior derivative d decomposes as the sum of bihomogeneous components $d = d_{0,1} + d_{1,0} + d_{2,-1}$ with the indicated bidegrees.

A natural idea is to try to find a relation between the cohomology $E_1 = H(\Omega, d_{0,1})$ (see e.g. [A][KT]) and the harmonic forms associated to the leafwise Laplacian Δ_0, the Laplacian canonically associated to $d_{0,1}$. In the spirit of Milgram and Rosenbloom [MR], one has then to study the heat flow associated to Δ_0, and its long-time behavior.

One constructs a chain of Hilbert space completions $\{H_{0,r}\}_{r\geq 0}$ of Ω_M with $H_{0,r} \supset H_{0,r+1}$, and self-adjoint extensions $\bar{\Delta}_{0,r} = \bar{D}^2_{0,r} : \mathrm{Dom}(\bar{\Delta}_{0,r}) \subset H_{0,r} \to H_{0,r}$ of $\Delta_0 : \Omega_M \to \Omega_M$. This induces a canonical continuous extension $\bar{\Delta}_{0,\infty} : H_{0,\infty} \to H_{0,\infty}$ to the Fréchet space

---

1980 AMS Subject Classification: 58E, 58G
Key words and phrases: Heat equation, Riemannian foliation.
[1]Work supported in part by a grant from the National Science Foundation.

(1.2) $$H_{0,\infty} = \bigcap_{r \geqq 0} H_{0,r}.$$

The heat flows associated to the operators $\bar{\Delta}_{0,r}$ give rise to the desired Hodge decomposition theorem of $H_{0,\infty}$ (see Theorem 2.24 below).

In the case of a closed oriented Riemannian manifold (viewed as a one leaf foliation) we have that each $H_{0,r}$ is the usual Sobolev space H_r, so $H_{0,\infty} = H_\infty \cong \Omega_M$ by the Sobolev Lemma.

A similar interpretation of $H_{0,\infty}$ can be given when the foliation is Riemannian and the metric bundle-like. In this case $H_{0,\infty}$ is formed by the space of elements in $L^2(\Omega_M)$, whose leafwise derivatives of any order exist and are also in $L^2(\Omega_M)$ (see Theorem 3.17 below).

We would like to thank I. D. Berg and F. Kamber for helpful comments.

2. LEAFWISE HODGE DECOMPOSITION. Let $\mathcal{F}$ be a smooth foliation on a closed oriented Riemannian manifold (M, g_M). Then we have the bigrading $\Omega^{u,v}$ of forms given by (1.1), and the decomposition $d = d_{0,1} + d_{1,0} + d_{2,-1}$. The formal adjoint δ of d decomposes as $\delta = \delta_{0,-1} + \delta_{-1,0} + \delta_{-2,1}$ with the indicated bidegrees. We have then the Dirac operator along the leaves

(2.1) $$D_0 = d_{0,1} + \delta_{0,-1} ,$$

and the Laplacian along the leaves

(2.2) $$\Delta_0 = D_0^2 = d_{0,1}\delta_{0,-1} + \delta_{0,-1}d_{0,1}.$$

For each integer $r \geqq 0$ let $H_{0,r}$ denote the Hilbert space completion of Ω_M with respect to the scalar product

(2.3) $$\langle \omega, \omega' \rangle_{0,r} = \sum_{i=0}^{r} \langle \Delta_0^i \omega, \omega' \rangle \quad \text{for} \quad \omega, \omega' \in \Omega_M.$$

For the corresponding norms $\| \ \|_{0,r}$, we have

(2.4) $$r \leqq r' \Rightarrow \|\omega\|_{0,r} \leqq \|\omega\|_{0,r'} \quad \text{for all} \quad \omega \in \Omega.$$

Thus we obtain the chain of continuous inclusions

$$(2.5)\qquad L^2(\Omega_M) = H_{0,0} \supset H_{0,1} \supset H_{0,2} \supset \cdots \supset H_{0,\infty},$$

where

$$(2.6)\qquad H_{0,\infty} = \bigcap_{r \geqq 0} H_{0,r}$$

is equipped with the obvious Fréchet topology.

D_0 is $\langle\ ,\ \rangle_{0,r}$-symmetric, so by Corollary 1.4 and Lemma 2.1 in [CH] it follows that any power of D_0 is essentially self-adjoint in $H_{0,r}$.

Let $\bar{D}_{0,r}$ and $\bar{\Delta}_{0,r}$ be the closures of D_0 and Δ_0 in $H_{0,r}$. By (2.3) and (2.4) we have

$$(2.7)\qquad r \leqq r' \Rightarrow \mathrm{Dom}(\bar{\Delta}_{0,r}) \supset \mathrm{Dom}(\bar{\Delta}_{0,r'}) \text{ and } \mathrm{Dom}(\bar{D}_{0,r}) = H_{0,r+1},$$

$\bar{D}_{0,r} : H_{0,r+1} \longrightarrow H_{0,r}$ is a bounded operator, and the diagrams

$$(2.8)\qquad \begin{array}{ccc} \mathrm{Dom}(\bar{\Delta}_{0,r}) & \xrightarrow{\bar{\Delta}_{0,r}} & H_{0,r} \\ \cup & & \cup \\ \mathrm{Dom}(\bar{\Delta}_{0,r'}) & \xrightarrow{\bar{\Delta}_{0,r'}} & H_{0,r'} \end{array}, \qquad \begin{array}{ccc} H_{0,r+1} & \xrightarrow{\bar{D}_{0,r}} & H_{0,r} \\ \cup & & \cup \\ H_{0,r'+1} & \xrightarrow{\bar{D}_{0,r'}} & H_{0,r'} \end{array}$$

are commutative.

From definition (2.3) it follows that

$$(2.9)\qquad H_{0,r+2} \subset \mathrm{Dom}(\bar{\Delta}_{0,r})$$

and the restriction is a bounded operator

$$(2.10)\qquad \bar{\Delta}_{0,r} : H_{0,r+2} \to H_{0,r}.$$

Hence the operators $\bar{\Delta}_{0,r}$ and $\bar{D}_{0,r}$ define continuous operators

$$(2.11)\qquad \bar{\Delta}_{0,\infty},\ \bar{D}_{0,\infty} : H_{0,\infty} \to H_{0,\infty}.$$

By the spectral theorem and the non-negativity of $\bar{\Delta}_{0,r}$, we have the semigroup of bounded operators

$$(2.12) \qquad e^{-t\bar{\Delta}_{0,r}} : H_{0,r} \to H_{0,r} \quad \text{for} \quad t \geqq 0.$$

The commutativity of the diagrams

$$(2.13) \qquad \begin{array}{ccc} H_{0,r} & \xrightarrow{e^{-t\bar{\Delta}_{0,r}}} & H_{0,r} \\ \cup & & \cup \\ H_{0,r+1} & \xrightarrow{e^{-t\bar{\Delta}_{0,r+1}}} & H_{0,r+1} \end{array}$$

follows by the commutativity of (2.8) and the density of the subspace of the space of analytic vectors s in $H_{0,r+1}$ (see [W], for example), for which the series

$$(2.14) \qquad \sum_{n=0}^{\infty} \frac{(-t\bar{\Delta}_{0,r+1})^n}{n!} s$$

is $\| \ \|_{0,r+1}$-convergent for all $t \geqq 0$.

Since $e^{-t\bar{\Delta}_{0,r}}$ converges strongly (as $t \to \infty$) to the orthogonal projection P_0^r of $H_{0,r}$ onto the kernel of $\bar{\Delta}_{0,r}$ (by the spectral theorem), from the commutativity of (2.13) we obtain the commutativity of the diagram

$$(2.15) \qquad \begin{array}{ccc} H_{0,r} & \xrightarrow{P_0^r} & H_{0,r} \\ \cup & & \cup \\ H_{0,r+1} & \xrightarrow{P_0^{r+1}} & H_{0,r+1} \end{array}$$

and hence

$$(2.16) \qquad \ker(\bar{\Delta}_{0,r}) \supset \ker(\bar{\Delta}_{0,r+1}).$$

On the other hand, with the same arguments as in Lemma A.4 of [HL] we obtain the following orthogonal direct sum decomposition

$$H_{0,r} \cong \ker \bar{\Delta}_{0,r} \oplus \overline{\operatorname{im} \bar{\Delta}_{0,r}}. \tag{2.17}$$

From (2.7) and (2.8) we obtain further

$$\operatorname{im}(\bar{\Delta}_{0,r}) \supset \operatorname{im}(\bar{\Delta}_{0,r+1}). \tag{2.18}$$

Then, since $H_{0,r+1}$ is $\| \ \|_{0,r}$-dense in $H_{0,r}$, from (2.16), (2.17) and (2.18) we obtain that $\ker(\bar{\Delta}_{0,r})$ and $\overline{\operatorname{im}(\bar{\Delta}_{0,r})}$ are the $\| \ \|_{0,r}$-completions of $\ker(\bar{\Delta}_{0,r+1})$ and $\operatorname{im}(\bar{\Delta}_{0,r+1})$ respectively. But it is clear that on $\ker(\bar{\Delta}_{0,r+1})$ we have $\| \ \|_{0,r+1} = \| \ \|_{0,r}$, and thus

$$\ker(\bar{\Delta}_{0,0}) = \ker(\bar{\Delta}_{0,1}) = \ker(\bar{\Delta}_{0,2}) = \cdots = \ker(\bar{\Delta}_{0,\infty}). \tag{2.19}$$

Therefore, since $\operatorname{im}(\bar{\Delta}_{0,\infty})$ is dense in each $\operatorname{im}(\bar{\Delta}_{0,r})$, the following orthogonal direct sum decomposition results:

$$H_{0,\infty} \cong \ker(\bar{\Delta}_{0,\infty}) \oplus \overline{\operatorname{im}(\bar{\Delta}_{0,\infty})}. \tag{2.20}$$

It is easy to check that

$$\ker(\bar{D}_{0,\infty}) = \ker(\bar{\Delta}_{0,\infty}), \quad \operatorname{im}(\bar{D}_{0,\infty}) \perp \ker(\bar{D}_{0,\infty}). \tag{2.21}$$

Therefore, (2.20) and (2.21) imply the orthogonal decomposition

$$H_{0,\infty} \cong \ker \bar{D}_{0,\infty} \oplus \overline{\operatorname{im} \bar{D}_{0,\infty}}. \tag{2.22}$$

Moreover, the spaces $\Omega^{u,v}$ are $\langle \ , \ \rangle_{0,r}$-orthogonal to each other for each integer $r \geqq 0$. It follows that $\bar{D}_{0,\infty}$ can be decomposed as the sum of the continuous operators

$$(2.23) \qquad \bar{d}_{0,1,\infty},\ \bar{\delta}_{0,-1,\infty} : H_{0,\infty} \longrightarrow H_{0,\infty} ,$$

which are extensions of $d_{0,1}$ and $\delta_{0,-1}$, respectively. Since $\operatorname{im} d_{0,1}$, and $\operatorname{im} \delta_{0,-1}$ are $\langle\ ,\ \rangle_{0,r}$-orthogonal for each integer $r \geqq 0$, from (2.22) we obtain the following leafwise Hodge decomposition theorem.

2.24 THEOREM. Let $\mathcal{F}$ be a smooth foliation on a closed oriented Riemannian manifold. Then we have an orthogonal direct sum decomposition

$$(2.25) \qquad H_{0,\infty} \cong \ker \bar{\Delta}_{0,\infty} \oplus \overline{\operatorname{im} \bar{d}_{0,1,\infty}} \oplus \overline{\operatorname{im} \bar{\delta}_{0,-1,\infty}}.$$

3. INTERPRETATION OF $H_{0,\infty}$. With the context of the preceding section we assume now in addition that $\mathcal{F}$ is Riemannian and g_M bundle-like. We define a new connection on M by

$$(3.1) \qquad \overset{\circ}{\nabla}_X Y = \pi_{\mathcal{F}}\, \nabla_X\, \pi_{\mathcal{F}} Y + \pi_Q\, \nabla_X\, \pi_Q\, Y$$

for vector fields X, Y, where $\nabla = \nabla^M$ is the Levi-Cività connection associated to g_M, and $\pi_{\mathcal{F}}$, π_Q the orthogonal projections of TM onto $T\mathcal{F}$, Q respectively. It is easy to check that

$$(3.2) \qquad \overset{\circ}{\nabla}_X = (\nabla_X)_{0,0} \quad \text{on} \quad \Omega_M,$$

where the double subscript indicates that we are considering the bihomogeneous component of bidegree $(0,0)$. Let $E_1,\dots,E_p$ be a (local) orthonormal frame of $T\mathcal{F}$ on a distinguished chart $U \subset M$, and $\alpha_1,\dots,\alpha_p$ the dual co-frame of $T\mathcal{F}^*|U$. Then, from Koszul's formula for d and δ (see e.g. [P, Ch. IV]), Reinhart's characterization of bundle-like metrics in [RE], and by (3.2) we have for any $\omega \in \Omega^{0,\cdot}(U)$

$$(3.3) \qquad d_{0,1}\omega = \sum_{i=1}^{p} \alpha_i \wedge \overset{\circ}{\nabla}_{E_i} \omega$$

(3.4) $$\delta_{0,-1}\omega = -\sum_{i=1}^{p} i_{E_i}\overset{\circ}{\nabla}_{E_i}\omega.$$

For the following we can assume without restricting generality that $\mathcal{F}$ is oriented. We choose an orientation of $T\mathcal{F}$, which with the given orientation of TM induces an orientation of Q. These orientations define a "leafwise star operator" and a "transverse star operator" respectively.

On a distinguished chart U we can find basic 1-forms $\beta_{p+1},\ldots,\beta_{p+q} \in \Omega^{1,0}(U)$, i.e. satisfying $d_{0,1}\beta_j = 0$, such that for each $x \in U$ the $(\beta_j)_x$ form an oriented orthonormal basis of Q_x^*. This implies the decomposition

(3.5) $$\Omega_c^{u,v}(U) = \Omega_c^{0,v}(U) \otimes \Lambda^u(\bigoplus_{j=p+1}^{p+q} \mathbb{R}\cdot\beta_j).$$

Then, using the expression of the star operator as a tensor product of the leafwise star operator with the transverse star operator, we obtain

(3.6) $$D_0 = D_0 \otimes \mathrm{id}$$

with respect to the decomposition (3.5).

Each $\beta \in \Lambda(\bigoplus_{j=p+1}^{p+q} \mathbb{R}\cdot\beta_j)$ is supposed to be extended by zero to the whole manifold M. In this way if $\beta \neq 0$ then it is not continuous on the boundary of U, but it still defines an element in $L^2(\Omega_M)$.

Now, for each integer $r \geq 0$ let $\overset{\circ}{\nabla}{}^r_{\mathcal{F}} : \Omega_M \longrightarrow \Gamma(\otimes^r T\mathcal{F}^* \otimes \Lambda T^*M)$ be the operator defined by $\overset{\circ}{\nabla}$ in the usual way:

(3.7) $$(\overset{\circ}{\nabla}{}^r_{\mathcal{F}}\,\omega)(X_1 \otimes\cdots\otimes X_r) = \overset{\circ}{\nabla}_{X_1}\cdots\overset{\circ}{\nabla}_{X_r}\omega \quad \text{for } X_i \in \Gamma T\mathcal{F}.$$

Let further $\overset{\circ}{\nabla}{}^*_{\mathcal{F}}$ denote the formal adjoint of $\overset{\circ}{\nabla}_{\mathcal{F}} = \overset{\circ}{\nabla}{}^1_{\mathcal{F}}$.

It is easy to check that on a distinguished chart we have

$$(3.8)\qquad \|\mathring{\nabla}^r_{\mathcal{F}}\omega\|^2 = \sum_{i_1,\dots,i_r=1}^{p} \|\mathring{\nabla}_{E_{i_1}}\cdots\mathring{\nabla}_{E_{i_r}}\omega\|^2 \quad \text{for all} \quad \omega\in\Omega_c(U),$$

where $\{E_i\}$ denotes an orthonormal frame of $T\mathcal{F}|U$. As in Chapter 2 of [RO 2] it can be proved that

$$(3.9)\qquad \mathring{\nabla}^*_{\mathcal{F}}\,\mathring{\nabla}_{\mathcal{F}} = -\sum_{i=1}^{p} \nabla^2_{E_i} \quad \text{on} \quad \Omega_c(U).$$

Using arguments similar to those in Chapter 2 of [RO 2], (3.3), (3.4) and (3.9) yield the following "leafwise Bochner-Weitzenböck formula"

$$(3.10)\qquad \Delta_0 = \mathring{\nabla}^*_{\mathcal{F}}\,\mathring{\nabla}_{\mathcal{F}} + \mathring{R}_{\mathcal{F}} \quad \text{on} \quad \Omega^{0,\cdot}_M,$$

where

$$(3.11)\qquad \mathring{R}_{\mathcal{F}}\omega = \sum_{i<j} E_i\cdot E_j\cdot \mathring{R}_{E_i,E_j}\omega$$

involving Clifford multiplication and the curvature $\mathring{R}$ of $\mathring{\nabla}$.

It can be seen easily that in general (3.10) does not hold on the entire Ω_M. Nevertheless, from (3.6), (3.8), and the finite-dimensional character of $\Lambda(\bigoplus_{j=p+1}^{p+q}\mathbb{R}\cdot\beta_j)$, we obtain that for each integer $r\geqq 0$ there is a constant $c>0$ such that if $\omega=\alpha\wedge\beta\in\Omega_c(U)$, with $\alpha\in\Omega^{0,\cdot}_c$ and $\beta\in\Lambda(\bigoplus_{j=p+1}^{p+q}\mathbb{R}\cdot\beta_j)$, then we have

$$(3.12)\qquad \|\omega\|_{0,r} \leqq c\cdot\|\alpha\|_{0,r}\cdot\|\beta\|,$$

$$(3.13)\qquad \|\alpha\|_{0,r}\cdot\|\beta\| \leqq c\cdot\|\omega\|_{0,r},$$

$$(3.14)\qquad \|\mathring{\nabla}^r_{\mathcal{F}}\omega\| \leqq c\cdot\sum_{i=0}^{r}\|\mathring{\nabla}^i_{\mathcal{F}}\alpha\|\cdot\|\beta\|, \quad \text{and}$$

$$(3.15)\qquad \|\overset{\circ}{\nabla}{}^{r}_{\mathcal{F}}\alpha\| \cdot \|\beta\| \leqq c \cdot \sum_{i=0}^{r} \|\overset{\circ}{\nabla}{}^{i}_{\mathcal{F}}\omega\|.$$

Now for each integer $r \geqq 0$ let $\| \|'_{0,r}$ be the norm on Ω_M defined by

$$(3.16)\qquad \|\omega\|'_{0,r} = (\sum_{i=0}^{r} \|\overset{\circ}{\nabla}{}^{i}_{\mathcal{F}}\omega\|^2)^{1/2}.$$

By (3.10) we obtain that each $\| \|_{0,r}$ is equivalent to $\| \|'_{0,r}$ on $\Omega^{0,\cdot}$, for $r \leqq 1$. For $r > 1$ this also follows by a standard recurrence argument using (3.7). Therefore, by (3.12) to (3.15), it follows that each $\| \|_{0,r}$ is equivalent to $\| \|'_{0,r}$ on the whole Ω_M, from which we obtain the following interpretation of $H_{0,\infty}$.

3.17 THEOREM. Let $\mathcal{F}$ be a Riemannian foliation on a oriented closed Riemannian manifold with a bundle-like metric. Then the space $H_{0,\infty}$ is the space of elements in $L^2(\Omega_M)$ defined by L^2-Cauchy sequences $(\omega_m)_{m\in\mathbb{N}}$ in Ω_M, so that any sequence of smooth forms obtained from (ω_m) by leafwise derivatives of any order of its local coefficient functions is also L^2-Cauchy.

REFERENCES

[A] J. A. Alvarez López, A finiteness theorem for the spectral sequence of a Riemannian foliation, Illinois J. of Math., to appear.

[CH] P. R. Chernoff, Essential self-adjointness of powers of generators of hyperbolic equations, J. of Functional Analysis, 12 (1973), 410–414.

[GL] M. Gromov and B. Lawson, Positive scalar curvature and the Dirac operator on complete Riemannian manifolds, Publ. Math. IHES, 58 (1983), 295–408.

[HL] J. Heitsch and C. Lazarov, A Lefschetz theorem for foliated manifolds, to appear.

[KT] F. Kamber and Ph. Tondeur, Foliations and metrics, Progr. in Math., 32 (1983), 103–152.

[MR] A. N. Milgram and P. C. Rosenbloom, Harmonic forms and heat conduction I: Closed Riemannian manifolds, Proc. Nat. Acad. Sc., 37(1951), 180–184.

[P] W. A. Poor, Differential geometric structures, McGraw–Hill, New York, 1981.

[RE] B. Reinhart, Foliated manifolds with bundle–like metrics, Ann. Math., 69 (1959), 119–132.

[RO] J. Roe, Elliptic operators, topology and asymptotic methods, Pitman Research Notes in Mathematics Series 179, Longman Scientific and Technical, 1988.

[W] J. Weidman, Linear operators on Hilbert spaces, Springer Verlag, 1980.

Jesús Antonio Alvarez López
Departamento de Xeometría e Topoloxía
Facultade de Matemáticas
Universidade de Santiago de Compostela
15705 Santiago de Compostela
SPAIN

Philippe Tondeur
Department of Mathematics
University of Illinois
1409 West Green Street
Urbana, Illinois 61801
USA

Contemporary Mathematics
Volume **105**, 1990

Aspects of the Novikov Conjecture

Shmuel Weinberger*

The Novikov conjecture has been recognized as one of the central problems in Mathematics. It is not only a very natural problem whose attempted solutions have led to much important work and many new ideas (although this is the case), but it is a problem that has many variants and refinements and implications. I therefore do not expect the solution of this problem to be the calamity of "killing the goose that lays the golden eggs"[1] (in the sense of H. C. Anderson). In some cases, more precise conjectures can be formulated, but they are known to be false in great generality. In this paper I would like to explain this circle of ideas and some of the problems involved.

Originally, my plan was to give a survey of the state of the art in the solution of the problem. Things have been moving too quickly for this to be realistic any more. I would like to thank Jerry Kaminker for encouraging that I write the early version (which only was half written before despair hit me) and the present version.

The paper is organized into many small paragraphs, each devoted to stating simply one idea. Roughly, the first 9 sections (except 4) are purely introductory stating and motivating the conjecture in its topological and operator theoretic settings, and deal as well with the interrelationships among these problems and the Borel rigidity conjecture. In 10 and 11, I deal with fiberring and splitting theorems and their relatives Wang and Mayer-Vietoris sequences. (§26 also deals with fiberring.) 12 deals with manifolds with boundary. The next four paragraphs deal with aspects of the equivariant versions of these problems. These results are given nonequivariant applications in 21 and 22. In 18-20 I deal with some quasiisometry invariant questions that are suggested by Novikov, and enter geometry very naturally. §23 is a brief discussion of how these problems actually involve the eta invariant as well as L-classes. The paper ends with some discussion of nonmanifolds. The reader might choose to (or have to) peruse the references to learn the idea in depth. If so, I feel this paper will have done its task.

1. The original statement of the Novikov conjecture concerns "higher signatures". The starting point for this is Hirzebruch's signature formula [Hi] which identifies the signature of an oriented 4k dimensional manifold M (the difference of the dimensions of the maximal positive and negative definite subspaces of the inner product space given by cup product on middle dimensional cohomology) as a universal polynomial L in the Pontrjagin classes of M:

$$\operatorname{sign}(M) = \langle L(M), [M] \rangle$$

Note that L is a graded polynomial and has pieces every fourth dimension, but only the top piece is given a homotopy theoretic interpretation. In fact, for simply connected manifolds this is the

1980 *Mathematics Subject Classification* (1985 *Revision*). Primary 57R20, 57R67, 58G12.

*Partially supported by an NSF grant, a Presidential Young Investigator Award, and a Sloan Foundation Fellowship.

[1] For this reason I am a fan of partial solutions even that do not yield any "new groups"; if they contain new ideas, that's good enough for me!

0271-4132/90 $1.00 + $.25 per page

only homotopy invariant piece of $L(M)$. Novikov's conjecture is the statement that for nonsimply connected manifolds the conceivably homotopy invariant part of $L(M)$ is in fact homotopy invariant.

Let me be more precise. Suppose the fundmental group of M is π. There is a universal space for principal π bundles, $B\pi$ (it is a space of $K(\pi,1)$ type, that is, it is the unique homotopy type with fundamental group π and contractible universal cover). The universal cover of M is such a bundle, and therefore gives rise to a well defined homotopy class of maps

$$f : M \to B\pi .$$

Suppose that one has α in $H^*(B\pi)$, then one can form the quantity

$$\operatorname{sign}_\alpha(M) = \langle (f^*\alpha) \cup L(M), [M] \rangle$$

The Novikov conjecture is the statement that this element is a homotopy invariant. The simply connected case is of course a consequence of Hirzebruch's formula.

It is sometimes more convenient to deal with a Poincare dual form of the conjecture, which is that

$$f_*(L(M) \cap [M]) \in H_*(B\pi)$$

is a homotopy invariant. This version has the advantage of being integral rather than just rational, but is false! We shall discuss better integral versions below. More importantly, it enables one to work with all of the cohomology of $B\pi$ simultaneously.

Let me mention a few small variants: One can consider arbitrary maps to $B\pi$ rather than just ones that come from the classification of fundamental group. This makes no difference. However, it enables one to make sense of the conjecture for infinitely presented groups. It also makes no difference to consider smooth or topological manifolds. Also, one can deal with rational homotopy (or better, twisted homology) equivalence rather than ordinary homotopy equivalence. In that case, because of technicalities involving fundamental classes it is best to deal with the homological version. This is also equivalent to the standard version. (See [KaM] for a discussion of the degree one case.) In fact, one can also work with $\mathbb{Q}$-homology manifolds equally well (they have L-classes by work of Thom and Milnor) and, using intersection homology, even certain types of singular spaces that include algebraic varieties, see [CW, We1 §9], all with no change. We shall also discuss variants that do not seem to follow from the conjecture later.

2. I would like to explain two things in this paragraph. First, why one only considers maps into aspherical spaces, and second, why only the L-classes and not some other genera. (All of this will be seen much more formally, and technically, in §7 below.)

The reasons are rooted in surgery theory which gives a method of classifying manifolds within a fixed homotopy type (see the first three sections of [We1] for a survey that is more than adequate for the purposes of this paper). There are two important theorems that force this.

The first is the $\pi - \pi$ theorem which reduces surgery theory by bordisms over the fundamental group. Let me be precise:

Suppose that $f : M \to X$ is a map which one would like to cobord to a homotopy equivalence to X. (This is the basic problem that surgery theory attacks, classification follows from a relative version of this existence problem and the h-cobordism theorem.) The $\pi - \pi$ theorem says that solving this problem up to cobordism where X is allowed to change by a bordism preserving its fundmental group, implies that one can solve the problem where X is left fixed! (In this formulation, it should be viewed as a cobordism invariance of intersection number type statement.)

As a consequence of this, and some elementary bordism arguments, any element of the kernel of $H_*(X) \to H_*(B\pi)$ occurs as the difference of pushed forward homology L-classes for manifolds with maps to X.

The second general theorem that bears on which characteristic classes can be homotopy invariant is the product formula/periodicity. If one crosses a surgery problem with a simply connected manifold of dimension $4k$, the obstruction is multiplied by the signature. This implies that signature 0 things $\times$ arbitrary bordism classes arise as differences of bordism classes of homotopy equivalent manifolds. The only characteristic class that can survive such a radical vanishing statement is the L-class.

In short the only conceivable rational invariant of cobordism and homotopy invariance is the homology higher signature. (It is also of some interest to solve this integrally, even for the trivial group!)

3. One might ask "what is the meaning of this conjecture?" I think that it is useful to consider the case of $\pi = \mathbb{Z}$, so that $B\pi = S^1$, the circle. In that case there is but one interesting cohomology class, in dimension 1, and it is Poincare dual to the homology class represented by a point. The higher signature associated to this cohomology class can be identified, via the Hirzebruch formula, with $\mathrm{sign}(f^{-1}(*))$. the Novikov conjecture is that this number is homotopy invariant.

There does not seem to be any good reason why a homotopy equivalence should preserve submanifolds in any good way. If one could make the inverse image homotopy equivalent, of course that would suffice. We will return to this later. In general the conjecture seems to involve a correspondence between carefully chosen submanifolds for homotopy equivalent manifolds.

In this case Novikov himself was able to prove the conjecture by identifying this signature with a $\mathbb{Q}$-homological invariant of the infinite cyclic cover of M. One can view this as prototypical for a certain type of topological approach to the problem: one looks for an a priori homotopy invariant of the cellular chain complex of the universal cover, and seeks to identify it with the signatures of "canonical submanifolds", or, equivalently, the higher signatures. However, even for the free abelian case, where one will be asking for signatures of transverse inverse images of subtori, it is quite difficult to extend this approach.

4. There are two approaches to the simply connected conjecture, the Hirzebruch formula. The first is Hirzebruch's; it is bordism theoretic, but does not seem to apply very often. The second is Atiyah-Singer's analytic, more conceptual argument, and this is generalizable. Index theory is of course intimately connected to the Novikov conjecture and its generalizations.

Lusztig [Lu] gave a beautiful method in his thesis for handling the free abelian case. He considered the family of signature operators on M twisted by a flat line bundle. This family is classified by $Hom(\pi, S^1)$, a torus. The index bundle of this family of operators is a homotopy invariant, and the index theorem for families identifies the chern character of this with the higher signature.

Later authors, such as Mischenko and Kasparov, have tried to do analogous things with families of infinite dimensional flat bundles, using C^*-algebra techniques.

5. There is a genuine connection between the analytic and topological approaches to the problem. Analytically, one examines variants of the signature operator (or better, the K-homology class determined by the signature operator) as the basic homotopy invariant object. Topologically, the basic homotopy invariant of an n-manifold is the cellular cahin complex with its Poincare duality structure; this is an element of a suitable Groethendieck group, $L^n(\mathbb{Z}\pi)$, and is called the symmetric signature [Ra]. (More precisely, one deals with the π-equivariant version of the universal cover.)

It is a basic fact that the topological invariant determines the analytic one: there is a natural map that takes the symmetric signature to the class of the signature operator ([Ka, KaM]).

The basic strategy followed by many papers is to deal with a natural map (often called the assembly map) from rational group homology to L-theory, or $K(B\pi) \to K(C^*\pi)$, and try to show that it is injective, so that the higher signature which lives in the former group and is a priori homotopy invariant in the target group, is homotopy invariant.

6. The analytic version is useful for other reasons beyond homotopy invariance of higher signatures. Rosenberg [Ro1] has pointed out that the Lichnerowicz method for showing that certain spin manifolds have no metric with positive scalar curvature (basically Atiyah-Singer theorem + Weitzenboch formula) actually implies the vanishing of the index of Dirac in $K(C^*\pi)$. Consequently, whenever Novikov can be proven in the C^* context, one sees that a higher $\hat{A}$-genus vanishes as well[1].

There are presumably holomorphic versions of this phenomenon for some problems involving the Dolbeault operator.

More generally, one can take any theorem for characteristic numbers of simply connected manifolds that is proven index theoretically, and ask about "higher" analogues. We shall indicate some transformation group examples of this type of theorem below. (§16)

7. The topological significance of the conjecture can be seen most clearly using surgery, and using this one can obtain some more precise geometric (although, as we've seen, less general) information from the topological form.

I have presented elsewhere a detailed review of surgery theory [We1] Here I would like to give a minimal statement of the "main theorem", the surgery exact sequence[2], although in later paragraphs I have to give various extensions.

For simplicity, I will assume that M is an oriented manifold[3]. If X is an arbitrary CW complex one defines a group $S_n(X)$ and a homology theory $H_n(X;L)$ for a spectrum L, and a 4-periodic sequence of abelian groups, functorial in $\pi = \pi_1(X)$, $L_n(\pi)$. It has the property that if M is an oriented n-manifold, $n \geq 5$, then there is a one to one correspondence between $S_n(M)$ and $\{(W,f)$ such that $f : W \to M$ is a proper homotopy equivalence which restricts to a homeomorphism in the complement of some compact subset of the interior of $M\}$/homotopy rel infinity to a homeomorphism. Thus $S_n(X)$ is a worthwhile object to study: it encodes the difference between homotopy and homeomorphism classification of manifolds. The main result is that there is a covariantly functorial 4-periodic[4] exact sequence:

$$\cdots \to S_n(X) \to H_n(X;L) \to L_n(\pi) \to S_{n-1}(X) \to \cdots$$

Recall that the Poincare conjecture (theorem in high dimensions) asserts that homotopy equivalences to the sphere are homotopy to homeomorphisms. This implies that $\pi_n(L) = L_n(e)$. In fact

1 There is some evidence that the vanishing, for spin manifolds, the real C^* analogue of this index is necessary and sufficient (Gromov-Lawson-Rosenberg conjecture; sometimes the vanishing of the higher $\hat{A}$-genus is given this name). I do not know what the correct conjecture should be for manifolds which are not spin, or even worse, manifolds with no spin cover.

2 For simplicity, I will deal with the topological version of surgery; the smooth version, while easier to set up does not have all of the same beautiful functorial properties. Consequently, to rigorize many of the claims in this paper one would ideally use Lipschitz manifolds and the Teleman signature operator in the philosophy of [SuT], see [Ro2] for a survey.

3 To be strictly correct, everything is a functor of X and an element of $H^1(X;\mathbb{Z}_2]$.

4 Here there is a slight lie involving one $\mathbb{Z}$; see [We1].

the homotopy type of L is known (due to Sullivan, see [MM]) and one has the calculation that

$$H_n(X;L)\oplus \mathbb{Z}_{(2)} = \oplus_{i=n \bmod 4} H_i(X;\mathbb{Z}_{(2)}) \oplus H_{i-2}(X;\mathbb{Z}_2)$$
$$H_n(X;L)\otimes \mathbb{Z}[1/2] = KO_n(X)\otimes \mathbb{Z}[1/2]\,.$$

Even more primitively

$$H_n(X;L)\otimes \mathbb{Q} = \oplus_{i=n \bmod 4} H_i(X;\mathbb{Q})$$

The functoralities involved are closely related to those involved in the index theorem (see [Wel §2]). The homology theory term is often called the normal invariant, and is a measure of the difference between the tangent bundles of the homotopy equivalent manifolds. The image of a homotopy equivalence in $H_n(M;L)\otimes \mathbb{Q}$[1] is just the difference between the Poincare duals of the L-classes of W and M[2]. In light of this and the diagram associated to the map $M \to B\pi$ one gets:

$$\begin{array}{ccccccccc} \cdots & \to & S_n(M) & \to & H_n(M;L) & \to & L_n(\pi) & \to & \cdots \\ & & \downarrow & & \downarrow & & \downarrow & & \\ \cdots & \to & S_n(B;\pi) & \to & H_n(B\pi;L) & \to & L_n(\pi) & \to & \end{array}$$

Novikov now becomes seen to be equivalent to the statement that $H_n(B\pi;L) \to L_n(\pi)$ is rationally injective.

Notice that if the Novikov conjecture were true canonically enough, one could produce a map (perhaps only rationally) $S : S_n(M) \to H_n(M, B\pi; L)$ which would encode the most precise type of tangential data that one can assign to a homotopy equivalence. We call the map S a solution to the Novikov conjecture. Most solutions of the Novikov conjecture actually produce solutions to the Novikov conjecture!

8. What is the connection between §5 and §7? there are commutative diagrams:

$$\begin{array}{ccc} H_n(B\pi;L) & \to & L_n(\pi) \\ \downarrow & & \downarrow \\ H_n(B\pi;L^*) & \to & L^n(\pi)\ ^3 \\ \downarrow & & \downarrow \\ K_n(B\pi) & \to & K_n(C^*\pi) \end{array}$$

The top map is the one of the previous section. The second map is the symmetric algebraic analogue. The maps $L_* \to L^*$ are isomorphisms away from 2. To a normal invariant one assigns the difference of symmetric signatures on the right[4]. On the left one has, at 2, a corrected version of the Morgan-Sullivan L-class (which is a variant of the Hirzebruch L-class defined for topological manifolds, see [MM]), and away from 2 the difference between the Sullivan orientations. (Rosenberg and I are planning a paper on this topic.[5]) The map from L-theory to K-theory is a

1 Away from 2 it is the difference between the classes of the signature operators. At 2 it is not the difference of two intrinsic invariants, and is rather more subtle.

2 This can be seen from the point of view of the product formula as in §2.

3 Weiss [Ws] has defined a very useful variant of symmetric L-theory for which this square is part of a ladder of long exact sequences with isomorphic third terms, and for which one can still use as a receptor for homotopy invariant algebraic data.

4 It also has the interpretation as the map that takes a the bordism class of a certain type of stratified space to its intersection homological symmetric signature.

5 It answers a few mysteries. For one, why is the class of the Teleman signature operator a topological invariant lying in integral K-theory? Topologists are most used to invariants lying in $K[1/2]$.

generalization of the signature of a quadratic form, and appears in work of Mischenko, Kasparov, and Kaminker-Miller. In $K_n(B\pi)$ one has the push forward of the class of the signature operator.

One can strengthen Novikov to the statement that these maps are integrally injections. This would imply strong homotopy invariance properties of indices, and the like. Unfortunately it is false in general; the reader can easily check this for any of the theories he is familiar with, for π finite. There are no known counterexamples for π torsion free.

9. The integral Novikov conjecture mentioned above has a nice geometric interpretation for $M = B\pi$. It says that any manifold homotopy equivalent to M is normally cobordant to M. This is equivalent to saying that the map $\times \mathbb{R}^3$ is properly homotopic to a homeomorphism. Thus the integral Novikov conjecture (for L-theory) is equivalent to a stable rigidity theorem for such manifolds. Note an important point about the topological version: if one analyzes a few specific manifolds (mainly to get it in all congruence classes of dimensions mod 4) with given fundmental group, then one has the result for all manifolds with that fundamental group.

In case M is complete and has nonpositive curvature, Kasparov has proven the integral $C^*\pi$ version, and Ferry and I have proven the L-theory version using rigidity ideas [FRW]. I believe that this proof can be extended to torsion free hyperbolic groups in Gromov's sense. (The rational version for these was proven by Connes and Moscovici [CM].)

A *much* stronger rigidity conjecture has been made by Borel. If $M = B\pi$ is finite dimensional, then $S(M) = 0$.[1] Farrell and Jones have recently announced this if M is a closed nonpositively curved manifold[2]. This translates to the conjecture that for such a group the assembly map is an isomorphism. One can conjecture the same thing for the $C^*\pi$ version, the best result being that of Kasparov (as far as I know) proving the result for discrete subgroups of $SO(n,1)$ and amenable groups. Hopefully, one can expect a proof of all the Borel conjectures in this paper for Hyperbolic groups before too long.

10. I would like to return to more old fashioned ideas on the problem for a moment. That is, let's return to the case of $\pi = \mathbb{Z}$ (or $\mathbb{Z}^n$) for a more careful look.

In §3 we discussed an interpretation of the Novikov conjecture as a statement about being able to make transverse inverse images homotopy equivalent. In the case of manifolds which fiber over a circle (the fiber being a suitable "canonically chosen submanifold") one can do this, and the result is intimately related to the second of the following three decompositions

$$K_i(R[T,T^{-1}] = K_i(R) \oplus K_{i-1}(R) \oplus Nil(R) \quad \text{([BHS] and [Q])}$$
$$L_i(\pi \times \mathbb{Z}) = L_i(\pi) \oplus L_{i-1}(\pi) \quad \text{([Sh])}^3$$
$$K_i(A \times \mathbb{Z}) = K_i(A) \oplus K_{i-1}(A) \quad \text{([PV])}$$

for rings, groups, and C^* algebras respectively (with fairly obvious extensions into long exact sequences for twisted Laurent extensions, extensions of $\mathbb{Z}$ by π, and crossed product algebras

1 He even conjectures that homotopy equivalence implies homeomorphism. This implies, via the h-cobordism theorem, that the algebraic K groups of π vanish in degree below 2. An algebraic K-theoretic analogue of the Novikov conjecture has been proven recently for groups with finitely generated integral homology by Bokstedt, Madsen, and Hsiang using an adaptation of some observations of Connes [Co] regarding Cyclic homology to Waldhausen's algebraic K-theory of spaces [Wald1].

2 They make much use of topology that is controlled with respect to a foliation. I believe that this is a very different sort of thing than is studied in the foliation index theorem. Presumably topologists and operator theorists have much to learn from each other in this foliation setting.

3 We are ignoring "decorations", a 2-torsion phenomenon, see [Wel §3].

respectively). The C^* algebra version could of course have been motivated by the commutative case, where the result would have been just the Kunneth formula, and then one might perspicaciously guess the others.

Let's note the Kunneth isomorphisms:

$$H_i(B(\pi \times \mathbb{Z}; L) = H_i(B\pi; L) \oplus H_{i-1}(B\pi, L)$$
$$K_i(B[\pi \times \mathbb{Z}]) = K_i(B\pi) \oplus K_{i-1}(B\pi)$$

and therefore, for instance:

$$S_i(X \times S^1) = S_i(X) \oplus S_{i-1}(X)$$

Consequently, the above decompositions contain K and L theoretic Borel conjectures inductively for $\mathbb{Z}$ extensions, and therefore, for example, for fundamental groups of nilmanifolds.

The formula for $S_i(X \times S^1)$ actually is actually equivalent to the L theoretic form, by comparing to the Kunneth formula in homology above. (The Wang sequence in homology matches up with the corresponding sequence for extensions of $\mathbb{Z}$.)

Let me try again to return to the geometry. Suppose we are given a manifold which we would like to know fibers over a circle. The first necessary condition is that the associated infinite cyclic cover, which would be a fiber $\times \mathbb{R}$ if the manifold fibered, must be homotopy equivalent to a finite complex. Assume this is so. Farrell, in his thesis (see [Fa]), showed that this is often sufficient. That is, an additional vanishing of an algebraic K-theoretic obstruction is necessary and sufficient[1].

Of course, this condition is homotopy invariant, so one sees that the appropriate codimension one submanifolds for such manifolds are homotopy equivalent. Notice also that the fiber is well defined up to h cobordism, so that an additional algebraic K theory obstruction suffices to kill this ambiguity. This yields, modulo the K theoretic decorations referred to above, the S theory statement, an therefore the L-theory statement.

This example suggests that one inflate the Novikov conjecture such that other signature type invariants associated to canonical submanifolds be homotopy invariant. Concretely (in the product case) one would want the maps (with the domains defined properly):

$$H_n(B\pi; L(\pi')) \to L_n(\pi \times \pi')$$
$$KK(C * (B\pi); C^*(\pi')) \to K(C^*(\pi \times \pi'))$$

to be rational injections in general, and isomorphisms in Borelian settings. These generalized conjectures[2] can be proven for many of the cases for which the usual Novikov conjecture can[3].

The L-theory conjecture is closely related to the issue of "approximately fiberring" a manifold over $B\pi$ (see §26). See [FRW] and [FJ] for some examples where L theoretic results of this sort are proven. I do not know of any analogous work on the C^* algebra side in this generality, although a little thought shows how to modify some of the proofs.[4]

---

1 Let me recommend the survey by Rosenberg [Ro3] to operator theorist interested in ways that algebraic K theory and topology interact.

2 They can be generalized further to allow more general rings and algebras that come from groups; they also can be twisted, as mentioned above.

3 Unfortunately the algebraic K theory proof of Bokstedt-Hsiang-Madsen does not extend gracefully to this generality, and there is much work to be done in this direction.

4 Since I'm not an expert I won't list which ones I see how to do lest I slight those other proofs which can just as easily be modified that I didn't mention.

11. Before leaving the "old fashioned conjecture", I'd like to mention a codimension one technique, extensively developed by Cappell. The idea is this: Thee was no need in the previous paragraph to produce every fiber over the circle; all that was necessary was a single fiber[1]. Therefore one should be able to get away with fibrations and only use two sided codimension one submanifolds that share some small amount of fundamental group data in common with fibrations. It turns out that the fruitful condition is that the submanifold's fundamental group inject in the fundamental group of the ambient manifold.[2] (This is because one then has a description of the universal cover as a union of fundamental domains glued together along universal covers of the submanifold; the fundamental group is a nontrivial amalgamated free product.)

One is led to consider the following question: Suppose $M \subset W$ is a locally two sided codimension one submanifold, and that $\pi_1 M \to \pi_1 W$ injects. Let $f : W' \to W$ be a homotopy equivalence, can one homotope f to be transverse to M and such that $f^{-1}(M) \to M$ is a homotopy equivalence?

There are algebraic K theory obstructions, like in the fiberring theorem, but they are fairly easily understood. When one is unobstructed one gets a Mayer-Vietoris sequence in L-theory, entirely analogous to the one one gets in homology (or K-homology) for the fundamental groups, by exactly the same reasoning as the Wang sequence above for extensions of $\mathbb{Z}$.[3]

Cappell shows that the obstruction is killed by 2 in three senses! First, the obstruction is quantified by a group UNil that is always 2-torsion. (This group is analogous to the Nil in [BHS].) Second, if $\pi_1 M \subset \pi_1 W$ is "square root closed", that is if $g^2 \in \pi_1 M$ then $g \in \pi_1 M$, then one is unobstructed (beyond the torsion conditions on the cohomotopy equivalence). Third, if one works with surgery to obtain a R-homology equivalence, and $1/2 \in R$, and thus tries to R-homology split, there is no obstruction.

This last result should not make it surprising that there is a Mayer-Vietoris sequence for all amalgamated free products for the K-theory of C^*-algebras.

A remarkable consequence of a nonvanishing UNil is the following [Ca]: a PL manifold homotopy equivalent to a connected sum, is necessarily a PL connected sum iff the fundamental group has no 2-torsion. However, this fails for all fundamental groups with 2-torsion. The primary source is an infinite number of homotopy $\mathbb{R}P^{4k+1}\#\mathbb{R}P^{4k+1}$ that are not themselves connected sums.

The splitting theorems establish the (C^*-algebra or L theory away from 2) Borel conjecture for groups that are accessible from the trivial group by HNN extensions and amalgamated free products along (not necessarily) square root closed subgroups.

One can deduce Novikov for yet a larger class of groups by making use of transfer tricks, see [Ca2]. Nonetheless, this class of groups do not contain $SL_3(\mathbb{Z})$ [4] or any subgroup of finite index in it.

Note, also, that the stronger versions of Novikov and Borel, discussed in the previous paragraph hold for thee classes of groups by keeping the group π' along in all of the MV sequences.

Perhaps the perspicacious point of view to have is that amalgamated free products and HNN

---

1 Of course, one has a single fiber there is only an h-cobordism obstruction to fiberring.

2 This is analogous to the standard method for analyzing irreducible sufficiently large three manifolds.

3 As above, the Mayer-Vietoris sequence is in fact equivalent to the splitting theorem; consequently, one can formally deduce the splitting theorem for the case of π in $\pi \times \mathbb{Z}$ from the fiberring theorem. Also, it should be pointed out that Waldhausen [Wald2] has proven the analogous Mayer Vietoris sequence in Algebraic K theory.

4 One should be able to compute the K and L theories of this using computations later in the paper and [Soule].

extensions act (not exactly properly discontinuously, but still not too badly, simplicially) by isometries on trees which have nonpositive curvature (at least intuitively). Therefore one is able to get information about the whole group in terms of the quotient and the fixed sets and their isotropies. Readers of later sections should be able to formulate their own conjectures.

12. What about manifolds with boundary?[1] There are various problems that one can consider:

1. What are the homotopy invariants of manifolds with boundary, where the boundaries are matched by a homeomorphism?

2. What are the homotopy invariants of manifolds with boundary?

3. What are the invariants between manifolds with boundary, where the map is not assumed to be a homotopy equivalence of pairs?

Similarly, one can consider the issue of positive scalar curvature for manifolds with boundary:

1′. Given a positive scalar curvature metric on a neighborhood of the boundary, when can its germ be extended to a positive scalar curvature metric on the whole manifold?

3′. Which manifolds with boundary have positive scalar curvature metrics with positive mean curvature boundary?

Both problems 1 and 1′ lead to obstruction living in absolute groups (because one is working relative to the boundary, the details are more complicated in 1') and the Novikov conjecture philosophy applies.

Some thought about examples and the surgery theorems that are a basic method in analyzing the positive scalar curvature problem [GL, SY, LM] leads one to conjecture that the obstruction is precisely the obstruction to putting a positive scalar curvature metric on the double. (This is in fact an obstruction). Analogously, one can make the same conjecture regarding the homology L classes for problem 3. Rosenberg and I have reduced this to the conjecture for 2 and statements about the Γ-surgery theory of Cappell and Shaneson [CS].

What is the conjecture for 2? It is as simple as one could imagine. The Poincare dual of the L class lies in relative homology, which can be mapped into relative group homology. Conjecturally, the image is a homotopy invariant. This is equivalent to the injectivity of a relative assembly map.

The evidence? If Borel conjectures held for the boundary components, then one cold deduce the result from the five lemma applied to the exact sequences of a pair in homology and L theory. The only way it cold really fail, assuming Novikov, is if Novikov were true for all groups, but there were no functorial splitting.[2]

13. Now I would like to discuss another variant, the equivariant Novikov conjecture [RsW]. Clearly it should have to do with the equivariant signature operator $\Delta(M) \in K^G(M)[1/2]$. (The topological invariance of this class is explained, by two very different approaches, in [RtW] and [CSW]; for odd order groups, this was first given a topological construction by [MR].) An equivariant $K(\pi, 1)$ according to homotopy theorists is an aspherical space all of whose fixed sets are unions of aspherical spaces. The natural conjecture would be that, rationally, the push forward of Δ into $K^G(B\pi M)$ is an equivariant homotopy invariant.

This is not true in general. The simplest case is the circle acting freely on the sphere. The variation of the Pontrjagin classes of homotopy complex projective spaces would contradict the conjecture. Thee have been two proposals "explaining" this failure. The first is some formal lack of finite dimensionality of the relevant K group, and one might only believe that the conjecture is

---

1 This material is joint work with Rosenberg.

2 Consequently, the algebraic K theory version of this is true in great generality because of Bokstedt-Hsiang-Madsen.

true for finitely generated homology groups. The second, that I prefer, is that there really is no equivariant Novikov conjecture, but only a stratified one, which I will get to soon. (It would be very interesting to understand this possibility analytically.)

In any case, if $B\pi M$ is a finite dimensional space, for instance, then the conjecture that the push forward is equivariant homotopy invariant is reasonable. Rosenberg and I have modified Kasparov's technique to show this when G acts by isometries on a nonpositively curved manifold. (This yields an equivariant C^*-algebra version, which therefore has applications to other elliptic operators as well.) We've also deduced this from an equivariant rigidity theorem of Ferry and myself.

14. Before returning to the equivariant NC, I would like to point out that there are many situations where one can produce maps to recognizable equivariant $B\pi$'s. For instance, elementary Hodge theory leads to a Jacobi map construction of the map promised by:

Proposition. If G acts by isometries on M with free abelian fundamental group, then there is a G action on the standard flat torus T by affine maps, and an equivariant map from M to T inducing the isomorphism of fundamental groups.

The method of [SY] implies that the same holds if the T is replaced by a finite volume hyperbolic manifold of dimension 3 and higher, and up to deformation, using the Nielson realization theorem [Ke], a hyperbolic surface. Also, if the action is by a trivial outer automorphism, one can dispense with the finiteness of volume, by relying on work of Corlette [C]. (I believe that this last case can also be done by pure group theoretic techniques!) I think that it is a very interesting problem in differential geometry to extend this to negative curvature. (I would doubt it in the case of nonpositive curvature.) Also, extensions of this phenomenon to Gromov's hyperbolic groups seems worthwhile.

15. Returning to ENC, it is interesting to note that, as for manifolds with boundary, one can ask for invariance under less than an equivariant homotopy equivalence (the counterpart of a homotopy equivalence of pairs). This is the notion of a "pseudoequivalence". The point is this: The generalization of signature to the equivariant case is G-signature. The definition can be found in [AS III]; one replaces dimensions by representations in the definition of signature. The G-signature is invariant under a larger class of maps than the equivariant homotopy equivalences: any equivariant map that happens to be a homotopy equivalence will preserve the G-signature. These are the pseudoequivalences.

Pseudoequivalence is not an equivalence relation.

Although the equivariant $B\pi$ is no pseudoequivalence invariant, the equivariant K theory of the C^*-algebra of the fundamental groupoid is! I do not think there is yet a very good theory of surgery up to pseudoequivalence (see [P] for a start).

16. One interesting feature of the G-signature formula is its combination with the localization theorem. One can do the same thing with respect to higher G-indices as well. This has a number of interesting corollaries. One is that if G acts trivially on $H(\ ;\mathbb{Q}\pi)$ (for this to make sense one needs the assumption that the G action lifts to a $G \times \pi$ action on the universal cover; this can often be phrased in easier to check conditions on M) then, assuming ENC,

$$f_*(L(M) \cap [M]) = (fi)_*((L(F) \cup D(\nu)) \cap [F])$$

where F is the fixed set of some element, and $D(\nu)$ is a characteristic class of the equivariant normal bundle to the fixed set. For instance for smooth involutions the right hand side is the higher signature of the self intersection of the fixed set.

For the choice of D (the average of all reasonable choices, in some sense, see [We2]) and semifree actions, the above formula is actually equivalent to the Novikov conjecture for π by purely topological arguments involving the assembly map and surgery. So is the special case asserting vanishing for free actions.

The above formula is true (assuming F nonempty), independent of the NC, without the $D(\nu)$ term, for circle actions. Browder and Hsiang [BH] have also established the vanishing of higher A-genera for spin manifolds with smooth circle actions. The key point is that one can readily do certain transversality arguments in this setting. For the homologically trivial actions, there is a nontrivial obstruction to "homologically trivial transversality" [We5].

17. ENC and EBC (equivariant rigidity of equivariant $K(\pi,1)$'s in the sense above) can both be translated into statements about a generalization of the assembly map using the stratified surgery developed in [We1]. The Borel conjecture's algebraic K theory concommitant[1] would be that a certain assembly map from a sheaf homology group is an isomorphism:

$$H(M/G;\ \mathbf{Wh}(G_m)) \to Wh(\Gamma)$$

where Γ is the orbifold fundamental group of M/G, i.e. the group of lifts of group elements to the universal cover of M, G_m denotes the isotropy subgroup of a point m, and $\mathbf{Wh}(G_m)$ denotes the Whitehead spectrum, which has homotopy groups in negative dimensions (the relevant ones!) the nonpositive K-groups. Note the case $V \times S^1$ for V a representation. The map is just the nonNil terms in the [BHS] formula, so the map split injects, but there is a deviation of Nil.

Connolly and Kosniewski have shown that if one makes the assumption of equivariant topological simplicity, then the EBC is true for odd order groups acting on tori. (Quinn [Qu] has proven the algebraic K theory analog, up to Nils, for virtually Poly $\mathbb{Z}$ groups.) However it is false even for involutions on tori!

The reason for this is that in the analogous L theoretic version there are UNil difficulties. See [We1] for a development of equivariant topological surgery[2]. Concretely, one can take Cappell's nonconnected sums in $S(\mathbb{R}P^{4k+1} \# \mathbb{R}P^{4k+1})$ and push them forward, using the naturality of S, into the structures of the nonsingular part of any orientation preserving involution on T^{4n+1} (so that the fundamental group split injects). It is not difficult to see, using well known facts about the topology of noncompact manifolds that the quotients of the nonsingular parts aren't homeomorphic, so the involution on the torus cannot be affine.

I will not state what the L theory version of Novikov and Borel are in this setting, but they contain, away from 2, the statements that the following map is injective or an isomorphism, respectively:

$$KO^G(B\pi)[1/2] \to L(\Gamma) \otimes \mathbb{Z}[1/2]$$

One can write the left hand side as a sheaf homology whose stalks are L-spectra, and make use of the connection between L-spectra and K-theory with coefficients. This makes the L theory more closely resemble the algebraic K-theory above. This morphism from the sheaf homology to KO theory was constructed in [CSW] to give a new proof a topological G-signature formula.

---

1 It's actually a consequence of the Borel conjecture; one should add similar statements for fixed sets of subgroups and the like.

2 Surprisingly, perhaps, it is in a much better shape than equivariant smooth surgery; for odd order groups Madsen and Rothenberg have established a topological equivariant surgery exact sequence by a philosophically very different path.

Of course, the operator theoretic ENC and EBC involve the same LHS (but perhaps integrally), and the RHS is $C^*\pi \times G$.

The reader should compare this to the Baum-Connes conjecture [BC].

18. Although we have seen that EBC is false in general, it (the simple homotopy version) is conceivably true for odd order group actions. I must confess some skepticism. I believed the case of even order group actions with only odd order isotropy to be just as likely, but conversations with Frank Connolly about the Klein bottle group have led to counterexamples to this. Conceivably the C^*-algebra EBC is always true!

19. Ferry and I have shown that ENC holds using ideas related to the EBC, for groups acting by isometries on manifolds of nonpositive curvature, we establish stable versions of Borel. Here is a theorem from [FRW] that is, I believe, suggestive of some interesting directions:

Theorem. Let W be a manifold of nonpositive curvature, and suppose that G acts by isometries on W. Suppose that $f : W \to W'$ is an equivariant proper homotopy equivalence such that there is a universal bound on the diameter of point inverse images, then W and W' are stably homeomorphic.

This theorem is very metric sensitive. There are metrics on almost any such noncompact W for which the action is still an equivariant $K(\pi, 1)$, but for which this theorem is false. I believe that the relevant condition should be that the universal cover has bounded contractibility property; every ball of radius n contracts inside a ball of radius $f(n)$ for some function $f(n)$. (Presumably linear functions are the most interesting. For manifolds with nonpositive curvature $f(n) = n$.)

20. Hopefully, the theorem from the previous paragraph will call attention to the importance of "bounded considerations". Here is a specific conjecture, which if proven sufficiently canonically, would imply Novikov for finite $K(\pi, 1)$ complexes by an elaboration of the previous method.

Conjecture: Suppose that W is an n-manifold with the bounded contractibility property, and $p : V \to W$ is a proper map. Suppose $f : V' \to V$ has the property that the homology of each ball inverse image is contractible inside a uniformly larger ball inverse image, then p^{-1} (point) and $(pf)^{-1}$ (point) have the same signature.

The example to keep in mind is the universal cover a homotopy equivalence between manifolds with free abelian fundamental groups, with W Euclidean space. The proof of this conjecture in the Euclidean case uses some ideas of bounded surgery [FP] to reduce to the case of $W \times N$, a simply connected closed manifold N (analogous to the way surgery enables one to make use of nice manifolds with given π to understand all manifolds with fundamental group π); there one can deform the map $V' \to W \times N$ to one with a good hold on the (shape of the) inverse image of each $\{w\} \times N$. (The technique for doing this is in the memoir of chapman [Ch].) Then one makes use of Novikov's proof of the topological invariance of rational Pontrjagin classes.

The Euclidean case is sufficient for the theorem of §19 above (although that was not the original argument given).

It would be very interesting to recast this type of argument in the operator theoretic context using appropriate types of not quite fiber preserving families of elliptic operators. Then one would have a very useful tool for studying differential geometric problems on manifolds with bounded geometry.

I should emphasize that this perspective is very different from what Atiyah [At] and Roe [Roe] do. Their indices for a universal cover correspond to the index downstairs; we are seeking to understand how to see the highest (codimensional) index.

21. The statements in §17 show that one can get more information about ordinary L groups for groups with torsion (or the K theory of their C^*-algebras) from ENC than from the ordinary NC. This then should lead to new information about manifolds with that bigger fundamental group. Here is an example where this is possible (see [We3]).

Let Γ be a discrete subgroup of a real semisimple Lie group G. $V = G/K$ then has a metric of nonpositive curvature. One can analyze the surgery exact sequence for $S(B\Gamma)$ using Selberg's observation that there is a normal subgroup Γ' of Γ which is torsion free and finite index. The double coset space associated Γ' is a manifold which has a Γ/Γ' action. The solution of ENC yields a rational split surjection

$$S_n(B\Gamma) \to KO^{\Gamma/\Gamma'}_{n+1}(V/\Gamma' \times E(\Gamma/\Gamma') \to V/\Gamma') \otimes \mathbb{Q}$$

The RHS of the above is supported on the singular set. We will investigate its significance shortly. If M has fundamental group Γ one then gets a map $S(M) \to S(B\Gamma)$. If one combines with a solution to NC for Γ as in §7 one would get rationally a split surjectve map

$$S_n(M) \to KO_n(M, B\Gamma) \times KO^{\Gamma/\Gamma'}_{n+1}(V/\Gamma' \times E(\Gamma/\Gamma') \to V/\Gamma') \otimes \mathbb{Q}$$

If EBC were true, this would be a rational isomorphism.

22. I would like to discuss the significance of the peculiar term $KO^{\Gamma/\Gamma'}(V/\Gamma' \times E(\Gamma/\Gamma') \to V(\Gamma')$ and its relation to invariants of [APS] (or, relatedly) those of [AB II], [AS III].

Consider $\mathbb{Z} \times G$ inside an affine group, via an orthogonal representation of a finite group G and a translation through an orthogonal line to the representation space. Then the relevant relative KO^G is isomorphic to O in odd dimensions and $RO(G)/i * RO(e)$ in even dimensions. So for manifolds with fundamental group $\mathbb{Z} \times G$ one is reading off a reduced peripheral invariant of a codimension one submanifold.

To be more precise, for an arbitrary codimension one separating submanifold with fundamental group G one (rationally) codimension one splits, as in §11, and takes the difference between the ρ-invariants, or twisted η invariants of the signature operator, between the "corresponding" submanifolds.

For other groups one is digging deeper and in more complicated ways into the peripheral invariants of the submanifolds.

It would be very interesting to understand this process better analytically. It is clearly related to the idea that eta invariants of manifolds occur as pieces in signatures of singular spaces. (See [Che], especially his formula for L-classes, i.e. the homology class associated under the chern character to the signature operator, in terms of eta of links.) One would presumably have to implicate the more analytic approach of Kasparov to the NC.

In fact, the analytic work of Cheeger combines with the ideas of the previous section, and the topological results in §16, to allow the definition of pieces of these higher ρ invariants under some acyclicity hypotheses (see [We3]). This is perhaps analogous to the use of acyclicity to define Reidemeister or analytic torsions [Mi, RS] which makes intrinsic for special manifolds the invariant (Whitehead torsion) of homotopy equivalences.

23. It should hardly be a surprise that work on the NC and BC lead to information about the conventional η-invariant as well. (This possibility was first raised by Neumann [Ne].)

Given a unitary representation ν of Γ one can twist the signature operator on an odd dimension manifold by ν and subtract from η with coefficients in this operator, η with coefficients in the trivial bundle of the same dimension. This produces an invariant that is independent of the metric according to [APS]. They observed that it is not in general a homotopy invariant by computing for the lens spaces. In fact if Γ has torsion and is virtually torsion free or residually finite, then this quantity cannot be a homotopy invariant for every representation. (There might be some nontrivial representation for which it is.)

If the BC, or the C^* version of it, is true for Γ then this invariant is a homotopy invariant. Assuming the NC, this difference is always rational, and is constant on the components of the representation space of Γ. (Either statement quickly yields a proof for free abelian groups, a case due to [Neu]; he goes much further and proves a homotopy invariant formula for this invariant. Unfortunately the group rings $\mathbf{R}\Gamma$ that occur here have many fewer nice properties than the Noetherian $\mathbf{R}[\mathbf{Z}^n]$. Still, one might again venture that operator theoretic techniques might be very useful for this purpose.)

A consequence of all of this material and Cappell's work verifying Borel often is the invariance of this type of η for fiber homotopy equivalences if the representation factors through the base of the fibration. This feels related to the holonomy phenomenon [BC].

I will not repeat the arguments from [We4] here; they are brief enough for the reader to simply go and read!

24. I would like to close by suggesting that more and more attention be paid to nonmanifolds. The phenomenon of intersection homology and in [Che] builds, in some sense, theories that manage to smooth over the difficulties related to the singularities. However, one also wants theories that for orbifolds are related to equivariant theory. (See the thesis of Farsi [Fs].) In the topological setting there is now a classification theory that extends surgery theory to fairly general stratified spaces. There is an extension of the L-classes for theses spaces that lives in a sheaf homology whose local stalk is related to the stratified local homotopy properties of the space. (It can be viewed as the normal invariant of some standard simply surgery problem $\times X$ for a stratified space X.) For all of this, see [We1]. It could be interesting to have a general index theory that parallels this theory.

25. In [We1] I proposed the following stratified Borel conjecture. (The stable version is Novikov!)

Conjecture[1]. Suppose the pure strata of a manifold stratified space X have the property that the map $(X^i, \epsilon) \to (B\pi_1 X^i, B\pi_1 \epsilon)$, where ϵ is the end of the stratum, is an isomorphism of $\mathbf{Z}$-homology. Then X is rigid (i.e. simple homotopy equivalence implies homeomorphism).

(It is, of course, false because of UNil. One might conjecture, but I'm skeptical, that if all the groups involved are torsion free, then the conjecture is actually true.)

If the aspherical pairs entering in this conjecture are hyperbolic, flat, torsion free Poly-(finite or infinite cyclic), square-root-closed accessible from the trivial group, then methods of [FJ2], [FH], [Ca4] combine with the ideas in [We1], [FRW] to verify this conjecture.

The previous paragraph can be viewed as minimally asking to find the operator theoretic analog of this conjecture.

26. I would like to go full circle with a part of this discussion, and point out that the stratified BC above, when verified, is intimately related to fibering problems, so that the use of Farrell's theorem is more than just an artifact in §10[2].

---

1 Cappell's work on the codimension one splitting problem can be viewed as the verification of a particular 4 stratum Borel conjecture.

2 Farrell and Jones consider a stronger question: "block fiberring" and only answer it in the absence of algebraic K-theory. They treat this as an analog of their rigidity theorem, rather than

Actually, as in §20, one realizes that one does not need or want fibrations, but rather maps which are sort of "shape fibrations" (these are technically called "approximate fibrations because of another interpretation of them) i.e. such that the inverse image of each ball has the proper homotopy type of $R\times$ the ball, in view of the ideas in Novikov's proof of the topological invariance of L-classes.

The way this comes to pass is this. (For simplicity this argument shows uniqueness rather than existence, but there are tricks to relate one to the other.) One takes the open mapping cylinder of a projection to a map that has a chance of fibering over an aspherical manifold. The fact that any stratified space simple homotopy equivalent rel the end is homeomorphic to this implies that the end determines the neighborhood. In [HTWW] neighborhoods in stratified spaces are classified in terms of approximate fibrations.

In the end, one sees that the whole obstruction to doing this as $\mathrm{Wh}^{\mathrm{top}} = \oplus$ Nils for the aspherical manifolds listed in §25.

References

[At] M. Atiyah, Elliptic operators, discrete groups, and von Neumann algebras, Asterisque 32 (1976) 43–72.

[AB] M. Atiyah and R. Bott, A Lefshetz fixed point formula for elliptic complexes, II: Applications, Annals of Math 88 (1968) 451–491.

[APS] M. Atiyah, V. Patodi, and I. Singer, Spectral asymmetry and Riemannian geometry I II III, Math. Proc. Cam. Phil. Soc. 77 (1975) 43-69, 78 (1975) 405–432, 79 (1976) 71–99.

[AS] M. Atiyah and I. Singer, The index of elliptic operators I III, Annals of Math. 87 (1968) 484–530, 546–604.

[BHS] H. Bass, A. Heller, and R. Swan, Publ. Math. d'IHES.

[BC] P. Baum and A. Connes, Geometric K theory for Lie groups and foliations, (preprint).

[BiC] M. Bismut and J. Cheeger, (preprint).

[BHM] M. Bokstedt, W.C. Hsiang, and I. Madsen, (in preparation).

[BH] W. Browder and W.C. Hsiang, G–actions and the fundamental group, Inven. Math. 65 (1982) 411–424.

[Ca1] S. Cappell, A splitting theorem for manifolds, Inven. Math. 33 (19876) 69–170.

[Ca2] S. Cappell, On connected sums of manifolds, Topology 13 (1974) 395–400.

[Ca3] S. Cappell, Unitary nilpotent groups and Hermitian K–theory, BAMS 80 (1974) 1117–1122.

[Ca4] S. Cappell, Mayer-Vietoris sequences in Hermitian K-theory, LNM 343.

[CS] S. Cappell and J. Shaneson, The codimension two placement problem and homology equivalent manifolds, Ann. of Math. 99 (1974) 27–346.

[CSW] S. Cappell, J. Shaneson, and S. Weinberger, Topological characteristic classes for group actions on Witt spaces, (preprint).

[CW] S. Cappell and S. Weinberger, The classification of certain stratified spaces, (preprint).

[Ch] T. Chapman, Approximation results in topological manifolds, Memoir AMS 251 (1981).

[Che] J. Cheeger, Spectral geometry of singular Riemannian spaces, J. Diff. Geo. 18 (1983) 575–657.

[Co] A. Connes, Noncommutative differential geometry, Publ. Math. D'IHES 62 (1986) 257–360 + more to come ...

[CK] F. Connolly and T. Kosniewski, Rigidity of crystollagraphic actions I, II (Notre Dame preprints).

[CM] A. Connes and H. Moscovici, Cyclic cohomology, the Novikov conjecture, and hyperbolic groups, preprint.

viewing it as another rigidity theorem.

[Co] K. Corlette, Flat G-bundles with canonical metrics, J. Diff. Geo. 28 (1988) 361–382.

[Fa1] F.T. Farrell, The obstruction to fibering a manifold over the circle, Indiana Math. J. 21 (1971) 315–346.

[FH] F.T. Farrell and W.C. Hsiang, Topological characterisationof flat and almost flat Riemannian manifolds $M^n (n \neq 3,4)$ Amer. J. of Math. 105 (1983) 641–672.

[FJ2] F.T. Farrell and L. Jones, A topological analogue of Mostow's rigidity theorem, JAMS 2 (1989) 257–370.

[Farsi] C. Farsi, 1989 Ph.D. Thesis (University of Maryland).

[FP] S. Ferry and E. Pederson, Controlled surgery theory, (preprint).

[FRW] S. Ferry, J. Rosenberg, and S. Weinberger, Equivariant topological rigidity phenomena, Comptes Rendus 1988.

[GL] M. Gromov and H.B. Lawson, Positive scalar curvature and the Dirac operator on complete Riemannian manifolds, Publ. Math. d'IHES 58 (1983) 83–196 (see also their references).

[HTW] B. Hughes, L. Taylor, and B. Williams, Controlled topology over manifolds of nonpositive curvature (preprint).

[HTWW] B. Hughes, L. Taylor, S. Weinberger, and B. Williams, The classification of neighborhoods in a stratified space (in preparation).

[KaM1] J. Kaminker and J.C. Miller, Homotopy invariance of the analytic index of signature operators over C^*-algebras, J. of Operator Theory 14 (1985) 113-127.

[KaM2] J. Kaminker and J.C. Miller, A comment on the Novikov conjecture, Proc. AMS 83 (1981) 656–658.

[Ka] G.G. Kasparov, Equivariant KK theory and the Novikov conjecture, Inven. Math. 91 (1988) 147–201.

[Ke] S. Kerkhoff, The Nielson realisation problem, Ann. of Math. 117 (1983) 235–265.

[M] J. Milnor, Whitehead Torsion, BAMS 72 (1966) 358-426.

[MM] I. Madsen and J. Milgram, The classifying spaces for surgery and cobordism of manifolds, Princeton University Press.

[MR] I. Madsen and M. Rothenberg, On the classification of G–spheres I II III Acta Math. 160 65–104 and several Aarhus preprints. See in particular Cont. Math. 19 (1983) 193–226 for a summary.

[No] S. Novikov, The algebraic construction and properties of hermitian analogues of K–theory for rings with involution from the point of view of the Hamiltonian formalism. Some applications to differential topology and the theory of characteristic classes, Izv. Nauk. SSSR ser. mat. 34 (1970) 253–288 478–500.

[PV] M. Pimsner and D. Voiculescu, Exact sequences for K–groups and Ext. groups for certain cross products of C^*-algebras, J. Operator Theory 4 (1980) 93–118.

[P] T. Petrie, Pseudoequivalences of G–manifolds, Proc. Symp. Pure Math. 32 (1978) 169–210.

[Q] D. Quillen, Higher algebraic K-theory, LNM 341.

[Qu] F. Quinn, Algebraic K-theory of poly-(finite or cyclic) groups, BAMS 12 (1985) 221–226.

[Roe] J. Roe, Index theory for open manifolds I, II, J. Diff. Geo. 1988.

[Ro1] J. Rosenberg, C^* algebras, positive scalar curvature, and the Novikov conjecture, Publ. Math. d'IHES 58 (1983) 197–212.

[Ro2] J. Rosenberg, Applications of analysis on Lipschitz manifolds, Proc. Special Year in C^* Algebras at the Centre for Mathematical Analysis in Australia, 269–283.

[Ro3] J. Rosenberg, K and KK, Topology and operator algebras, (preprint).

[RoW1] J. Rosenberg and S. Weinberger, with an appendix by J.P. May, An equivariant Novikov conjecture, submitted to K-theory.

[RoW2] J. Rosenberg and S. Weinberger, Higher signatures and scalar curvature for manifolds with boundary, (in preparation).

[RoW3] J. Rosenberg and S. Weinberger, Higher G–indices and their applications, Ann. Sci. Ec. Norm. Sup. 21 (1988) 479–495.

[RS] D. Ray and I. Singer, R-torsion and the Laplacian on Riemannian manifold, Adv. in Math. 7 (1971) 145-210.

[RtW] M. Rothenberg and S. Weinberger, Group actions and equivariant Lipschitz analysis, BAMS 1988.

[SY1] R. Schoen and S.T. Yau,

[SY2] R. Schoen and S.T. Yau, Compact group actions and the topology of manifolds with nonpositive curvature, Topology 18 (1979) 361–380.

[Sh] J. Shaneson, Wall's surgery obstruction groups for $\mathbb{Z} \times G$, Ann. Math. 90 (1969) 296–334.

[SuT] D. Sullivan and N. Teleman, An analytic proof of Novikov's theorem on rational Pontrjagin classes, Publ. Math. d'IHES 58 (1983) 79–81.

[Wald1] F. Waldhausen, Algebraic K-theory of spaces, Proc. Symp. Pure Math. 17 (1978) Stanford conference.

[Wald2] F. Waldhausen, Algebraic K-theory of amalgamated free products, Ann. of Math. 108 (1978) 135–256.

[We1] S. Weinberger, The topological classification of stratified spaces, submitted to BAMS.

[We2] S. Weinberger, Group actions and higher signatures II, Comm. Pure Appl. Math. 40 (1987) 179–187.

[We3] S. Weinberger, Higher ρ-invariants, submitted to K-theory.

[We4] S. Weinberger, Homotopy invariance of η-invariants, Proc. Nat. Acad. Sci. USA 1988.

[We5] S. Weinberger, Class numbers, the Novikov conjecture, and transformation groups, Topology 27 (1988) 353–365.

[Ws] M. Weiss, On the definitions of the symmetric L-groups, (preprint).

Department of Mathematics
University of Chicago